Theory construction and selection
in modern physics: the S matrix

Theory construction and selection in modern physics

THE S MATRIX

James T. Cushing

Professor of Physics
University of Notre Dame

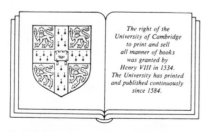

The right of the
University of Cambridge
to print and sell
all manner of books
was granted by
Henry VIII in 1534.
The University has printed
and published continuously
since 1584.

CAMBRIDGE UNIVERSITY PRESS

Cambridge

New York   Port Chester

Melbourne   Sydney

Published by the Press Syndicate of the University of Cambridge
The Pitt Building, Trumpington Street, Cambridge CB2 1RP
40 West 20th Street, New York NY 10011, USA
10 Stamford Road, Oakleigh, Melbourne 3166, Australia

© Cambridge University Press 1990

First published 1990

Printed in Great Britain at the University Press, Cambridge

British Library cataloguing in publication data

Cushing, James T. (James Thomas) 1937–
Theory construction and selection in modern physics:
the S matrix
1. Physics. Theories. Development
I. Title
530.01

Library of Congress cataloguing in publication data

Cushing, James T., 1937–
Theory construction and selection in modern physics: the S Matrix
James T. Cushing
      p.   cm.
Includes bibliographical references
ISBN 0-521-38181-9
1. S-matrix theory. 2. Physics–philosophy. 3. Physics–
Methodology. I. Title
QC174.35.S2C87 1990
530. 1′22–dc20

ISBN 0 521 38181 9

MCP

TO NHC

Contents

Preface   xi
Acknowledgments   xviii

1   Introduction and background   1
1.1   Internal history of recent science   3
1.2   Philosophical issues and the Forman thesis   6
1.3   The purview of this case study   7
1.4   Quantum field theory (QFT) background   8
1.5   Renormalized quantum electrodynamics (QED)   12
1.6   Feynman diagrams   19
1.7   Gauge field theories   22
1.8   Summary   26

2   Origin of the *S* matrix: Heisenberg's program as a
background to dispersion theory   28
2.1   The *S* Matrix: Wheeler and Heisenberg   30
2.2   Discussion of Heisenberg's *S*-matrix papers   34
2.3   Heisenberg's subsequent role in the program   39
2.4   Work on the program just after WWII   42
2.5   The *S* matrix and nuclear theory   51
2.6   Causality and dispersion relations   57
2.7   A problem background for the 1950s   63
2.8   Summary   65

3   Dispersion relations   67
3.1   Goldberger, Gell-Mann, Thirring and microcausality   69
3.2   'Proofs' of dispersion relations   70
3.3   Phenomenological use of dispersion relations   77

3.4   Pragmatic attitude of many practitioners   80
3.5   More general proofs   81
3.6   Other applications of dispersion relations   83
3.7   Summary   87

4   Another route to a theory based on analytic reaction amplitudes   89
4.1   Some technical preliminaries   91
4.2   Fermi and the impulse approximation   94
4.3   The Chew model and phenomenology   97
4.4   A digression: crossing symmetry   100
4.5   The Chew–Low-model   104
4.6   Chew's particle-pole conjecture   109
4.7   Summary   113

5   The analytic S matrix   115
5.1   Gell-Mann: a new approach to QFT   116
5.2   The Mandelstam representation   118
5.3   Proofs of double dispersion relations   123
5.4   Regge's innovation   126
5.5   A bootstrap mechanism   127
5.6   An independent S-matrix program   129
5.7   Summary   132

6   The bootstrap and Regge poles   134
6.1   The Pomeranchuk theorems   136
6.2   The Chew–Mandelstam calculation   139
6.3   Regge poles and asymptotic boundary conditions   141
6.4   The LaJolla conference and Chew's rejection of QFT   142
6.5   Early successful predictions   145
6.6   Degenerating Regge phenomenology   152
6.7   Applications of the bootstrap   155
6.8   Too much complexity   161
6.9   Summary   165

7   An autonomous S-matrix program   167
7.1   QFT versus SMT   169
7.2   A conceptual framework for SMT   173
7.3   A bootstrapped world   176
7.4   A shift to 'higher' philosophical ground   179
7.5   Axiomatic SMT and its offshoots   182

7.6	Enormous complexity again	185
7.7	Summary	187
8	**The duality program**	**189**
8.1	*S*-matrix origin of duality	191
8.2	The Veneziano model	193
8.3	A topological expansion	196
8.4	The topological *S*-matrix program	200
8.5	The emergence of quantized string theories	203
8.6	Superstrings–the ultimate bootstrap	207
8.7	Summary	208
9	**'Data' for a methodological study**	**209**
9.1	An overview of this case study	210
9.2	Hallmarks of this episode	213
9.3	A review of several developments in modern physics	217
9.4	An illustration: the compound nucleus model	223
9.5	Causal quantum theory	231
9.6	Summary	236
10	**Methodological lessons**	**238**
10.1	Some general characteristics of science	239
10.2	The role of sociological factors in high-energy physics	243
10.3	Structures and dynamics in methodology	249
10.4	Changes in methodological rules	256
10.5	A uniqueness in our theories?	262
10.6	Convergence of scientific opinion	271
10.7	*A* view of science	281
	Appendix	291
	Notes	316
	References	330
	Glossary of technical terms (from physics and from philosophy)	372
	Some key figures and their positions	393
	Index	399

Preface

A major, overarching cluster of problems central to the philosophy of science and certainly underlying much of the debate in the recent literature is how scientific theories are constructed, how they are judged or selected, and what type of knowledge they give us. There are two aspects of answers to any of these three questions: what has actually occurred according to the historical record and what is the rational status of each of these activities or of the knowledge produced. A simple schema, that is based on induction and the hypothetical-deductive method and that provides answers to the above queries, is the sequence: observation, hypothesis, prediction, confirmation. This model or picture of science has a long tradition. We can see its roots already in Bacon's (1620 (1960, pp. 43–4 and 98-100)) advocating a slow and careful ascent from particulars to generalities (Aphorisms, Bk. I, XIV, XXII, CIII–CVII). He urged use of a combination of induction and deduction in arriving at knowledge. In Bacon's ladder of axiom, one is to make modest generalizations based on specific observations and data, check these modest theories by comparing their predictions with facts once again, then combine these generalizations into more general ones, check their predictions against observations, and in this way carefully proceed to the most general axioms, theories or laws. Whewell* (1857, Vol. I, p. 146) speaks of the epochs of induction, development, verification, application and extension. This is often taken as the hallmark of the scientific method that results in truth, true knowledge or true theories about the world. While the proverbial 'man

---

* At the end of this book there are both a glossary of technical terms (used in physics and in philosophy) and a list of key figures, along with their major positions. Since not *every* reader will need *all* of this information repeated, I have not put it into the text proper, where it might interfere with the flow of the narrative. If in doubt, check!

in the street' may subscribe to this representation of science, as many working scientists seem still to do in general outline, few philosophers of science would accept so simplistic a response. It would be nice if the world were as simple as this, but such is not the case. Let us refer to this as the *simple model* of science. Although we shall expand at length in the text proper on several developments of this model in the philosophy of science, let us sketch here the evolution of the position we shall follow.

The logical positivists sought to elaborate and formalize this model by attempting to base an explanation of the rationality of science upon its empirical foundation and even to develop a logic of induction. The program did not succeed, both because its foundationist assumptions led to a description of science (or, actually, what science should be like) that just did not accord with actual historical scientific practice and because of internal problems that have become evident in retrospect (Friedman, 1988). One of the difficulties with a reliance on the straightforward inductive-hypothetical-deductive method is the so-called Duhem–Quine thesis according to which any theory is underdetermined by the empirical facts upon which it is based and which serve to confirm the theory. A reaction to this is instrumentalism which sees the goal of science as constructing laws and theories that provide a means for calculating and correlating empirical results, but which need not give us a true picture (at the level of theoretical entities) of the world. Furthermore, another difficulty for any fixed truth claims made on behalf of science is that the historical record of the development of science provides ample evidence of laws, concepts and theories, once held to be true, that have later been abandoned as false at a foundational level (e.g., classical mechanics being replaced by relativity and quantum theory).

To cope with these and other shortcomings of the simple model, philosophers of science have come to view the functioning scientific enterprise in terms of a three-level scheme: practice (theory), methods and goals. Theories function to explain, correlate and organize phenomena. The heading 'practice' also includes experiments and the activity of judging the empirical adequacy of a proposed theory. For many scientists, and perhaps for most people in general, this may seem to be just about the whole of science. However, the question remains of *how* theories are to be evaluated. What standards or rules are to be applied in testing and selecting theories? This is the level of method. Some examples of such criteria would be predictive accuracy, simplicity, coherence and fertility. Finally, there is the metalevel of the goals or aims of science (such as giving true explanations of phenomena versus merely providing rules that allow us to calculate and predict without necessarily providing literally true pictures of the world; or, a goal for science could be the control of nature). A fallback position from the simple model is to admit the obvious corrigibility of our theories (i.e., they *do* change and develop), but to hold out for stability or 'fixedness' at the (meta) levels of methods and of goals in order to underpin an invariant rationality that characterizes science and the knowledge it obtains. Even here though, a concession is usually made that the goal of science is approximate truth about nature (in some sense of a correspondence with reality). It is also typical to make a distinction between the *discovery* of a scientific theory and the *justification* of an already formulated theory. Postpostivist philosophy of science tends to bracket the problem of the means by which scientific theories and hypotheses are discovered or constructed (leaving these to some other area such as psychology, luck, inspired guess, etc.) and to concentrate on the justification of an articulated theory (whatever its origin). The rationality of science is to be located in the logic of justification of its theories. It is to this aspect of the scientific enterprise that the three-tiered schema of practice, methods, goals is to apply. But, one has now to decide the status of these allegedly fixed methods and goals (that are to underpin the *rationality* of science). Are these to be argued for and justified on the basis of some logically necessary first principles (the foundationist approach) or on the basis of (contingent) historical fact? Examples of methodologies of the first type, the foundationist or rationalist school, are those of Popper (1963), Lakatos (1970, 1976) and Watkins (1984), while the best-known proponent of the historicist school is Kuhn (1970).

To set the scene between the opposing views, let us outline, as representative of each view, the positions of Kuhn and of Lakatos. Kuhn (1970) sees two essential components in science. Normal science, which is the activity the majority of scientists engage in most of the time, is guided by paradigms and consists largely in puzzle-solving within a fairly well-articulated set of ground rules. This phase of science defines problems to be solved. The pressure, rather than being on the theories themselves, is on the individual scientist to apply successfully the currently accepted paradigm in solving a problem that has been set (Kuhn 1970; pp. 4–5). In a crisis situation, the paradigm (or 'rules of the game') become loosened. While normal science recognizes anomalies and crises, it cannot, of itself, change the paradigm. It is then that a transition to revolutionary science takes place. By such revolutions, new paradigms are generated and science advances (or evolves). A successful revolution is followed by another period of normal science.

For Lakatos (1970, 1976), the proper unit for theory appraisal is the research program, an entity that establishes a tradition within which scientists work. This methodology of scientific research programs (MSRP) is intended to be a development of Popper's view of science according to which *the* hallmark of scientific theories is that they are (in principle) refutable or falsifiable. That is, bold hypotheses are exposed to the hazard of refutation (by observation and experiment) and the successful ones survive (for a longer or shorter time). For Lakatos, the process of refutation alone is not sufficient to represent actual science, but research programs are seen as essential for comprehending scientific practice. In Lakatos' theory of science, the various components of a research program are (i) the 'hard core' of assumptions which are kept unfalsifiable by methodological decision, (ii) the auxiliary hypotheses (or assumptions) to which the falsifiability criterion is directed when anomalies arise, and the (iii) the (positive) heuristic, a set of suggestions for modifying the auxiliary hypotheses. Roughly speaking, the negative heuristic (or hard core) tells one what *not* to do in the sense that certain assumptions are to be left largely untouched. The positive heuristic (often referred to hereafter simply as the heuristic) is a partially articulated research policy to guide one through all the possibilities allowed (or not forbidden) by the negative heuristic. The positive heuristic provides a direction for research and the evolution of a program. If MSRP is correct, then applications of the heuristic to specific problems should generate a sequence of theories by which a research program develops. These changes should then be classifiable as degenerating or progressive problem shifts, respectively, in terms of their *ad hocness* (or contrived nature) to meet anomaly or their fertility for further research and confirmation by experiment. Popper and Lakatos, like others in the rationalist school, see science as having its own distinctive *internal* logic by which it tests and selects theories. This exercise of theory justification is seen as (logically, even if not always temporally) distinct from the means by which theories are discovered, conjectured or constructed.

However, it has been demonstrated by case studies that discovery and justification are *not* disjoint enterprises. Galison (1983b) has shown that this distinction is not meaningful in modern experimental high-energy physics. We provide further examples of such blurring in the present book. The more closely one looks at the historical record of science, the more difficult it is to find *the* hallmark of science (valid for all science in all ages). As we shall see later, the writings of Fine (1984,

1986), Laudan (1984a), Nickles (1987), Nersessian (1984) and Shapere (1984) are particularly important in this regard. Their work has shown a trend toward the naturalization of the philosophy of science (i.e., to base work upon the actual historical record of science and to stress the use of the methods of science in studying the scientific enterprise itself). An important element introduced into the discussion has been the influence of factors (e.g., social ones) *external* to science proper upon the form and content of science, a point already made by Kuhn (1970), but more recently emphasized by Bloor (1976) and by Pickering (1984). These sociologists of knowledge have stressed important and previously undervalued (by philosophers of science) factors of scientific practice, but they go too far when they suggest that such factors account for *all* of science. Shapere (1986) has argued that the demarcation between internal and external factors is not static, but that science internalizes once-external factors as it develops and as it finds it useful to do so. A concurrent debate, which has paralleled and been interwoven with the evolution of views on methodology, is that of realism versus anti-realism in the philosophy of science. This is basically an argument over the degree of uniqueness or the tightness of constraint on theories and their worldviews as provided by empirical results. There is, of course, a whole spectrum of views on realism. Fine (1986) has recently argued that this is not really too fruitful a debate.

The purpose of the present monograph is to use an extensive and detailed case study of a research program in modern theoretical physics to examine how theories are constructed, selected and justified in actual scientific practice. The book is intended both for philosophers of science and for interested physicists. Since the text mixes physics and philosophy, each group will probably find parts of it difficult. Hence, there is an extensive glossary of terms and a list of the main players collected at the end of the book so as not to clutter up the narrative text too much (because 'half' of the intended audience will already be familiar with 'half' the material, and vice versa). My claim is that the origin of methodologically interesting ideas and questions, at least in modern physics, lies in the (highly) *technical* details of practice. For that reason, I present the details on key developments for those who can and wish to follow them. Extended instances of this are set off in smaller type (to allow the more casual reader to pass over them easily). Each chapter begins with an extensive discussion of the historical relevance, to the overall program, of the detailed developments to be presented. There are verbal, nontechnical summaries of such material in the text proper and at the end of the chapters. The entire case study is summarized early in Chapter 9. The first eight chapters are devoted to the case study itself, with philosophically relevant comments interspersed throughout. The concluding Chapter 10 assesses the import of the case study and of other episodes in modern physics for these basic philosophical issues, illustrates several themes from 'naturalized' philosophy of science and suggests a useful framework within which to view scientific practice. These last two chapters can be taken on their own, provided one is willing to accept my summaries of the details of the case study of the previous chapters. It is essential to realize that these conclusions are based mainly on *one* case study so that no claim to universality would be warranted. My final position is not a wholly skeptical one that claims we have not learned anything, but rather one that asks how much and what part of that is peculiar to *science*. If science is to have any universal characteristics (as some claim), then it is fair to examine a field (e.g., modern theoretical physics) that is commonly acknowledged to be scientific to see whether it conforms to this general characterization. It is true that the case study proper (Chapters 1 through 8) and the general discussion of methodology (Chapter 10) form two separate parts, but ones which are, I claim, connected in an important way. A detailed case study of a 'failed' research program (*S*-matrix theory) is perhaps of relatively little interest unless it is connected to some larger philosophical or methodological issues. The analysis of Chapter 10 is based on history – including not just the direction in which things finally went ('good' science), but also a consideration of how things *might* (consistently) have gone a very different way and why they did not.

Perhaps it is of some relevance to mention that when, as a practicing scientist, I became interested in the philosophy of science some ten years or so ago, I had no particular methodological ax to grind. Rather naively I attempted to apply some currently fashionable methodologies from the philosophy of science to an area of high-energy physics I was familiar with. Things didn't mesh too well and I have been brought to a rather skeptical position, as is evident from the tenor of this book. Being an autodidact in this business, I am certain the following pages contain many statements that philosophers of science will find outrageous. I hope that some of the material may be useful, nevertheless.

I wish to thank all of those scientists who contributed their recollections and comments to this case study. The list is too large to reproduce here. They are acknowledged in appropriate footnotes. Several colleagues in the philosophy of science have been supportive of my work over the last several years, most notably Professor Ernan McMullin of the University of Notre Dame. Professor J. W. N. Watkins of the London School of Economics has welcomed me as an Academic Visitor and Professor M. L. G. Redhead of Cambridge University extended his hospitality to me as a Visiting Scholar. The appointment as a Visiting Fellow at St. Edmund's College, Cambridge, furnished an atmosphere conducive to writing a complete draft of the manuscript. The National Science Foundation, through its Program in the History and Philosophy of Science, provided partial support for several years while the research for this work was being done (grants Nos. SES-8318884, SES-8606472 and SES-8705469).[1] Ernan McMullin, John Polkinghorne and Michael Redhead contributed useful comments and criticisms on a late draft of the manuscript. Over the years my colleague, Professor Gerald L. Jones, has provided trenchant and commonsense observations to keep me in touch with the real world of physics. These acknowledgments are not meant to imply that the people cited necessarily agree with my conclusions or that they are in any way responsible for the errors that remain. The comments and criticisms of an anonymous referee were especially helpful for the last revisions prior to publication. Neal Nash generously offered to redraw the figures for the text. Finally, Susan Varnak has done a remarkable job of coping with an typing several successive versions of a difficult manuscript.

University of Notre Dame   James T. Cushing

Acknowledgments

The following have given permission to include in this book previously copyrighted material. The section, chapter and figure references given below indicate where excerpted, often modified, versions of this material appear in the present publication.

American Journal of Physics: Cushing (1986b) (Section 10.2).
Centaurus: Cushing (1986a) (Chapter 2).
Foundations of Physics: Cushing (1987b) (Section 9.4); Cushing (1989e) (Section 10.2).
Physical Review Letters: Figure 6.5.
Studies in History and Philosophy of Science: Cushing (1985) (Section 7.3).
Synthese: Cushing (1982) (Sections 1.4 and 1.5); Cushing (1989f) (Section 10.3).
Academic Press: Figure 8.1.
Benjamin-Cummings Publishing Co.: Figure 6.4.
Philosophy of Science Association: Cushing (1983a) (Section 10.5); Cushing (1984) (Section 10.6); Cushing (1986c) (Section 9.4).
Plenum Press: Cushing (1989a) (Chapters 2 and 3).
Princeton University Press: Figures 4.2 (a) and 4.2 (b).
Springer–Verlag: Figures 6.6 (a) and 6.6 (b).
John Wiley & Sons: Figures 6.1 and 9.2.

1

Introduction and background

In recent years there has been a move to 'naturalize' the philosophy of science. This has meant basing work in the philosophy of science upon the actual historical record of real scientific practice and stressing (in varying degrees) the use of the methods of science in studying the scientific enterprise. This attention to actual scientific practice has been supported by traditional realists, an example being Ernan McMullin (1984) who early on (1976a) argued for a central role for the history of science in the philosophy of science; by philosophers of various anti-theoretical bents, such as Nancy Cartwright (1983) and Ian Hacking (1983); and by empiricists, like Bas van Fraassen (1980, 1985). In the early 1960s, it was attention to the historical record that led Thomas Kuhn (1970), in his *The Structure of Scientific Revolutions*, to stress the importance of social factors in the practice of real science. The spirit of the present work is that careful and detailed study must be made of the actual development of science *before* conclusions are drawn about the appropriateness of any particular methodology of science. Ours is mainly a story about theory, but not one uncoupled from its relation to experiment.

We claim that many of the philosophically interesting questions in science, especially in regard to possible changes in the methodology and goals of science, can be seen and appreciated only upon examination of the technical details of that practice. So, in this chapter we discuss some motivations for studying *current* scientific practice and then set the problem background out of which the S-matrix program arose. This sketch is somewhat ahistorical since we mention some field theory developments that occurred *after* 1943 (when Heisenberg introduced his S-matrix program). The formalism of classical particle mechanics and of wave phenomena have been, in this century, successively reinterpreted to yield (nonrelativistic) quantum mechanics, relativistic quantum field theory for electromagnetic phenomena and, finally, relativistic quantum field theory for strong (or 'nuclear') phenomena. The prototypical experimental arrangement (of either the real or *gedanken* type) for studying the fundamental interactions of nature has been a scattering process in which a target (such as a nucleus) is probed by a projectile (such as an electron or a proton).

However, the development of these quantum field theories has not always been a smooth one. In particular, in the 1930s and early 1940s, the quantum field theory program had run into considerable technical and experimental difficulties (Cassidy, 1981; Galison, 1983a). Mathematical inconsistencies, most notably divergences or infinities produced by calculations with the formalism, occurred for quantum electrodynamics (QED) at short distances (or, equivalently, at high energies) of the order of the size of the electron. Similar problems plagued Fermi's theory of β decay and Yukawa's meson theory of nuclear forces. Here we have examples of three of the four basic forces in nature: the electromagnetic, which is responsible for the atomic phenomena producing those features of the world we commonly encounter; the weak, which accounts for the spontaneous decay or conversion of a free neutron into a proton; and the strong, which predominates at very short distances for nuclear processes. The last is the gravitational force, which plays no essential role in our story here. During the same period of confusion in the arena of theory, experimental results from cosmic ray data seemed to contradict expectations based on quantum field theory (QFT). It appeared as though cosmic ray showers, or 'explosive' events, occurred (in contrast to the cascades built up from many essentially pairwise events, which could be readily accounted for by Dirac's hole theory). Heisenberg took the existence of these multiple processes to signal a breakdown of conventional quantum field theory and to require the introduction of a fundamental length into the theory. The field theory situation was further complicated by the confusion (in the 1930s) caused when the mesons observed in cosmic-ray interactions were at first identified with Yukawa's nuclear-interaction π meson (pion), before they were finally identified as μ mesons (muons), which are essentially 'heavy' electrons. These difficulties encountered by the quantum field theory program provided a significant part of the motivation for Heisenberg's proposing his *S*-matrix theory (SMT).

By the late 1940s, a mathematical technique (renormalization) had been formulated which allowed one to circumvent the divergences of QED and to make accurate predictions confirmed by experiment. This is the first cycle of the oscillation of theory between QFT and SMT. Others occurred when QFT was stymied by the strong interactions, from which it subsequently recovered with gauge field theories. This back and forth between formalisms, with their corresponding paradigms, is an important feature of the episodes we present. It will be especially relevant for our evaluation of methodology in science (in Chapter 10).

1.1 Internal history of recent science

This work is largely, but not exclusively, an *internal* history of an extended episode in modern high-energy physics. That is, the published physics literature is a major source for the technical developments we present. Nevertheless, interviews and correspondence are also employed. The primary interest in and motivation for doing the research necessary for this case study are philosophical. Some obvious questions, that arise about the value and wisdom of doing an internal history of a *current* (and hence not completed) episode in a *highly technical* (or specialized) subject area, must be addressed.

Schweber (1984, p. 41), in his history of the early developments of quantum field theory, has stated one of the problems of internal history as follows:

> [I]nternal history faces the problem common to all good history: how to avoid the pitfalls of Whiggish history, that is, the writing of history with the final, culminating event or set of events in focus, with all prior events selected and polarized so as to lead to that climax.

So, while the philosophy of science must be based on history (i.e., events as they actually occurred), it can be important not to focus exclusively on the form and content of 'successful' scientific theories alone. The arguments and contingent events that inform the course of development and selection of theories are essential. That is, how things might have gone a very different way at certain crucial junctures and why they did not may be as important as the reasons for the 'right' choices that science has made. The present case study focuses on a 'failed' program that has never been proven to be incorrect. It is evident that one cannot explain its rejection just in terms of falsification. Perhaps there is something to be learned from the history of such a dead-end theory. The relevance of such ('sociological') factors as the previous interests and expertise of the participants becomes apparent enough.

Historians may extol the virtues (in fact, necessity) of doing the history of an episode only long after the clamor of the day has settled. They can argue that once time and events have produced a stable picture of the past, one feels some confidence that one may be able to find 'the objective truth' in those long-dead events (or 'corpse') (Burckhardt, 1963, pp. 74–76).[1] There is an old tradition of this attitude in the history of science. Thus, in Whewell's *History of the Inductive Sciences* (1857, Vol. II, p. 434) we find: 'It is only at an interval of time after such events have taken place that their history and character can be fully understood, so as to suggest lessons in the Philosophy of Science.' Even if one accepts that thesis (and it *can* be debated), he can still feel that something (perhaps important, Burckhardt and Whewell to the contrary notwithstanding) has slipped away. The detailed dynamics of the events and the motivations of the protagonists have been lost (in large measure, at least) behind the veil of time. Now if one believes that all final scientific positions are reached ultimately through rational judgments *alone*, then there is probably even virtue in waiting until the flotsam has been swept away by time to leave a residue of objective truth. But it is not clear that science operates (*even* in the long run) quite as objectively as we might like to think. An examination of the record of actual scientific practice may shed some light on that question. The goal is not to clear up the rules and mechanisms that regulate the eternal ups and downs of fashions and fads in theoretical physics. It is not certain that there is *a* set of rules and mechanisms, but we can learn what some of them might plausibly be.

There are inherent dangers in studying fairly recent episodes in physics (or in anything else) (Brinkley, 1984). But, since the final 'verdict' is not yet in and since many of the participants are still alive, there are opportunities here that are not available in more traditional 'corpse dissections'. Most obviously, one can ask the major figures involved what their motivations were, how they saw events at the time, and what they recall about the interactions of other scientists. An obvious danger in gathering such recollections is that people sometimes feel (rightly at times, but often not) that their own contributions have been slighted. This 'interview' approach can be taken too far, as when sociologists of science monitor the day-to-day routine activities of scientists. An additional useful dimension can be added by examining a recent episode in science.

Many of the most interesting questions in the philosophy of science come from studying actual scientific practice, rather than from armchair *a priori* reasoning that philosophers sometimes engage in. I have chosen the dispersion-theory and S-matrix theory program of theoretical physics because I had some familiarity with the technical literature of that program in the 1960s and 1970s, because the major activity in that area was confined to a time period of several years and that activity was reasonably localized (around relatively few central theorists), and because several philosophically significant issues, such as the origin, development and selection of theories, can be illuminated with specific instances from a history of that program.

A difficulty in doing a case study of a major episode in modern theoretical physics is that one of the traditional sources of corroboration – an extensive personal correspondence among the major creators of the theories – is by and large no longer available. That is, historians of science are wary of taking at face value and relying solely upon the personal recollections of individuals. While such recollections are an invaluable source for leads about what actually went on behind the 'story' as reconstructed from the published physics literature, those recollections must be checked for support against other documents, usually the published literature and the private correspondence among key theorists. While the leading theorists of an earlier era (e.g., Einstein, Bohr, Schrödinger, Heisenberg, Pauli) did correspond frequently and extensively with one another (and much of that correspondence survives), markedly increased use of the telephone and relatively easily-available travel to many topical conferences have obviated the need for such correspondence among already busy individuals. The situation for the history of recent experimental physics is not so bad since laboratory notebooks and, more often, research proposals to funding agencies and the internal memoranda of large groups give details of what was going on in the major experiments (cf., Galison, 1987). However, research proposals for theoretical work provide a less reliable guide to what a theorist actually ends up doing.

This problem of a missing record of corrspondence among theorists is especially bad with those generations of theorists who have begun working since the end of the Second World War. Lacking a large body of such correspondence, the only resource available appears to be getting as many independent recollections as possible of key episodes in the development of the disperison-theory and S-matrix programs and then looking for the common overlap among these.

One can also question the value of studying a frontier area involving the creation of new physical theories, since this may be a singular exercise in science. Rigden (1987) has characterized such creative developments as follows.

When first-rate minds are engaged in the intellectual activity called physics, as was the case in February 1927 when Heisenberg was struggling with the 'pq–qp swindle', it is an activity with no equivalent in any other natural science. In fact, there is no equivalent in any intellectual arena except, possibly, first-rate theological thinking. These special times in physics do not come often, but when they do, physicists must often create new constructs for which neither previous experience nor previous thought patterns provide guidance. New words representing entirely new concepts must be created, words whose meaning cannot be rendered even by the most deliberate use of older words. The new meaning takes form slowly, but with a groping awkwardness. Soon the new ideas become the basis for empirical predictions and, in the process, a 'sense of understanding' emerges. However, in the end, the basic concepts of physics are aloof, they remain outside our ability to convey their meaning.

Rather than taking this to mean that such activities in theoretical physics are largely irrelevant to the philosophy of science, we can see in these episodes a unique opportunity to examine how foundational theories are created – perhaps at a time of singular flexibility and underdetermination of the outcome.

1.2   Philosophical issues and the Forman thesis

Since a primary interest of ours here is certain philosophical questions, references are not given to *every* technical development in S-matrix and dispersion theory. By examining in detail a major episode in contemporary physics, we hope to illuminate somewhat the processes by which theories are generated and selected by the scientific community. A question of central interest for us is the relative importance of internal versus external factors in the development of a scientific theory. Forman (1971) initially raised this issue with regard to the origin and acceptance of the concept of acausality in physics in Germany after the First World War when modern quantum mechanics was being formulated. We shall often use the expression 'Forman thesis' to refer more generally to the role of social and sociological influences in the development and acceptance of a scientific concept or theory. It does seem evident that, once we 'buy' into a set of starting assumptions, then the 'internal' logic of a formalism can largely take over (Raine and Heller, 1981). However, the origin of hypotheses central to a theory often lies in very specific and technical developments, having little, if anything, to do with overarching philosophical schemes. For that reason some fairly extended discussion of technical details is necessary. (The reader can get an overview of the philosophical issues and conclusions from Chapters 1 and 10 alone, aided perhaps by the introductions and brief summaries at the end of each intervening chapter.) Retrospectively, the central tenets of a theory may be put into or associated with a particular philosophical world view. Furthermore, the acceptance (or the effective infectivity) of a theory can be greatly influenced by the social environment and by generally accepted overarching principles. This case study does not support the radical Forman (1971) thesis that the social milieu plays a central role in the *creation* of scientific theories, but it is consonant with the more modest Forman-type thesis (Forman, 1979; Hendry, 1980) that social factors are relevant for the *acceptance* of a theory. There does remain an important distinction for science between internal factors (such as formalism, logic and experiment) and external ones (such as group interests and social influences).

Another set of issues to be discussed in the context of this episode in physics is the interplay between the discovery and the justification of a scientific theory (both initially and later in the program) and the symbiosis between theory and experiment in the development of a program. This is just *one* case study and its conclusions may or may not have any general applicability to the way other scientific theories have developed. It is by no means clear that there is *a* (i.e., one) scientific rationality that applies usefully to all science in all eras (in spite of some claims made, for example, by Popper, with his emphasis on falsification, and Lakatos, with his representation of the dynamics of science in terms of progressive and degenerating research programs).

1.3 The purview of this case study

Because this case study is intended mainly for historians and philosophers of science who have an interest in the modern scientific enterprise, I have attempted to give an essentially accurate representation on technical matters, but have usually avoided telling the *whole* truth (i.e., giving all the technical details). Rather, the central concepts and techniques are often illustrated with simple mathematical examples. Although I do not want to reconstruct past developments from the biased vantage of today's state of knowledge, I have nevertheless employed a unified notation in these mathematical examples in order to make the line of argument more accessible to a wider audience. Along the way I do point out important notational and conceptual differences between my illustrative examples and the original presentations found in the physics literature.

The last introductory comment concerns my use of the expression 'S-matrix program'. I do not mean to equate the dispersion-theory program and the S-matrix theory (SMT) program[2] nor do I wish to obfuscate the distinction between the 'bootstrap condition' as a uniqueness criterion and the much broader implications that term has in the program associated with Geoffrey Chew and his collaborators. This case study should make it clear (1) that at any given time the term 'bootstrap' has not had a unique, universally accepted meaning among theoretical physicists (if, indeed, it has any specific meaning at all) and (2) that within a given group or school of theorists the meaning of that term has evolved over the years. To respect this caveat, I shall use the designation 'autonomous S-matrix program' to distinguish the radical or fundamentally revisionary conjecture from the more general S-matrix and dispersion-theory program[3].

Much of the S-matrix program discussed in this study will appear as a largely American project and this may give the entire project too much of an American, even 'Chewian', flavor. It is true that major developments in several areas took place in Europe or have been made by Europeans. I attempt to point this out in the narration that follows. Nevertheless, it does remain that much of the major activity of the S-matrix program was centered around Geoffrey Chew and his collaborators in Berkeley.

1.4   Quantum field theory (QFT) background

A brief sketch of the development of quantum field theory (QFT) is necessary in order to place S-matrix theory (SMT) in some historical context for the reader. More complete discussions of the history of QFT can be found in Cushing (1982), Darrigol (1984, 1986) and Schweber (1984, 1990). Philosophical problems associated with QFT have been addressed by Stöckler (1984). In fact, the following is largely a summary from my (1982) paper on high-energy theoretical physics. Detailed references to the relevant physics literature can be found there, so that we shall not repeat those references here. This condensed outline is essentially ahistorical in that it presents only the central ideas of QFT without any pretense at maintaining a strict historical sequence. Let us begin by reproducing (Cushing, 1987a) a chronological outline of the sequence of developments (Schweber, 1987) we shall take as background for the following chapters in this case study. Some of these topics will be elaborated here, others in subsequent chapters.

1925–1927: formulation of nonrelativistic quantum mechanics
1927–1947: formulation of relativistic quantum field theory (QFT)
1947–1950: renormalization program for quantum electrodynamics (QED)
late 1950s–1970: a period of serious problems for perturbative QFT, with various alternative avenues pursued
    (*a*) axiomatic QFT (Wightman school)
    (*b*) local, asymptotic QFT (LSZ formalism)
    (*c*) dispersion relations and *S*-matrix theory (SMT)

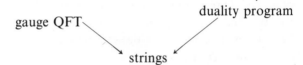

1970s–present: gauge QFT

         duality program

         strings

The transition from classical mechanics to (nonrelativistic) quantum mechanics in the period 1925–27 can be seen (at least now, retrospectively) as a reinterpretation of the equations and of the formalism of classical mechanics to represent phenomena in the atomic domain. In the Hamiltonian formulation of classical mechanics (for a single particle here), the time evolution of the canonical variables $q(t)$ ('position') and $p(t)$ ('momentum') is governed by Hamilton's equations of motion

$$\dot{q} = \frac{\partial H}{\partial p},\tag{1.1a}$$

$$\dot{p} = \frac{-\partial H}{\partial q}.\tag{1.1b}$$

Here $H(q, p)$ is the Hamiltonian and is (in our case) just the total mechanical energy of the system. For example, a particle of mass m moving in a conservative force field $F(q) = -\partial V/\partial q$ has the Hamiltonian

$$H(q, p) = \frac{p^2}{2m} + V(q)\tag{1.2}$$

in terms of the potential energy function $V(q)$. For a classical system, $q(t)$ and $p(t)$ are simply ordinary functions of the independent time variable t. They are solutions to the coupled set of differential equations (1.1) subject to the initial conditions $q_0 = q(t_0)$, $p_0 = p(t_0)$ at some (arbitrary but definite) initial time $t = t_0$. In this simple example, Eqs. (1.1) are nothing more or less than (equivalent to) Newton's second law of motion

$$m\ddot{q} \equiv ma = \frac{-\partial V}{\partial q} = F(q). \tag{1.3}$$

In the early part of the present century, it became evident that for atomic systems not all of the solutions (or 'orbits' for particle motion) are in fact allowed or realized in nature. For example, only certain orbits, or energy levels, for a bound electron in a hydrogen atom are permitted (as evidenced by the discrete spectrum of the light emitted or absorbed by a hydrogen atom). The program of the old quantum theory (say, 1913–1925) was to find a set of rules that would allow one to select from the (continuous) infinity of classically-allowed solutions (or 'orbits' or energy levels) those actually realized in nature. Bohr's classic 1913 paper gave one such rule in terms of the quantization of the orbital angular momentum l,

$$l = n\hbar, \quad n = 0, 1, 2, \ldots \tag{1.4}$$

Here $\hbar$ is $h/2\pi$, where h is Planck's constant. The old quantum 'theory' amounted in essence to a set of quantization rules, that were generalizations of Eq. (1.4). It consisted of a set of *ad hoc* guesses guided by Bohr's correspondence principle, which was initially a requirement that certain quantities derived in the (old) quantum theory should pass over into their classical counterparts in a suitable limit.

Heisenberg's 1925 paper laid the foundations of a systematic quantum mechanics by reinterpreting the classical q and p variables as quantities satisfying the commutator relation (in units with $\hbar = 1$)

$$qp - pq \equiv [q, p] = i. \tag{1.5}$$

(We make *no* claim that Heisenberg, Schrödinger or Dirac originally presented their ideas in the form we represent them here. This is 'Whiggish' history, which we avoid in our study proper.) Hamilton's equations (1.1) and the Hamiltonian (1.2) were to be retained, but the (operators) q and p were now required to satisfy the (commutator) condition of Eq. (1.5). That is, one must seek solutions to the eigenvalue problem

$$H(q, p)\Psi = E\Psi \tag{1.6}$$

for the allowed eigenvalues E. Here (to make a long story short) Ψ is the (Schrödinger) eigenfunction (or eigenvector). One typically finds a representation for q and p (e.g., $q \to q$, $p \to -i\partial/\partial q$) which satisfies Eq. (1.5) and then uses this in the $H(q,p)$ of Eq. (1.2) to re-express Eq. (1.6) as (in units with $2m = 1$)

$$\left[-\frac{\partial^2}{\partial q^2} + V(q) \right] \Psi(\theta) = E\Psi(q). \tag{1.7}$$

This particular representation of Eq. (1.7) is usually referred to as the Schrödinger equation (which Schrödinger in 1925 arrived at independently and by a route different from that indicated here). Dirac in 1925 produced an elegant general set of rules for passing directly from the classical Hamiltonian formulation of a problem to the corresponding quantum-mechanical equations by replacing the classical Poisson brackets $\{q,p\}$ with the commutator $[q,p]$ as

$$\{q,p\} \to \frac{1}{i}[q,p]. \tag{1.8}$$

Schrödinger also established the formal equivalence of his wave mechanics and of Heisenberg's matrix mechanics. We use the term 'quantum mechanics' to refer to either of these representations without distinction.

Quantum field theory – really, quantum electrodynamics (QED) – was born in 1927 when Dirac applied perturbation theory to an atomic system (such as a hydrogen atom) in a radiation field (such as the electromagnetic field of light). The stationary state problem (e.g., the energy levels in the hydrogen atom) had been solved by quantum mechanics, but the mechanics or details of light emission and absorption remained to be handled. The quantum-mechanical problem considered was

$$i\frac{\partial \Psi}{\partial t} = (H_0 + V)\Psi, \tag{1.9}$$

which is just the standard time-dependent Schrödinger equation for an atomic system (whose unperturbed Hamiltonian is H_0) acted upon by an external potential V (here the radiation field due to light). (Similar applications of Schrödinger theory are discussed in more detail in the Appendix which should be consulted if the reader feels at a loss for specifics.) The equations of motion (i.e., Maxwell's equations) for the free electromagnetic field can be transformed into an equivalent set having the form of a denumerably infinite set of uncoupled harmonic oscillators, which can easily be quantized in terms of a set of creation (a†) and annihilation (a) operators satisfying the commutation relation

$$[a, a^\dagger] = 1. \tag{1.10}$$

That is, if Ψ is a state having n quanta (say, photons), then $a^\dagger\Psi$ is a state having $(n+1)$ quanta and $a\Psi$ is a state having $(n-1)$ quanta. Straightforward perturbation theory (think, say, in terms of the strength or intensity of the radiation field as the expansion or 'smallness' parameter) can be applied to compute the transition probability from one atomic (or hydrogen atom) level to another (under the stimulus of the radiation field). Calculations to *lowest* order in this expansion (or perturbation) turned out to be finite and reasonable. However, when this formalism was applied to the problem of the dispersion (or scattering) of light by a collection of atoms, divergent terms were present. During the period from the late 1920s throughout the 1930s, theoretical physicists were occupied with these and other serious consistency problems of quantum field theory, as well as with attempts at formulating a satisfactory relativistic quantum field theory. For our present telescoped 'history' of QFT, we simply pass over this period and turn to the major developments that occurred immediately after the Second World War.

1.5   Renormalized quantum electrodynamics (QED)

Lamb and Retherford in 1947 determined experimentally that two energy levels of hydrogen (the $2s_{\frac{1}{2}}$ and the $2p_{\frac{1}{2}}$) which should have been degenerate in one-electron Dirac theory are in fact separated by a small but finite energy difference. For here and for reference later, it may be worth pointing out that the energy-level notation is nl_j, where n is the principal quantum number of the Bohr theory (that is, the n of Eq. (1.4)) and takes on values (in units of $\hbar$)$n = 1, 2, 3, \ldots$; l is the orbital angular momentum and takes on values $0 \leqslant l \leqslant n-1$ (with the old spectroscopic notation $l = 0 \leftrightarrow s$, $l = 1 \leftrightarrow p$, $l = 2 \leftrightarrow d, \ldots$); and j is the total angular momentum (l plus $s = \frac{1}{2}$, the spin of the electron). Thus, $2s_{\frac{1}{2}}$ stands for $n = 2, l = 0, j = \frac{1}{2}$. In Dirac's relativistic theory of the hydrogen atom, the energy levels E_{nj}, depend only upon n and j, but not explicitly upon l. For $l = 1$ (and $s = \frac{1}{2}$, of course), there are two possible values for j (where $\mathbf{j} = \mathbf{l} + \mathbf{s}$): $j = \frac{1}{2}$ and $j = \frac{3}{2}$. Hence, the $2s_{\frac{1}{2}} (n = 2, j = \frac{1}{2})$ and $2p_{\frac{1}{2}} (n = 2, j = \frac{1}{2})$ energy levels should be the same (or 'degenerate'). In Figure 1.1 we show the Dirac-theory predictions for the energy level E_{nj} with solid lines and the actually observed $2p_{\frac{1}{2}}$ level with a dotted line. The experimentally observed split between the $2s_{\frac{1}{2}}$ level and the (dotted line)

$2p_{\frac{1}{2}}$ level is the Lamb shift. This difference is observed as a splitting or displacement of the spectral lines emitted when electrons make transitions from one level to a lower one.

In a remarkable calculation that same year, Bethe made a *nonrelativistic* perturbation calculation of the self-energy of an electron in a bound state of a hydrogen atom. The most singular term was a linearly divergent integral that Bethe realized was present for a free electron. (By linearly divergent we simply mean that the calculation yielded an expression involving an integral of the form $\int^{\infty} dk$, which diverges linearly (i.e., as a first power of k) as $\lim_{k \to \infty} k \to \infty$.) Arguing that this should be included in the physically observable mass of the electron, he discarded it as a not-separately observable effect. There still remained a logarithmic divergence (i.e., of the form $\int^{\infty} dk/k$) of the type already known from Dirac's hole theory which he also discarded in the hope that a relativistic calculation would produce the same logarithmic term plus, possibly, an additional small finite correction. Bethe took the finite remainder of his calculation to be *the* Lamb shift. His calculation produced a result

$$\Delta W/h = 1040 \text{ megacycles } s^{-1} \tag{1.11}$$

to be compared with the early experimental value of 1000 Mc s^{-1}. The extreme accuracy of these measurements becomes clear when we appreciate that $\Delta W_{\text{Lamb}}/W_{\text{energy level}} \approx 10^{-7}$.

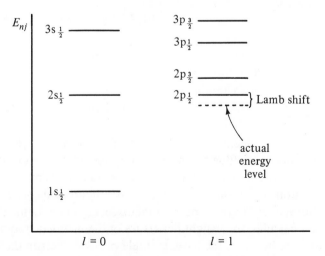

Figure 1.1 Hydrogen atom energy levels in one-electron Dirac theory and the Lamb shift (not drawn to scale).

Kramers pointed out at the 1948 Solvay Conference that observable effects calculated in quantum electrodynamics depend upon e and m only as measured experimentally (that is, upon these *structure-independent* properties of the electron). At this time people were able to do some higher-order corrections to electron scattering cross sections by subtracting out (that is, really, just throwing away) these divergences as they occurred in calculations. However, the answers obtained, while finite, were not unique and the entire procedure was a delicate and ambiguous one. Kroll and Lamb in 1949 made a relativistic quantum-electrodynamics calculation of the level splitting for $2s_{\frac{1}{2}} - 2p_{\frac{1}{2}}$ and obtained 1051 Mc s^{-1}. The results were still somewhat ambiguous since subtraction of infinite terms to yield finite ones was not unique because there was no subtraction procedure independent of a specified Lorentz frame. Also, these calculations were carried out to order e^2 in these so-called radiative corrections and were not yet known to remain finite in higher order. 'Radiative correction' is the term used to refer generically to any virtual process in which a photon, say, is radiated (perhaps from an electron) and then reabsorbed (by the same emitter) before it can be observed directly. Virtual processes are essentially those that are allowed by the QED formalism but that cannot be seen directly. We shall give illustrations of several of these below.

Since the Lamb shift remains one of the showpieces of QED, let us give some indication of recent experimental and calculational results.

$$\textit{experimental:}\ \ \Delta E \equiv \Delta E(2s_{\frac{1}{2}}) - \Delta E(2p_{\frac{1}{2}})$$
$$= 1057.77 \pm 0.01\ \text{Mc s}^{-1}. \quad (1.12)$$

$$\begin{array}{lr}
\textit{theoretical:}\ \text{self-energy} & 1011.45 \\
\text{vacuum polarization} & -27.13 \\
\text{vertex modification} & \underline{67.82} \\
& 1052.14
\end{array} \quad (1.13)$$

Bethe in his initial calculations had estimated only the electron self-energy modification (due to the emission and reabsorption of a virtual photon) which, it turns out, gives the overwhelmingly major part of the contribution. The result in Eq. (1.13) is the theoretical correction to order e^2. Subsequent to the experimental discovery of the Lamb shift and to Bethe's calculation, a slight departure of the magnetic moment from its value, predicted on the basis of single-particle electron theory, was detected experimentally by Foley and Kusch in 1948. That is, according to Dirac (single-electron) theory, the electron should have an intrinsic magnetic moment of

$$\mu = \frac{eh}{2mc},\tag{1.14}$$

whereas the experimental value is

$$\mu_{exp} = \mu + \delta\mu \tag{1.15}$$

with

$$\delta\mu = (0.001\,165 \pm 0.000\,011)\mu.\tag{1.16}$$

The QED prediction is to order e^2

$$\delta\mu = \frac{\alpha\mu}{2\pi} = 0.0\,011\,614\mu \tag{1.17}$$

as shown by Schwinger prior to the experimental measurement.

In his report to the 1948 Solvay Conference, Oppenheimer gave a masterful summary of the state of QED at the time. The first problem was to find a manifestly Lorentz-invariant and gauge-invariant method (or prescription) for subtracting (or neglecting) infinite quantities that arise in corrections made to finite order in $\alpha = e^2/\hbar c (=1/137)$, the fine-structure constant (*beyond* the already-finite terms that are first-order in α). The basis for this was provided by Schwinger and by Tomonaga and applied by Schwinger to calculating, to various orders in α, corrections to the magnetic moment of the electron and to the scattering of electrons by a Coulomb field. Several of these results were simply 'announced' by Schwinger in a 1947 work with no calculational details given. In these formulations a perturbation series expansion in α was essential and the renormalization was proved to remove the infinities only in lowest order. In contrast to the highly formal theory of Tomonaga and of Schwinger, which stressed the wave aspect of the problem, Feynman, whose early papers involved a modification of divergent integrals that ultimately disappeared from the final result, developed diagrammatic techniques that were based on concepts emphasizing the particle-like aspect of the interactions of electrons and photons. Subsequently, Dyson in 1949 showed the equivalence of the Tomonaga–Schwinger and Feynman formalisms. Feynman's diagrammatic expansion has the great virtue that it makes QED relatively simple and straightforward for calculations. Dyson also established the validity of the renormalization program to all orders in α. That is, he proved that, after mass and charge renormalization, every term in the perturbation expansion is *finite*, although the convergence of the series itself has never been established.

The renormalization program in QED consists basically of the following. In doing a perturbation correction to the scattering amplitude to first order in $\alpha = e^2/\hbar c (= 1/137)$, one obtains two types of divergent integrals, one of which can be associated with an electromagnetic mass δm which modifies the 'bare' mass m_0. One simply takes the (finite) physically observed mass m to be

$$m = m_0 + \delta m. \tag{1.18}$$

Only m appears in *calculated quantities* which correspond to physically observable processes, but never m_0 or δm *separately*. Similarly, to first order *corrections* in α another divergent term occurs which again appears only in a suitable linear combination with the 'bare' charge e_0 of the electron. This can be interpreted as a modification of the charge due to (vacuum) polarization effects and the combination is taken to be the physically observed charge e of the electron. Here, too, neither e_0 nor the polarization term appears *separately*, but only the combination making e. Therefore, up through and including all terms of order α and α^2 in the scattering amplitude, charge and mass renormalization may be employed to make the calculated theory finite. It is important to realize that this procedure can be carried out in a completely covariant, and therefore unambiguous, fashion and that these results agree extremely well with experiment (see, for example, Eqs. $(1.12)-(1.13)$ and $(1.16)-(1.17)$ above). The crucial question now becomes what happens when higher-order terms in α are calculated in the perturbation expansion. Divergent terms again appear, but they can *all* be written in terms of those of mass and charge renormalization, and *no* others. In fact, to *every* order α^n, the same procedure of mass and charge renormalization will remove all infinite quantities, leaving finite corrections. This is a remarkable circumstance and is necessary for QED to be useful as a calculational tool. If new types of divergences were to occur in each successive order of perturbation, then, even if we could associate each of these infinities with a renormalization or redefinition of various physically observable quantities, the theory would have *no* predictive power. Any field theory for which a *finite* number of redefinitions is sufficient to remove all the divergences to all orders of perturbation theory is termed *renormalizable*.

The renormalized QED program can be summarized as follows. The time evolution of the state vector $\Phi(t)$ is governed by the interaction Hamiltonian H_I as (in units with $\hbar = 1$)

$$i \frac{\partial \Phi(t)}{\partial t} = H_I \Phi(t) \tag{1.19}$$

which has the formal solution

$$\Phi(t) = \Phi(-\infty) - i \int_{-\infty}^{t} dt_1 H_I(t_1)\Phi(t_1) \tag{1.20}$$

satisfying the initial condition

$$\Phi(t) \xrightarrow[t \to -\infty]{} \Phi(-\infty). \tag{1.21}$$

Here $\Phi(-\infty)$ is the initial state in which the system was prepared. For a scattering process we are interested in the state vector in the remote future $\Phi(+\infty)$ and this is connected to the initial state vector $\Phi(-\infty)$ *via* the S operator (or S matrix) as

$$\Phi(+\infty) = S\Phi(-\infty). \tag{1.22}$$

Once the S matrix (or scattering amplitude) is known, the scattering cross section (or probability) for a given reaction can immediately be calculated (essentially as the square of the modulus of S). If we recursively iterate Eq. (1.20), we obtain the formal expression for S in terms of the Hamiltonian density $H_I(x)$ as

$$S = \sum_{n=0}^{\infty} \frac{(-i)^n}{n!} \int d^4x_1 \ldots \int d^4x_n P\{H_I(x_1) \ldots H_I(x_n)\}. \tag{1.23}$$

Even though we have not specified the precise meaning of all the operations on the right-hand side of Eq. (1.23), we have stated it to indicate both that, once the interaction has been specified in terms of $H_I(x)$, then S can be computed and that the (infinite) series is an expansion in ascending powers of $H_I(x)$ (the perturbation expansion so often referred to in this section). In QED, $H_I(x)$ contains an overall multiplicative constant e, the electron charge.

It is basically an historical accident that the first quantum field theory (QED) compared with experiment turned out to be renormalizable. Salam, in 1952, proved the renormalizability of certain other field theories (in which more than just two renormalization constants appeared), but many (in some loose sense, 'most') quantum field theories are nonrenormalizable. In fact, Fermi's 1933 β-decay theory (the paradigm process of which is the spontaneous decay of a free neutron into a proton with the emission of an electron (or 'β ray')) provides an example of a nonrenormalizable theory. A concise statement of any renormalization program has been given by Matthews and Salam as consisting of three steps:

1. the number of types of infinities must be shown to be finite;
2. a subtraction procedure must be found to remove the infinities;

3. a theoretical justification for this procedure must be found (i.e., it must be shown to hold to all orders in perturbation theory).

Even if a theory is renormalizable, though, this does not necessarily make it useful for calculations since the renormalization program and the calculations themselves can be carried out explicitly only within the framework of perturbation theory. For QED this is fine since the expansion parameter (or coupling constant) $\alpha = 1/137$ is small and the first few terms in the series might reasonably be expected to give a good approximation (even if the series itself should only be an asymptotic one). However, in field theories relevant to nuclear and elementary-particle physics, the coupling constant (which measures the strength or rate of a process) is of the order $g \approx 15$, so that a perturbation expansion is useless to obtain numerical results. These difficulties for strong-interaction theory were amply apparent by the early 1950s. The numerical results were in terrible disagreement with experiment. This was one of the motivations for theorists' turning to other programs, such as dispersion theory and S-matrix theory in the late 1950s and throughout the 1960s. These alternative avenues of research are the subject of the following chapters in this case study.

However, it remained clear that even QED was at base a mathematically inconsistent theory. There had been some hope that the infinities that arose were a result of the approximate nature of the perturbation calculations rather than of the theory itself. As a rather naive analogy, consider the following power series expansion for $x > 0$

$$e^{-1/x} = 1 - \frac{1}{x} + \frac{1}{2!x^2} + \cdots . \tag{1.24}$$

In the limit $x \to 0^+$, every term in the series (except the first) diverges, but the exact result is zero. However, that such is not the case for the renormalization constants of QED was indicated in 1953 by Källen. Independent of any perturbation theory, he argued that in an exact formulation of QED not all of the renormalization constants of QED can be finite. Dyson (1952) and Edwards (1953) have questioned the convergence of the perturbative expansion. Landau (1955) and his coworkers (Landau, Abrikosov and Khalatnikov, 1954a, 1954b, 1954c, 1954d; Landau, 1965) stressed the importance (at high energies) of singularities that are not handled by renormalization. Schwinger has pointed out the fundamental inconsistency of assigning operators for definite-mass fields at localized space-time points (e.g., $\psi(x, t)$) since a precise measurement of these properties would result in arbitrarily large field fluctuations via the uncertainty principle (i.e., the interaction

energy cannot be limited while using an exact space-time description).
Also, the actual magnitude of the electron's charge is explainable only
once we understand the strong interaction because charge renormalization must, in principle, be linked to *every* charged field with which the
electron can interact. Dirac has discussed similar difficulties with QED.

1.6   Feynman diagrams

For reference in later chapters, we now introduce some elementary
concepts associated with Feynman diagrams. Before that, though, we
define some terminology appropriate to describing scattering phenomena. Some of the best-known and most precise predictions of QED are
quantities like the Lamb shift and the anomalous magnetic moment of
the electron and these would seem to have little to do with scattering of
one particle by another. Nevertheless, most of our basic information
about fundamental interactions is gained through scattering processes,
as we shall see throughout this book. Figure 1.2 represents a typical
scattering event in which an incoming projectile (with initial velocity v_0)
is scattered (or deviated) by a target, which will itself recoil (through an
angle ϕ with a final velocity V) during the interaction. How the
projectile is scattered (as indicated by the angle θ and the final velocity v'
in the figure) is determined by the interaction or force between the target
and the projectile. And, conversely, it is through a study of such
scattering reactions that we learn something about the details of the
forces acting between the particles. In Chapter 2 we discuss scattering
formalism in some detail. Feynman diagrams will prove useful for
treating scattering reactions.

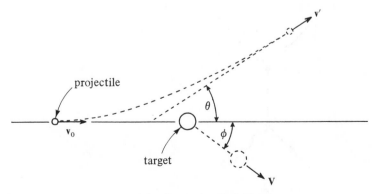

Figure 1.2   The scattering of a projectile by a target.

Basically, Feynman constructed a diagrammatic representation of the perturbation expansion of Eq. (1.23) for the S matrix (or for the scattering amplitude) and a set of rules that allow one to assign a definite mathematical expression to each of these diagrams. That is, there is a one-to-one correspondence between the terms in the perturbation expansion of Eq. (1.23) and the Feynman diagrams (or 'pictures') for any given scattering process. A few simple examples will illustrate the points we shall need for future reference. Figure 1.3 shows the lowest-order Feynman diagram for (free) electron–photon scattering (also known as Compton scattering). This can be pictured as (taking time to flow upward in the figure) the absorption of a photon (γ) by the initial electron (e_i^-), the propagation of a virtual electron (e^-) in the intermediate state and then the subsequent emission of a photon (γ) leaving the final electron (e_f^-). The complete Compton scattering amplitude is obtained by summing *all* possible Feynman diagrams (of which there are infinitely many), allowing for arbitrarily many virtual electrons and photons in the intermediate states. At each $e^-e^-\gamma$ vertex one picks up a factor e in the perturbation expansion. Thus, the Feynman diagram of Figure 1.3 would make a contribution of order e^2 (or α) to the scattering amplitude. We shall not state in any detail the specific Feynman rules for recovering the mathematical form of the contribution to the perturbation expansion from the corresponding diagram. A great virtue of the Feynman diagram technique is that it allows one to 'picture' a scattering process in terms of the exchange of virtual particles.

Figure 1.4 shows two lowest-order Feynman diagrams for electron–electron (or Møller) scattering. If we denote by $p_j, j = 1, 2, 3, 4$, the four momenta of the incident and scattered electrons, then (neglecting

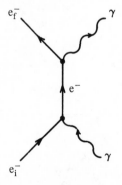

Figure 1.3   Electron–photon (Compton) scattering.

electron spin and photon polarization complications here) the contribution to the scattering amplitude corresponding to Figure 1.4(*a*) is proportional to

$$\frac{e^2}{(p_1 - p_3)^2} \tag{1.25}$$

and that of Figure 1.4(*b*) to

$$\frac{e^2}{(p_1 - p_4)^2}. \tag{1.26}$$

Figure 1.5 is an example of a higher-order contribution to $e^- - e^-$ scattering (of order e^4 or α^2). Finally, Figure 1.6 is a renormalization contribution (a 'vertex' radiative correction) that contributes to charge renormalization. The integral corresponding to this diagram diverges, but the renormalization procedure allows one to assign a *finite* contribution to the scattering amplitude.

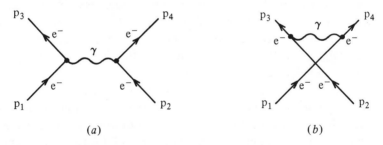

Figure 1.4   Electron–electron (Møller) scattering.

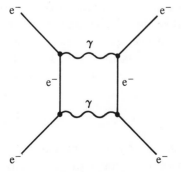

Figure 1.5   A higher-order diagram for electron–electron scattering.

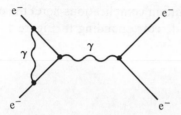

Figure 1.6   A renormalization contribution.

1.7   Gauge field theories

We have already indicated that by the late 1950s even renormalized QFT was unable to cope with strong-interaction phenomena, thus producing interest in the *S*-matrix program. However, QFT did make a strong comeback in the 1970s, in the form of gauge field theories. To complete our brief historical sketch, we now mention some of the key developments in that program. The basic idea used in modern gauge field theories was put forward in 1954 by Yang and Mills. Although their argument concerned a symmetry between protons (p) and neutrons (n) in the strong interactions, the line of reasoning is most directly explained for a simpler case. Just as in classical particle mechanics any invariance or symmetry of the Lagrangian implies the existence of a conserved quantity (or generalized momentum), so in a field theory. (Think, for example, of rotational symmetry implying conservation of angular momentum or of translational symmetry yielding conservation of linear momentum.) Noether's theorem guarantees that, corresponding to any transformation or invariance group that leaves the Lagrangian density (representing the interactions) for the theory invariant, there exists a conserved quantity or constant of the motion (that is, a quantity whose time derivative vanishes by virtue of the equations of motion). As an example, the Lagrangian, which yields the nonrelativistic Schrödinger equation *via* the Euler–Lagrange variational equations, is left invariant under the phase transformation of the wave function $\psi(x)$

$$\psi(x) \rightarrow \psi'(x) = e^{i\alpha}\psi(x) \tag{1.27}$$

where α is a *constant*. (This α should not be confused with the fine structure constant.) This is known as a global gauge transformation since, once α has been chosen, the phase is fixed at every space-time point (i.e., globally). Essentially, what Yang and Mills argued was that such a 'nonlocal' fixing of the phase (in their case, the fixing of the proton p relative to the neutron n) should be avoided in a local field theory by requiring *local* gauge invariance as

$$\psi(x)\rightarrow\psi'(x)=e^{i\alpha(x)}\psi(x) \tag{1.28}$$

where $\alpha(x)$ is now an arbitrary function of the space-time variable x. However, the Lagrangian for the coupled electromagnetic field $(A_\mu(x))$ and the electron wave function $(\psi(x))$ is no longer invariant under Eq. (1.28) unless the electromagnetic four-potential $A_\mu(x)$ is simultaneously subjected to the transformation

$$A_\mu(x)\rightarrow A'_\mu(x)=A_\mu(x)+\frac{1}{e}\frac{\partial\alpha(x)}{\partial x_\mu}. \tag{1.29}$$

In the context of electromagnetic theory, this is just Weyl's gauge invariance.

Yang and Mills were able to show how to implement a general symmetry group by means of such local gauge fields. The only problem was that each gauge field had associated with it a massless particle, just as $A_\mu(x)$ has the photon. This simply produced too many massless particles that were not found experimentally. For well over a decade, the Yang–Mills theory was considered an interesting but essentially useless curiosity. It is worth noting that their motivation for introducing such local gauge transformations was an abstract or purely theoretical one, not really required by any empirical evidence.

The other problem that provided impetus for the current gauge theories of weak interactions was the fact that the original 1933 Fermi theory of weak interactions is nonrenormalizable and its lowest-order term eventually violates unitarity (or conservation of probability) at high enough energy, a fact realized long ago by Heisenberg in 1936. It was conjectured that Fermi's form of the interaction in which the neutron n decays directly into a proton p and electron e and a neutrino ν as

$$n\rightarrow p+e^-+\nu \tag{1.30}$$

might be only an approximation to the actual two-step process

$$n\rightarrow p+W$$
$$\mathrel{\raise2pt\hbox{$\llcorner$}}\!\!\rightarrow e^-+\nu \tag{1.31}$$

where W is the so-called (massive) intermediate vector meson. Salam in 1960 had shown that neutral (i.e., uncharged) vector meson theory was renormalizable. Unfortunately, the (charged) W meson theories are known to be nonrenormalizable unless certain very specific cancellations of divergent terms happen to take place. The problem was to find a renormalizable charged vector meson theory that could describe the weak interactions.

The resolution began when Goldstone in 1961 showed that, even though a Lagrangian (or, equivalently, the equation of motion) for a field theory might possess a certain symmetry, this symmetry could be spontaneously broken so that the solution had a lower degree of symmetry than the Lagrangian. (Spontaneous magnetization of a solid, which does pick out a preferred direction in space, is an example of a solution, or physical situation, having a lower degree of symmetry than the fundamental equations describing the system. The physical solution has only axial symmetry, whereas the equations themselves have complete spherical symmetry (i.e., complete rotational invariance).) Goldstone proved that this can occur whenever the Lagrangian is invariant under a continuous symmetry group but *only* if zero-mass bosons (that is, integer-spin particles) are present. This spontaneous symmetry breaking seemed to offer a plausible mechanism for explaining the broken symmetries (such as SU(3) of Gell-Mann and of Ne'eman) that exist in strong-interaction dynamics, except for the fact that massless particles would again be present, and none were observed experimentally. The proof of this theorem was generalized by Goldstone, Salam, and Weinberg in 1962. In a series of papers Higgs not only established that the proof of the Goldstone theorem breaks down in theories which couple a conserved current to *gauge* fields but also exhibited a particular model field theory in which the Goldstone boson acquires mass and becomes part of a vector meson field (which had initially been massless) so that no massless particles exist in the (broken) solutions to the model.

In the late 1960s Weinberg, and independently Salam, combined the Yang–Mills gauge fields with the Higgs mechanism to produce a unified theory of the weak and electromagnetic phenomena. An analogy between many properties of weak and electromagnetic interactions was used as motivation. In this Weinberg–Salam model the Yang–Mills vector boson field responsible for the weak interactions is put on an equal footing with the photon field (i.e., the electromagnetic four-potential) and both are initially massless, as are the electron and neutrino fields. A Higgs field breaks this symmetry, giving the electron its mass as well as the intermediate vector boson its mass. The massless Yang–Mills vector bosons and the Goldstone bosons conspire so that the physically observed 'broken' theory has a massive charged spin-1 field W_μ (the charged intermediate vector boson), a neutral massive spin-1 field Z_μ (the neutral intermediate vector boson), and the massless spin-1 photon field A_μ, in addition to a massive electron and a massless neutrino. Initially it was a hope that this theory would be renormalizable. In 1971 't Hooft proved that such is indeed the case. It is precisely the gauge vector–meson theories that are the renormalizable ones. The Higgs meson plays a peculiar role in the Salam–Weinberg model since it is essential to provide the spontaneous symmetry-breaking mechanism and yet, since these Higgs bosons have not to date been observed experimentally, its mass must be made extremely high so that it has essentially no effect on many of the predictions of the theory.

Renormalization assumes a distinctly different status in gauge theories from what it had in previous QFT. In QED, for instance, one *proved* renormalizability. In gauge theories, as is particularly clear in Salam's writing, renormalizability is used as a *criterion* for constructing an acceptable theory. It provides a guide to writing down a suitable Lagrangian just as Lorentz invariance and conservation laws (*via* Noether's theorem) restrict the possible forms of the Lagrangian. This constraint severely limits the choice of Lagrangians.

This fairly complete and well-confirmed Salam–Weinberg model for the unification of the weak and electromagnetic interactions has prompted an attempt to unify these with the strong interactions in a grand unified field theory (or GUT in the trade). One first attempts to build a gauge theory of strong interactions. The basic ingredients in such a theory are quarks (that is, fractionally charged, spin $\frac{1}{2}$ elementary particles), implementation of SU(3) symmetry *via* Yang–Mills gauge fields, the spontaneous breaking of this symmetry by the Higgs mechanism, and the requirement of renormalizability. At first sight one might expect such gauge theories of strong interactions to be of little value for making numerical calculations since ordinary QFT perturbation series diverge for the large coupling constants of the strong interaction. However, a remarkable property of such gauge theories (often referred to as quantum chromodynamics, QCD), known as asymptotic freedom, was proved by Politzer. This means that for certain high-energy scattering processes the lowest-order perturbation calculations provide the major contribution to the cross sections (which are a measure of scattering reactions). The structure, or identifying characteristics, of scattering processes governed by QCD have been calculated using asymptotic freedom and have been confirmed by experiment. One can further attempt to unify these two gauge theories

(Salam–Weinberg and QCD) into one overarching grand unified theory of weak, electro-magnetic and strong interactions (GUT). Discussion of that work is not necessary for our present purposes. There has recently been a merging of ideas from the duality program (from S-matrix theory) and from gauge QFT in the superstring theories of current interest. String theories and duality are part of the material of Chapter 8.

1.8   Summary

To give the general reader an overall picture of the problem background out of which the S-matrix program emerged, we have presented a sketch of the highlights in the developments of quantum field theory (QFT) in this century. The central theme in that story is one of successive reinterpretations of the formalism (or of the equations) of classical (Hamiltonian and Lagrangian) mechanics: first (Heisenberg in 1925) by promoting the canonical variables q, p of classical mechanics to the noncommuting operators, $[q, p] = i$, of nonrelativistic quantum mechanics; then (Dirac in 1927) by considering the (Schrödinger) wave function Ψ itself to be a quantized field operator. The long struggle (1927–47) with mathematical inconsistencies in QFT culminated (1947–50) in the highly successful renormalization program of quantum electrodynamics (QED). As we shall see, Heisenberg (in 1943) proposed his initial S-matrix theory (SMT) as a possible means of circumventing the (pre-renormalization) divergences of QFT. Subsequently, renormalization procedures proved adequate for the electromagnetic interactions. The Feynman diagram technique (of 1949) provides a useful pictorial representation of the physical scattering processes corresponding to the mathematical expansion of the perturbation series of quantum field theory. When the perturbation-series expansion method of calculation proved inadequate for the strong interactions (mid to late 1950s), theorists turned to dispersion relations and again to the S-matrix program. After some initial successes, SMT encountered calculational difficulties at a time (1970s) when QFT made a dramatic comeback in the guise of gauge field theories. Today's (1980s) superstring theories can be seen as a merging of ideas from the duality program of SMT and from gauge field theories.

In this chapter we have defined the scope and context of an historical case study of an episode in modern theoretical physics. The sources of this largely internal history will be the published scientific literature and correspondence from the participants themselves. Cognizance is also taken of the larger social context, such as the structure of the scientific community, within which the theoretical activity took place. The relevance of historical developments to pertinent philosophical issues will be pointed out in the narrative itself. Results of this study will then be used to address questions about the construction, selection and justification of theories in science. The central subject of this book is the S-matrix program, from its beginnings in the late 1930s and early 1940s with Wheeler and Heisenberg, as an interim program from 1943 to 1955, as dispersion theory and mass-shell quantum field theory from the mid 1950s to the 1960s, then as the bootstrap and duality models of the 1970s, leading finally to the current superstring theories. We shall analyze the role of the S-matrix program played in generating these theories of current interest.

2

Origin of the S matrix: Heisenberg's program as a background to dispersion theory[1]

A common perception of Heisenberg's S-matrix program of the 1940s is that it encountered difficulties quite early on and then quickly died out. One can easily get the impression that the original Heisenberg program was irrelevant for the theoretical developments that provided the background out of which the dispersion-theory and later S-matrix theory program emerged. In this chapter we wish to show that Heisenberg's original program posed a set of questions the criticism of and response to which led to the dispersion-theory program of Goldenberger and Gell-Mann. It is not our purpose to review all of the elementary particle physics of the 1950s, or even all of what today, in retrospect, is judged to have been the 'best' or most important physics of that period.

In correspondence or in direct conversations, Geoffrey Chew[2], Marvin Goldberger[3], and Murray Gell-Mann[4] all recall that Heisenberg's old S-matrix program had essentially no direct influence on their own work which led to the dispersion-theory and S-matrix theory programs of the late 1950s and of the 1960s. All had known of Heisenberg's general ideas from some lectures given at the University of Chicago by Gregor Wentzel or possibly from reading Heisenberg's papers. It was only later that they became aware of any relevance of their work to Heisenberg's S-matrix program.

We begin by reviewing Heisenberg's program (Cushing, 1982, 1986a; Grythe, 1982; Oehme, 1989; Rechenberg, 1989) and the considerable theoretical activity related to it during the period from the mid-1940s to the early 1950s. A summary of the theoretical background available in the mid to late 1930s is given in the Appendix. There the reader can find an elementary discussion of the main results we draw upon, as well as several simple examples that we use later for illustration of subsequent developments. In the main body of this work there are many references to the equations of the Appendix. Readers with little background in theoretical physics, but with a general knowledge of undergraduate physics, may find it helpful to begin with the Appendix. The ground rules, here and in the rest of this work, for the technical level of the presentation will be roughly the following. Fairly detailed technical arguments will be confined to subjects requiring as background undergraduate physics or what can be gleaned from the Appendix. The discussion of more advanced topics will be less quantitative, although some nodding or 'popular' acquaintance with the 'pictures' (or diagrams) of Feynman will be assumed. The discussion in the latter part of Chapter 1 will suffice for this. Furthermore, several fairly technical developments in the text are set off in smaller type. The essentials of the argument can be followed without studying those sections in detail.

Aside from the difficulties presented for quantum field theory by the confusing status of cosmic ray experiments (in the late 1930s), the early developments, pro and con, in the S-matrix program were theory motivated. That is, the divergences present in the QFT formalism and Heisenberg's belief in the need for a fundamental length in such a formalism (coupled with his or anyone else's inability to construct such a theory) led to the formulation of the S-matrix program. That program depended in an essential fashion on one's ability to abstract hoped-for general results (as principles of the program) from specific models and incomplete theories. The importance of the early (1940s) S-matrix program for subsequent developments in theoretical elementary-particle physics was a series of questions or problems that it gave rise to, such as how causality (essentially as a first-signal principle) was to be incorporated into a scattering formalism, what type of restrictions causality could place on the mathematical form of the scattering amplitudes (which in turn determine the predicted scattering cross sections of experiment) and specifically how the results of scattering experiments would determine the form of the interactions between the scattered particles (at least in model situations governed by the nonrelativistic Schrödinger equation).

An historically important and extremely fruitful interplay between the general theoretical approach to scattering phenomena and experimental practice in the 1930s and 1940's was the study of resonance scattering in nuclear physics. Just as a mechanical system, such as a bridge or even a child's swing, will exhibit exceptionally large responses, vibrations or oscillations when energy is fed into the system at the proper frequency or rate (as by marching over a bridge or by rhythmically pushing the swing, in our two examples), so atomic and nuclear systems also react particularly strongly to certain stimuli. Hence, the blue color of the sky is due to the preferential (or strong) scattering of light in the blue part of the visible spectrum by atomic energy levels of air molecules. That is, the difference in energy between a pair of bound-state energy levels of an electron in a molecule corresponds to a frequency in the blue part of the spectrum. This produces an especially strong interaction (here, absorption and subsequent reemission of 'blue' light) between the molecule and the light of a certain frequency. In fact, a plot of the scattering cross section (or 'rate' of scattering) versus the frequency of the scattered light would show a peak or 'bump' in the cross section at such a resonance frequency. These resonance phenomena had long been known and understood (both classically and quantum-mechanically) for the interactions between light and atoms. When similar bumps or resonances were observed in the scattering of, say, neutrons from nuclei, it was natural to exploit the analogy with known resonance phenomena. At first the *form* of the mathematical equation used for the atomic resonance formula was essentially just carried over to provide a fit to the nuclear resonance scattering data. This became known as the Breit–Wigner formula. Subsequently, a profound connection was established among causality (as an upper limit on the speed of propagation of a 'cause' to produce an effect at a distant point), analyticity (as a mathematical statement of and constraint upon the smoothness and singularity structure of the function describing the scattering of a projectile by a nucleus) and the resonance or bump structure of these nuclear cross sections. Causality finally provided an underpinning for nuclear resonance formulas. This relation between causality and analyticity would prove a key insight for subsequent work in dispersion theory and S-matrix theory (a theme we develop in later chapters).

   The continuity in problems and the chain of workers from the early S-matrix program to other advances in dispersion theory and quantum field theory are major themes of our story. A fertile problem background was set by this early S-matrix theory.

2.1   The *S* matrix: Wheeler and Heisenberg

Wheeler (1937a, b), in the context of a theoretical description of the scattering of light nuclei, introduced the concept of a scattering matrix.[5] His motivation appears to have been pure nuclear physics and his

S matrix was a tool toward that end. In modern notation, the S matrix consists of elements $S_{\beta\alpha}$, α, $\beta = 1, 2, \ldots, N$, that give the relative strengths of the asymptotic forms of the wave functions for various channels (or types of reactions) as (cf. Eq. (A.33))

$$\psi_{\alpha\beta}^{+}(r_\beta) \xrightarrow[r_\beta \to \infty]{} \frac{i}{2k_\beta r_\beta} [\delta_{\alpha\beta} e^{-ik_\alpha r_\alpha} - S_{\alpha\beta} e^{ik_\beta r_\beta}]. \tag{2.1}$$

Here $e^{-ik_\alpha r_\alpha}$ represents the incident, or incoming, (spherical) wave and $e^{ik_\beta r_\beta}$ the scattered, or outgoing, (spherical) wave. It is the square of such a wave function that is related to the probability of the corresponding reaction taking place (as we show below). In order to keep notational complications to a minimum, we outline the situation only for the $l = 0$ partial wave (or angular momentum state) in Eq. (2.1) and later. There are N possible entrance channels and also N possible exit channels. In Wheeler's paper the scattering matrix is simply an N^2 array of elements connecting the incident wave to possible exit channels. There is no sense of using this S matrix as the central entity in an independent theory. Even though Wheeler presents his scattering matrix essentially as a convenient and powerful calculational tool, he does point out that the elements of this scattering matrix are completely determined by the (vanishing of the) Fredholm determinant of the linear integral equations (which are equivalent to the Schrödinger equation) so that the asymptotic form of these wavefunctions themselves need not be found explicitly.[6] Once these S-matrix elements are known, all the relevant transition probabilities and cross sections can be calculated directly in terms of them. This can be seen as follows.*

By the same type of argument that led from Eqs. (A.20) to Eqs. (A.25) and (A.26), one can show (here for $l = 0$) that, for scattering in the channel $\beta = \alpha$ (i.e., *elastic* scattering), the cross section is

$$\sigma_{sc} \equiv \sigma_{\alpha\alpha} = \frac{\pi}{k_\alpha^2} |S_{\alpha\alpha} - 1|^2, \qquad \beta = \alpha$$

and for transfer from channel α to channel β (i.e., *inelastic* scattering)

$$\sigma_{\alpha\beta} = \frac{\pi}{k_\alpha^2} |S_{\alpha\beta}|^2. \qquad \beta \neq \alpha.$$

---

* Here and throughout the text, we set off in smaller type technical material that can be passed over without interrupting the continuity of the development.

Both of these can be summarized as

$$\sigma_{\alpha\beta} = \frac{\pi}{k_\alpha^2} |S_{\alpha\beta} - \delta_{\alpha\beta}|^2.$$ (2.2)

The (net) reaction cross section (which takes flux out of the entrance channel α) is

$$\sigma_r = \sum_{\beta \neq \alpha} \sigma_{\alpha\beta} = \frac{\pi}{k_\alpha^2} \sum_{\beta \neq \alpha} |S_{\alpha\beta}|^2.$$ (2.3)

As we shall discuss in more detail in the next section, conservation of flux (or probability) and the orthogonality of the ψ_α,

$$\langle \psi_\alpha, \psi_\beta \rangle = \delta_{\alpha\beta},$$ (2.4)

imply that

$$\sum_{\gamma=1}^{N} S_{\alpha\gamma} S_{\beta\gamma}^* = \delta_{\alpha\beta}$$ (2.5)

or, as a matrix equation,

$$SS^\dagger = 1,$$ (2.6)

which is a statement of the unitarity of the S matrix. Equations (2.3) and (2.5) together imply that

$$\sigma_r = \frac{\pi}{k_\alpha^2} (1 - |S_{\alpha\alpha}|^2),$$ (2.7)

which is similar to Eq. (A.26). The total cross section is

$$\sigma_t = \sigma_{sc} + \sigma_r = \frac{2\pi}{k_\alpha^2} (1 - \mathrm{Re}\, S_{\alpha\alpha}) = \frac{4\pi}{k_\alpha} \mathrm{Im}\, f_{\alpha\alpha}(\theta = 0),$$ (2.8)

since the $l = 0$ part of the elastic scattering amplitude $f_{\alpha\alpha}(\theta)$ is just $(S_{\alpha\alpha} - 1)/2ik_\alpha$. Here we have argued for the optical theorem (Feenberg, 1932) of Eq. (2.8) only in s-wave approximation. However, the proof goes through in general so that Eq. (A.27) becomes

$$\sigma_t(k_\alpha) = \frac{4\pi}{k_\alpha} \mathrm{Im}\, f_{\alpha\alpha}(\theta = 0).$$ (2.9)

This remarkable and important theorem states that the *total* cross section is completely fixed once the (imaginary part of the) forward scattering amplitude is known.

The point to be emphasized is that once the scattering matrix elements $S_{\alpha\beta}$ are given, then all the cross sections, or observables of interest here, can be calculated in terms of them. This gives some indication of the power or importance of the S matrix.

In a series of papers, Heisenberg (1943a, 1943b, 1944, 1946) proposed as an alternative to quantum field theory (QFT) a program whose central entity was a matrix he denoted by S and termed the 'characteristic matrix' of the scattering problem. Cassidy (1981) has shown how Heisenberg interpreted cosmic ray showers or 'explosions' as indicating the existence of a fundamental length, Oehme (1989) has briefly outlined some results contained in Heisenberg's S-matrix papers, and I have previously discussed some of the difficulties with quantum field theories in the 1930s which motivated Heisenberg to seek an alternative to quantum field theory (Cushing, 1982, especially pp. 45–48).

In the period from the late 1930s to the late 1940s, the divergences present in quantum electrodynamics indicated to some that quantum field theory may have been a mistake. Both the theoretical and experimental situations left the future of QED in doubt (Galison, 1983a). Cosmic ray showers (or 'explosions') and the divergence of cross sections beyond a certain energy in a classical (nonlinear) field theory version of Fermi's β-decay formalism were taken by Heisenberg (1936, 1938b) to indicate the existence of a fundamental length and the need for a profound revision of elementary-particle dynamics. Not knowing what that future theory would be, he proposed the S-matrix theory as an interim program. Heisenberg wanted to avoid any reference to a Hamiltonian or to an equation of motion and to base his theory only on observable quantities. The point here is not that Heisenberg felt that the Hamiltonian (in the sense of the total energy) is not a quantum-mechanical observable. Rather, the calculational route from the basic Hamiltonian and the equations of motion of quantum field theory to many of the important experimentally observable quantities (such as bound-state energy levels and scattering cross sections) was an ill-defined one, often producing infinities.[7] Heisenberg wanted a theory based directly on finite, experimentally meaningful quantities. Since Heisenberg wanted to base his theory on general properties that were independent of any particular model, we can see why the S matrix appeared a reasonable place to begin. The formalism outlined in the Appendix indicates that, within the framework of nonrealativistic potential theory, the potential $V(r)$ determines the S matrix and hence the experimentally measurable cross sections. However, if one were given directly the S-matrix elements $S_{\alpha\beta}$, then one would be able to compute the cross sections. In quantum field theory, just as in nonrelativistic potential theory, once the Hamiltonian H for the system has been given, the scattering matrix is (at least formally) determined. Heisenberg's desire to base a theory on observable quantities only was a return to an idea that had proven useful in his earlier successful formulation of matrix mechanics (Cushing, 1982, p. 19). That is, in his 1925 paper he set out to construct a quantum mechanics wholly in terms of observable quantities. It worked pretty well once – why not try it again!

2.2 Discussion of Heisenberg's *S*-matrix papers

Heisenberg's stated purpose in his seminal paper (1943a), 'The observable quantities in the theory of elementary particles', was to abstract as many general, model-independent features of *S* as possible. In the abstract and introduction to that paper we read (Heisenberg, 1943a, p. 513):

> The known divergence problems in the theory of elementary particles indicates that the future theory will contain in its foundation a universal constant of the dimension of a length, which in the existing form of the theory cannot be built in in any obvious manner without a contradiction. In consideration of such a later modification of the theory, the present work attempts to extract from the foundation of quantum field theory those concepts which are not likely to be discarded from that future, improved theory and which, therefore, will be contained in such a future theory.
>
> In recent years, the difficulty, which still stands in the way of a theory of elementary particles, has been pointed up in many ways. This difficulty manifests itself surprisingly in the appearance of divergences (infinite self energy of the electron, infinite polarization of the vacuum, and the like), which hinders the development of a mathematically consistent theory and must probably be perceived as an expression of the fact that, in one manner of speaking, a new universal constant of the dimension of a length plays a decisive role, which has not been considered in the existing theory.'

This paper is remarkable for the number of new ideas it introduces, many of which would be put on a firm mathematical basis only years later. We outline below his proof of a key property of the *S* matrix – namely, unitarity. This property would remain central to the *S*-matrix program.

Heisenberg proved the unitarity of the S matrix for any system governed by a Hamiltonian. Unitarity, as an independent principle, would remain a key ingredient in the S-matrix program. Some details of his proof are given here. Heisenberg worked in a momentum-space representation outlined by Dirac (1935, pp. 195–200)[8] and defined the S matrix as the matrix coefficient of the outgoing waves.

Let us write this somewhat symbolically[9] (in analogy with Eq. (2.1)) as

$$\psi_{i,k}^{+} = \phi_{i,k}^{-} + S_{ik}\phi_{i,k}^{+}. \tag{2.10}$$

Here the plane wave $\phi_{i,k}$ has been decomposed into its ingoing and outgoing components as

$$\phi_{i,k} = \phi_{i,k}^{-} + \phi_{i,k}^{+}. \tag{2.11}$$

In this representation $\phi_{i,k}$ is real and

$$(\phi_{i,k}^{+})^{*} = \phi_{i,k}^{-}. \tag{2.12}$$

Having defined the S matrix in general, Heisenberg then established the unitarity of S within the framework of a Schrödinger equation

$$\sum_{j} <k|H|j> \psi_{i,j}^{+} = E_{i}\psi_{i,k}^{+}. \tag{2.13}$$

Heisenberg expressed these solutions ψ^{+} as a formal operator limit of the solutions to the time-dependent Schrödinger equation (cf. Eq. (A.1)) as

$$\psi_{i,k}^{+} = \lim_{t \to \infty} e^{iE_{i}t}\langle k|e^{-iHt}|i\rangle. \tag{2.14}$$

This paper contains (in an often symbolic and certainly nonrigorous fashion) the essential elements of formal time-dependent scattering theory, which would later be further developed, for example, by Lippmann and Schwinger (1950), by Gell-Mann and Goldberger (1953) and by Brenig and Haag (1959). Heisenberg's ψ^{+} and ψ^{-} are, respectively, the in and out scattering states of later theory.[10] Since another solution ψ^{-} to (2.13) can be obtained from (2.14) by letting $t \to -\infty$ and since H is hermitian ($H^{\dagger} = H$), it follows that

$$\psi_{i,k}^{-} = (\psi_{k,i}^{+})^{*} = \phi_{i,k}^{+} + S_{ki}^{*}\phi_{i,k}^{-}. \tag{2.15}$$

Heisenberg then observed that any linear combination of these ψ^{-} must also be a solution, in particular, $\sum_{j} S_{ij}\psi_{j,k}^{-}$, whose outgoing part is just $S_{ik}\phi_{i,k}^{+}$. Because the boundary conditions make the solutions to Eq. (2.13) unique, this must be ψ^{+}, from which it follows that

$$SS^{\dagger} = 1. \tag{2.16}$$

The essence of this formal proof is that hermicity ($H^{\dagger} = H$) implies unitarity for S. We shall later see that unitarity is also related to conservation of probability.

Heisenberg's development of the scattering matrix concept appears to have been independent of Wheeler's, although Heisenberg does indicate in a footnote (1943a, p. 533) that Wick had mentioned to him a paper by Breit on the use of the scattering matrix. In this footnote, Heisenberg refers to this paper only as 'Breit (Phys. Rev. 1941)'. This is almost certainly a reference to Breit's 'Scattering matrix of radioactive states' that appeared in the December 15, 1940, issue of *The Physical Review*. There Breit applied Wheeler's (1937b) scattering matrix formalism to unstable systems that have a resonance structure. (The 'Breit–Wigner' (1936) formula, Eq. (A.96), is used in this paper by Breit.) Heisenberg thanks Wick for the information but also states that he had not been able to see Breit's paper. (*The Physical Review* did not regularly reach Germany during the war.) He appears to have had no prior or direct knowledge of Wheeler's ideas on the scattering matrix. Dr. Helmut Rechenberg states[11] that, when he asked Heisenberg directly about this, Heisenberg said 'no' to knowing about Wheeler's S-matrix at the time he wrote his own paper on that subject. Professor John Wheeler is also of the opinion[12] that Heisenberg was unaware of Wheeler's work at the time.

But, it does seem implausible that Heisenberg had not even *seen* Wheeler's 1937 *Physical Review* paper on the S matrix. As indicated above, Heisenberg was certainly definite in claiming not to know about (or to have been aware of) Wheeler's work in 1941–42. However, it is not wholly implausible that Heisenberg might not have been aware of Wheeler's scattering matrix concept even if he had come across Wheeler's 1937 paper since the title and context of Wheeler's work indicated a concern with models for light nuclei. Furthermore, Heisenberg's main interests in 1937–40 appear to have been cosmic ray showers and the problems they presented for quantum field theory as then understood. Interestingly, though, Heisenberg (1938a) did publish an article in 1938 in which he referenced a paper by Feenberg in Volume 52 (1937) of *The Physical Review*. This is the same volume that contained Wheeler's S-matrix paper. So, it appears unlikely that

Heisenberg did not at least (literally) 'see' Wheeler's paper, although its content may not have attracted his attention.[13] In any event, Heisenberg certainly brought the concept of the S matrix to the attention of theoretical physicists. It has remained one of the central tools of modern physics.

A reader today looking at Heisenberg's original article (1943a) is amazed at what he 'sees' there (sometimes clearly only in retrospect, though.) There are the removal of the energy-momentum δ-function singularities to yield a true 'function' for S, the derivation of expressions for scattering and production cross sections in terms of the elements of S (cf. Eqs. (2.2) and (2.3) above), the proof of the unitarity of S, the Lorentz invariance of S, the separation of S as[14]

$$S = 1 + 2iT \tag{2.17}$$

with T representing the effects produced by interactions, discussion of the fact that S respects the usual connection between spin and statistics, and expression of S as

$$S = e^{i\eta} \tag{2.18}$$

where η is an hermitian phase matrix. With the decomposition of Eq. (2.17), the unitarity condition of Eq. (2.16) becomes

$$-\frac{i}{2}(T - T^\dagger) = TT^\dagger. \tag{2.19}$$

The diagonal element of this equation is just

$$\operatorname{Im} T_{ii} = \sum_j |T_{ij}|^2. \tag{2.20}$$

This relation was derived by Heisenberg and is essentially the optical theorem (Eq. (2.9)) relating the total cross section σ_t to the imaginary part of the forward elastic scattering amplitude. It is an elegant and quite 'modern' derivation. The form of Eq. (2.20) makes the meaning of unitarity and the origin of the optical theorem particularly transparent. The quantum-mechanical probability amplitude T_{ij} represents the effects of scattering from channel i to channel j (e.g., particles a and b in channel i going to or scattering into particles c and d in channel j: $a + b \rightarrow c + d$; or, more compactly, $i \rightarrow j$). So, $|T_{ij}|^2$ is the probability ('rate', 'chance' or cross section) for the reaction $i \rightarrow j$. The expression $\sum_j |T_{ij}|^2$, summed over *all possible* outcomes (or final states) j, is the total reaction rate (or cross section). Equation (2.20) relates this total cross section to the imaginary part of T_{ii}, the elastic (e.g., $a + b \rightarrow a + b$)

scattering amplitude. That is the optical theorem. Furthermore, Eq. (2.20) shows that a particular amplitude, here T_{ii}, is related to *all* other amplitudes T_{ij}, representing channels (j) that can be connected to channel i. This coupling, or 'entanglement', of reactions with each other will be an important feature of S-matrix dynamics.

The hermitian phase matrix η was the primary quantity to be determined in the theory (Cushing, 1982; Grythe, 1982), essentially by guessing at that time. Its determination would replace equations of motion, such as the Schrödinger equation or the Hamiltonian formalism of quantum field theory. Heisenberg also indicated (1943a) how the matrix η could actually be calculated in those cases where a Hamiltonian H was known, as in quantum mechanics and field theory. In his second paper (1943b) he computed S for particular models of η (i.e., essentially for certain enlightened guesses for the form of η).

An important point for the subsequent development of the S-matrix program is the following. Heisenberg (1944, p. 94, footnote) thanks Kramers for the suggestion that S be considered as an analytic function of the energy variable. (Heisenberg (1946, p. 612) later acknowledged discussing this with Kramers in Leiden in 1942. Grythe (1982) also references some correspondence between Heisenberg and Kramers on this point.) This analyticity suggestion was related to earlier unpublished work by Kramers and Wouthuysen. Here the stationary (or bound) states correspond to the zeros of the S matrix on the (negative) imaginary k axis, where the energy is $E = k^2$. (We shall discuss below poles versus zeros of the S matrix and their connection with bound states.) Wouthuysen, whose association with Kramers began as a graduate student in 1937, recalls:[15]

> In 1940, if I remember well, he [Kramers] proposed to me a subject of research: the study of nuclear resonances, in relation with papers by Breit and Wigner, Kalckar, Oppenheimer and Serber, Kapur and Peierls. A later paper by Seigert (1939) inspired me very much: he pointed out the relations between 'radioactive states' (in the sense of Gamow) and resonance scattering. The Gamow states appeared as singularities for complex energy of the scattering amplitude as well. I proceeded to show the analyticity of the Schrödinger scattering amplitude and in this way found the bound states as singularities with negative energy. These properties I illustrated by simple examples, like square well potentials with square potential barriers. After a seminar I gave on the subject, Kramers' verdict was 'it is interesting and new'. Shortly afterwards, summer 1942, I had to abandon my studies, due to the war circumstances. Later, in 1943, I learned indirectly from Kramers 'that he made good use of my work during a visit of

Heisenberg in Leiden'. Clearly (now by hindsight!) Kramers had seen much wider implications of my work than I myself had dreamt of.

Heisenberg (1944) demonstrated this connection explicitly in specific model calculations. A simple example of this correspondence can be seen in the S matrix $S(k)$ for the scattering by a square well (Eqs. (A.39b), (A.42) and the comment following (A.42)). This important reflection by Kramers implied that the scattering matrix not only determined those observables related to cross sections but could also fix the bound-state energies of the system. However, the difficulty was that, since no Hamiltonian was assumed, there were no correspondence rules with classical theory to guide one in constructing (or guessing) S (or η).

2.3   Heisenberg's subsequent role in the program

Heisenberg did not remain a major player in S-matrix theory beyond the late 1940s. The program influenced theoretical physics through the work of others. So, before we detail the history of these developments, let us indicate the path taken by Heisenberg's subsequent research. In 1946 he (Heisenberg, 1946) summarized his S-matrix program and promised a fourth paper in this series. Interestingly enough, this paper appears never to have been published.[16] That manuscript discusses the many-body problem in the S-matrix framework. As we indicate below, the many-body problem posed difficulties for Heisenberg's program. Heisenberg himself soon lost interest in the S-matrix program and turned to the theory of turbulence (1947–48) and then to nonlinear field theory (Heisenberg 1949a, 1950a, 1950b, 1951, 1952, 1953, 1954, 1957, 1958; Heisenberg, Kortel, and Mitter, 1955; Ascoli and Heisenberg, 1957) in which there would be just *one* fundamental field that would underlie all of particle physics. That Heisenberg's interests had shifted back to quantum field theory by the late 1940s is not surprising since the renormalization program had shown how to cope with divergences. One of Heisenberg's earlier motivations for studying a *nonlinear* field theory was that he believed such a theory could account for the 'explosive' cosmic ray events we mentioned in the last chapter (see also Cassidy, 1981, pp. 19–21). That there was a connection between the turbulence and nonlinear quantum field research is already made plausible even by the title of Heisenberg's (1952) paper ('Meson production as a shock wave problem') in which he used previously proposed nonlinear equations to discuss meson production. His view (as summarized much later, Heisenberg, 1966, 1967) seems to have been that nonlinear problems are overwhelmingly more common in nature than are linear ones. Hence, a nonlinear equation was much more likely to describe a fundamental law of nature than would be a linear one.

Neither Heisenberg's *S*-matrix program (as a serious, independent program), nor his theory of superconductivity (Heisenberg, 1949c), nor his nonlinear spinor field theory was well received by the majority of theoretical physicists. His attempts at foundationally new theories no longer commanded the attention they once had. For example, Pauli was at first quite enthusiastic about Heisenberg's *S*-matrix program (as evidenced even in the present chapter by the number of people he described the program to; cf., also, Jost, 1984). But, Pauli was never convinced by the theory and eventually thought the program rather empty. Still, while Pauli remained at Princeton during the war, '... he engaged several of his collaborators at Princeton to explore the *S*-matrix' (Grythe, 1982, p. 200). Similarly, Pauli was initially interested in Heisenberg's nonlinear spinor theory but soon became an almost derisive critic of it. For example, in his comments on Heisenberg's (1958) paper given at the 1958 CERN Conference, Pauli was very negative (Ferretti, 1958, pp. 122–6).

> I reached the conclusion that [Heisenberg's papers on the spinor model] are mathematically objectionable.
>
> I disbelieve all the more in the possible excuses for such a contradiction.
>
> This I discussed already in April and I wonder that you again repeat it all.
>
> Well, I think that it is superfluous.

Møller also eventually became disenchanted with Heisenberg's *S*-matrix program (Grythe, 1982, p. 201). In this same vein, Jost states:[17]

> Heisenberg's reputation was of course enormous. He was trusted with an unfailing intuition and with the ability to be able to do almost anything ... His fame began to fade after 1945.
>
> ... Heisenberg's futile attempt at a theory of superconductivity ... dispelled once and for all the spell which he exerted on his friends (and even his enemies).
>
> I was present at H[eisenberg]'s seminar in Princeton (fall 1950) and witness to the violent attack against Heisenberg. Oppenheimer had warned ... [us] ... before the seminar, to treat the speaker kindly.

The lion had lost his claws.

Since we do not wish to pursue Heisenberg's nonlinear field theory here, let us indicate briefly his final position in which he had abandoned the S matrix as *the* fundamental entity in a new theory. In a review article on the quantum theory of fields, he stated (1957, p. 270):

> To avoid ... fundamental difficulties ... the efforts of many physicists have in recent years been concentrated on the S matrix. The S matrix is the quantity immediately given by the experiments.

> The S-matrix formalism does not by itself guarantee the requirements of relativistic causality.

> It is perhaps not exaggerated to say that the study of the S matrix is a very useful method for deriving relevant results for collision processes by going around the fundamental problems. But these problems must be solved some day and one will then have to look for a mathematical formalism that allows one to calculate the masses of the particles and the S matrix at the same time. The S matrix is an important but very complicated mathematical quantity that should be derived from the fundamental field equations; but it can scarcely serve for formulating these equations.

Some small comment is appropriate on the first two paragraphs of this quotation. First, it would be an overstatement to claim that *all* S-matrix elements are given directly by experiment. Some, such as those with two bodies in the initial state, have a reasonably direct relation to experimentally measurable quantities. However, there are many S-matrix elements that one would be extremely hard pressed to find any plausible way to measure. Second, Fierz (1950) had criticized Heisenberg's (1950a, 1950b) attempt to construct a 'convergent field theory' expression for S and had shown it to be not causal. This emphasized the difficulty of implementing causality without having a new Hamiltonian theory.

At the Solvay Conference (Stoops, 1961, pp. 174–5) Heisenberg also reflected upon his reasons for abandoning the S-matrix program:

> When I had worked on the S-matrix for a while in the years 1943 to 1948 I came away from the attempt of construction of a pure S-matrix theory for the following reason: when one constructs a unitary S-matrix from simple assumptions (like a hermitian η-matrix by assuming $S = e^{i\eta}$), such S-matrices always become non analytical at places where they ought to be analytical. But I found it very difficult to construct analytical S-matrices. The only simple way of getting (or guessing) the correct analytical behaviour seemed to be a deduction

from a Hamiltonian in the old-fashioned manner. ... My criticism [of the original program] comes only from the practical point of view. I cannot see how one could overcome the enormous complications of such a program.

Here, as later, complexity would be a recurrent theme of and problem for the S-matrix program.

2.4   Work on the program just after WWII

Let us now return to the historical sequence of developments. Some sense of the status of Heisenberg's early S-matrix program can be gotten from Kramers' remarks for a symposium at Utrecht in the spring of 1944 (Kramers, 1944, 139–40; 1956, p. 838):

> III. Heisenberg's recent investigations concerning the possibility of a relativistic description of the interaction that is not based on the use of a Hamiltonian with interaction terms in a Schrödinger equation. Heisenberg considers only free particles and introduces a formalism ('scattering matrix') by means of which the result of a short interaction (scattering) between these particles can be described. Formerly the scattering matrix could be derived from the Hamiltonian, but now we are to consider the scattering matrix as fundamental. We do not care whether a Schrödinger equation for particles in interaction exists; we do care which correspondence requirements exist and how the scattering matrix can obey them. It is interesting that the scattering matrix is also able in principle to answer the question in which stationary states the particles considered can be bound together. These are related to the existence and the position of zeros and poles of the eigenvalues of the scattering matrix, considered as a complex function of its arguments. Heisenberg could already give a (very simple) model of a two-particle system, in which a perfectly sharply relativistically determined stationary state occurs, while there are no divergence difficulties whatsoever.
>
> However promising, this is still only a beginning, and in particular with regard to a correct description of the electromagnetic fields of photons I expect difficulties, which the investigations in this direction will have to overcome. Fortunately, Heisenberg's program is still open in several respects, and one may perhaps expect a great deal from a fortunate combination with further ideas.

In two lengthy papers and in a briefer note, Møller (1945, 1946a, 1946b) studied the properties of Heisenberg's S matrix and fleshed out many of Heisenberg's original arguments. (Grythe (1982) has some interesting comments about the correspondence between Heisenberg and Møller during the period when these papers were written.) In the first paper, he considered the quantum-mechanical dynamics of an arbitrary number of identical interacting particles, gave a careful treatment of the unitarity of S, showed that the results of scattering processes (at real positive values of $E = k^2$) are independent of the bound-state energies (1945, p. 18), pointed out the completeness condition for the wave matrix states, proved the Lorentz invariance of S from cross-section invariants, and discussed the 'collision constants' (or conserved quantities) arising from the invariance of S under sets of transformations (such as the Lorentz transformations). In the second paper, Møller restricted himself mainly to a two-body system and developed extensively the analytic properties of the wave matrix to establish the connection between the zeros of S and the bound-state energies. There Møller proved (for a two-body system) that at a zero of S in the lower half k plane at $k_o = i\kappa$, $\kappa > 0$

$$
i(-1)^l \frac{dS_l}{dk}\bigg|_{k_0} > 0 \tag{2.21}
$$

for any given partial wave l (1946a, p. 29) and showed how resonance energies and half-lives were determined by the *poles* (or singularities) of the analytic S matrix in the lower half energy plane at $E - i\Gamma/2$, where Γ is the reciprocal of the lifetime of the state (cf. Eq. (A.101)). He established that for a given matrix S there might exist either no Hamiltonian H or many Hamiltonians that would yield the same S. In other words, even when the usual equations of motion exist, there would not necessarily be a unique correspondence between S and H. The square-well and Coulomb-potential S matrices were given explicitly. Møller was fairly optimistic about the general applicability of results that had been proved in special cases (1946a, p. 45):

> The results obtained for two particle systems in this paper may be supposed to hold also in the general case of a many particle system with possibilities for creation and annihilation processes, the only difference probably being that the number of collision constants ... is then larger than in the case of a simple two particle system.

In 1946 Heisenberg (1946) summarized the status of the S-matrix program as set forth in his three papers (1943a, 1943b, 1944) and in Møller's two papers (1945, 1946a).

There Heisenberg used the completeness of the $l=0$ wave functions

$$\int dk \, \psi^{+*}(k,r)\psi^{+}(k,r') + \sum_n \psi_n^*(r)\psi_n(r') = \pi\delta(r-r')$$

$$(2.22)$$

to express the constants c_n for the bound-state wave functions $(k_n = -i\kappa_n, \kappa_n > 0)$

$$\psi_n(r) \xrightarrow[r\to\infty]{} \frac{c_n}{\sqrt{2}} \, e^{-ik_n r}$$

$$(2.23)$$

in terms of the contour integral of the S matrix

$$|c_n|^2 = \oint dk \, S(k)$$

$$(2.24)$$

where the integral encloses only the pole at $k_n = i\kappa_n$. This was another example of using Hamiltonian-independent, general principles to obtain specific dynamical results about bound states from the S matrix. The point is that the completeness condition of Eq. (2.22) is just a statement that the states $\{\psi(k)\}$ can be used as a basis for expanding 'any' function. It is essentially Fourier's theorem that allows the expansion of an arbitrary function, $f(x)$, in terms of sines and cosines or of exponentials, e^{ikx}. It is a property the ψ's would be expected to have.

Heisenberg (1946, p. 613) acknowledged that Møller had essentially this result in the summer of 1944. Equation (2.24) can also be derived directly by an older method due to Kramers (1938; Ma, 1947). In this paper (1946, p. 609) he also promised the fourth paper (which we referred to above) on many-particle systems. This and his Cambridge lecture at the end of 1947 (Heisenberg, 1949b) appear to have been about the last of Heisenberg's publications on the S matrix. Even at this time, though, Møller (1947, p. 195) remained quite pessimistic about the prospects for a successful conventional quantum field theory.

> [I]t seems that divergencies are intrinsic difficulties of all relativistic quantum field theories of the Hamiltonian form and that the frame offered by the Hamiltonian scheme of quantum mechanics is too narrow.

> [I]t even seems doubtful that in any strictly relativistic theory of the future a Hamiltonian and a Schrödinger equation will exist at all in general.

Now that we have indicated what the central concepts of the Heisenberg S-matrix program were, let us consider the impact these ideas had. Since our chief interest in Heisenberg's early S-matrix theory lies in the proposed responses to a series of questions it provoked, we begin by focusing on the work of two theorists, Walter Heitler and Ernst C. G. Stückelberg.

As early as 1941 Heitler (1941) proposed a quantum theory of radiation damping (which refers to the effects caused by the radiation or emission of quanta by a particle) to do for the strong-interaction mesons what classical and quantum theory had done for light. By analogy, the mesons were seen as the quanta of the nuclear force just as the photon was the quantum of the electromagnetic field. The time-dependent theoretical treatment was very much in the spirit of that for the interaction of a quantized atomic system with the electromagnetic field as done by Weisskopf and Wigner (1930) and by Weisskopf (1931). (See Eqs. (A.86)–(A.95) and the discussion following them.) Heitler and Peng (1942) then proposed a scheme for omitting all the divergent terms in the perturbation expansion but still retaining effects of radiation damping. They (1943) used this procedure to make approximate calculations of the matrix elements for meson production in cosmic rays. In the late 1940s Pauli visited the United Kingdom and spoke on the difficulties in quantum field theories (Pauli, 1947) at the 1947 Cambridge Conference. This review included a summary of Heisenberg's S-matrix theory and of Heitler's *ad hoc* recipe for keeping only the finite terms in the perturbation expansion for S. Prior to Pauli's discussions with Heitler about Heisenberg's recent work, the S-matrix program seems to have been unknown at the Dublin Institute for Advanced Studies where Heitler was. Heitler and Hu (1947, p. 124) thank Pauli for having given them an outline of Heisenberg's work. After this exchange, Heitler and Hu (1947) used the approximate matrix elements of Heitler and Peng (1943) to calculate the S matrix and hence the bound states or the (isobar) particle spectrum for the meson–nucleon system. As Heitler and Hu (1947, p. 140) acknowledge, Pauli pointed out that an analytic continuation of an *approximate* expression for S from real k to imaginary values of k need not give reliable values for the zeros (and hence for the isobar masses). However, Heitler and Hu had indicated how bound-state masses could (in principle) be calculated even when a complete Hamiltonian theory is not known.

We can summarize Heitler's procedure as follows (Heitler, 1947; Wentzel, 1947). If the Hamiltonian H is written as $H = H_0 + H'$, where H' (representing the effects of radiation) causes a perturbation on the eigenstates of H_0, then a reduced matrix element of the operator H' is defined as[18]

$$\langle i|\overline{H'}|j\rangle = \frac{\langle i|H'|j\rangle}{E_j - E_i}. \tag{2.25}$$

The matrix K (sometimes termed the reaction matrix) is defined as

$$K = \sum_{n=0}^{\infty} (\overline{H'})^n H'. \tag{2.26}$$

The equation for the T matrix of Eq. (2.17) takes the form

$$K(1 + iT) = T \tag{2.27}$$

whose (formal) solution is

$$T = (1 - iK)^{-1} K. \tag{2.28}$$

Heisenberg's S matrix (cf. Eq. (2.17)) can be expressed as

$$S = (1 - iK)^{-1}(1 + iK). \tag{2.29}$$

If the interaction Hamiltonian H' is characterized by a coupling constant (or coupling strength) g, then Eq. (2.26) can be seen as an expansion in powers of g. The first non-vanishing term is finite, whereas higher-order ones diverge (i.e., are infinite). Heitler's *ad hoc* recipe consisted in keeping only this lowest-order, finite term for K and dropping (or 'subtracting') all the rest. He (1941) expressed the hope that these equations, which still included radiation damping, could be *exact*, although he later (1947, p. 189) relegated this to a (perhaps interim) working hypothesis:

> The attempt to be described below aims not at a final solution to this difficult problem, but rather at obtaining a preliminary working hypothesis from which can be derived physical results which shall be reasonable and, as far as possible, in agreement with the experiments.

Heitler had generated a relativistically invariant procedure for producing a finite K and hence a finite S matrix whose zeros should give the bound states of the system. Still, this was not a procedure based on some more general principle that eliminated the divergences, but rather a pragmatic rule for simply throwing away all the troublesome terms.

Stückelberg[19] had worked on a consistent classical model for a point electron (1938, 1939, 1941) and later (1942) extended this to a quantized theory. He applied (1944, 1945) Heisenberg's S-matrix formalism to these problems and proposed a series of correspondence rules (i.e., previously-known results to which the quantum expressions must pass over in suitable limits) to guide one in writing down Heisenberg's η matrix, or Heitler's K matrix. These matrices are related as

$$\tan\left(\frac{\eta}{2}\right) = K. \tag{2.30}$$

He also (1946) imposed the constraint that, in a scattering process, a particle cannot appear in the final asymptotic state before the incident particles collide. These correspondence and 'causality' constraints eliminated some of the arbitrariness and the divergences in K (or in η), but still did not determine uniquely the structure of the higher-order terms. This program was an attempt, however, to eliminate the divergences on the basis of independent, generally accepted principles. Stückelberg and his co-workers (Rivier and Stückelberg, 1948; Stückelberg and Rivier, 1950a, 1950b; Stückelberg and Green, 1951) continued in their efforts to develop a quantum field theory free of the divergences present in the Schwinger–Tomonaga–Feynman–Dyson theory, but without ultimate success. Wentzel (1947) gives a nice review of the status of strong-interaction field theory just after the Second World War and, in particular, of the Heisenberg–Stückelberg–Heitler theory. Wentzel's evaluation of Heisenberg's program at that date is that this scheme '... is very incomplete; it is like an empty frame for a picture yet to be painted' (p. 15). He saw Heitler's and Stückelberg's work as attempts to fill this frame.

However, before we continue the historical development of work directly related to Heisenberg's S-matrix program, let us make a few observations about how the S matrix had entered theoretical physics as a practical and important calculational tool by the late 1940s. Most important for theoretical elementary-particle physics was Schwinger's (1948) introduction of the unitary operator $U(t, t_0)$ that gives the time evolution of the state vector $\Psi(t_0)$ at some initial time t_0 to the $\Psi(t)$ at a later time as[20]

$$\Psi(t) = U(t, t_0)\Psi(t_0) \tag{2.31}$$

so that the Lorentz-invariant collision operator S connecting the initial and final states is[21]

$$S = U(\infty, -\infty). \tag{2.32}$$

Schwinger also expressed this S in terms of the hermitian reaction operator K by Eq. (2.29). Dyson (1949a, 1949b) made explicit the relation of Feynman's (1949a, 1949b) method for calculating the elements of Heisenberg's S matrix and also recalled (1949b, p. 1736) that Stückelberg had anticipated several of Feynman's results. In 1946–47 Dyson had studied under Nicholas Kemmer at Cambridge. Kemmer had Dyson read Heisenberg's S-matrix papers as well as Gregor Wentzel's (1943) *Einführung in die Quantentheorie der Wellenfelder* (published in Vienna during the War). When Dyson came to America in 1947, he was well acquainted both with S-matrix theory and with quantum field theory. Dyson recalls[22]:

> I was well prepared by Kemmer to put this knowledge [of the S matrix and of quantum field theory] to use in the reconstruction of quantum electrodynamics. I well remember the joy of recognition when I suddenly realized that Feynman's rules of calculation were just the practical fulfilment of Heisenberg's S-matrix program. This was in the Fall of 1948.

The fact that Dyson saw Feynman's theory as the *fulfillment* of Heisenberg's program is very clear from a letter to J. R. Oppenheimer that Dyson wrote in 1948[23]:

> I believe it to be probable that the Feynman theory will provide a complete fulfillment of Heisenberg's S-matrix program. The Feynman theory is essentially nothing more than a method of calculating the S-matrix for any physical system from the usual equations of electrodynamics. It appears as an experimental fact (not yet known for certain) that the S-matrix so calculated is always finite; the divergencies only appear in the part of the theory which Heisenberg would in any case reject as meaningless. This seems to me a strong indication that Heisenberg is really right, that the localisation of physical processes is the only cause of inconsistency in present physics, and that so long as all experiments are interpreted by means of the S-matrix the theory is correct.
>
> The Feynman theory exceeds the original Heisenberg program in that it does not involve any new arbitrary hypothesis such as a fundamental length.

Thus, Heisenberg's S matrix as a calculational tool, even if not as the central entity in an independent theory, was very much in the 'consciousness' of theoretical physics before 1950. This renormalization program was so computationally successful that the preponderant majority of theoretical high-energy physicists worked within its tradition beginning in the late 1940s. This success of renormalized quantum electrodynamics undercut, as pointed out earlier, Heisenberg's original motivation (i.e., the inability of quantum field theory to produce finite, unique results) for his S-matrix program. As we shall see later, in the late

1950s and early 1960s, the inability of even renormalized quantum field theory to calculate results for the strong interactions once more led to a new S-matrix program. And, again, once quantum field theory (in the form of gauge field theories) became calculationally adequate, relatively few theorists continued to work on the S-matrix program. In both instances, field theories won the contest because they could be cast into a form that allowed reliable perturbation calculations to be made (since no exact, closed-form solutions are known for realistic situations), while the S-matrix program could not be cast into a form useful for extensive (numerical) calculations. However, once again let us tie off this thread in our story and return to developments more closely related to Heisenberg's program.

A straightforward application of Heisenberg's S-matrix zero ⇔ bound-state energy rule soon encountered difficulties in the secure testing ground of nonrelativistic quantum mechanics. Ma (1946, 1947) showed by explicit calculation for an exponential potential in Schrödinger theory that there were 'false' or 'redundant' zeros of the S matrix which did not correspond to actual bound-state energies.[24] This discovery, Møller said, made him doubtful about the S-matrix theory and he left the new theory in the summer of 1946 (Grythe, 1982, p. 201). Ter Haar (1946) attempted to use Møller's inequality (2.21) to rule out these false zeros, but Jost (1946, 1947) criticized that argument and concluded that all false zeros could not necessarily be ruled out with that prescription.[25] Jost (1947) considered s-wave scattering and introduced the irregular solution $f(k,r)$ to the radial Schrödinger equation (Eq. (A.19))[26]

$$f(k,r) \xrightarrow[r \to \infty]{} e^{-ikr} \tag{2.33}$$

and an $f(k)$ (today known as 'Jost functions') as[27]

$$f(k) = f(k, r=0). \tag{2.34}$$

The solution regular at the origin is

$$\phi k,r) = \frac{1}{2ik} [f(k)f(-k,r) - f(-k)f(k,r)]. \tag{2.35}$$

$$\xrightarrow[r \to \infty]{} \frac{1}{2ik} [f(k)e^{ikr} - f(-k)e^{-ikr}]. \tag{2.36}$$

From Eq. (A.33) we see that

$$r\psi^+(k,r) = \frac{\phi(k,r)}{f(-k)} \tag{2.37}$$

and that

$$S(k) = \frac{f(k)}{f(-k)} = e^{2i\delta(k)}. \tag{2.38}$$

Here $\delta(k)$ is the phase shift (for s-wave scattering). Equation (2.36) makes it clear that $\phi(k,r)$ will be a bound-state wave function for $k = -i\kappa(\kappa > 0)$ when $f(-i\kappa) = 0$; that is, when $S(k)$ vanishes. (An explicit expression for $f(k)$ for the square well can be read off from Eq. (A.39b).) Jost discussed the analytic properties of these $f(k)$ and laid the foundation for a rigorous treatment of the analytic properties of the scattering amplitude. Unless one confines the discussion to strictly finite-range potentials, there can be redundant zeros in the S matrix found in potential theory. Meixner (1948) also studied the analytic properties of $S(k)$ in Schrödinger theory. Touschek (1948) and Wildermuth (1949a, 1949b) examined many-body effects within the framework of potential theory and Wildermuth (1949c) attempted unsuccessfully (van Kampen, 1951) to resolve the redundant zero problem. These redundant zeros remained a difficulty for the S-matrix program.[28]

Fröberg (1947, 1948a, 1948b), Hylleraas (1948) and Bargmann (1949a, 1949b), all motivated by Heisenberg's program, considered the possible equivalence, in nonrelativistic Schrödinger theory, of the S matrix (or of phase shift $\delta(k)$) and the potential $V(r)$. That is, the question was whether the $V(r)$ could be reconstructed from the $\delta(k)$. In fact, Fröberg (1948a) acknowledged Heisenberg's S-matrix program and thanked Pauli for suggesting the inversion problem to him in Zürich. Hylleraas (1948) also pointed out the importance of the inversion problem for Heisenberg's S-matrix program. Bargmann (1949a, 1949b) cited Møller's (1946a) proof that many Hamiltonians could correspond to one S matrix (cf., comment following out Eq. (2.21) above). Fröberg gave an incomplete, *formal* solution of the inversion problem. Hylleraas somewhat cautiously commented upon and extended Fröberg's work. Then Bargmann gave explicit examples of one phase shift $\delta(k)$ corresponding to several potentials $V(r)$ in Schrödinger theory and thus showed that Fröberg's and Hylleraas' inversion procedure could not be valid in all cases. These results led to Levinson's (1949a, 1949b) interest in the inversion problem, as indicated by

Levinson's (1949a) reference to Fröberg's work. He found conditions on the asymptotic value of the phase shift $(\delta(\infty) - \delta(0))$ which would guarantee that only *one* potential $V(r)$ could have produced this $\delta(k)$. Holmberg (1952) pinpointed the flaw in Fröberg's and Hylleraas' uniqueness agrument for the inversion problem. Gel'fand and Levitan (1951a, 1951b) and Jost and Kohn (1952a, 1925b, 1953) discovered a constructive procedure for finding $V(r)$, given $\delta(k)$ as well as the energies and normalization factors for the bound-state wave functions. Even today, after over three decades of work on the question, a complete solution to the inversion problem is not a simple matter to state. A simplified (and hence incomplete) answer is the following. If $\delta(k)$ is given for all real positive k for $l = 0$ and if there are no bound states, then (for suitable restrictions on the potential) there is a unique $V(r)$ for that problem. For each bound state (where the number n of bound states is given by $\delta(0) - \delta(\infty) = n\pi$), there is an arbitrary parameter in the potential (and, hence, no uniqueness). (See Newton, 1966, Chapter 20, and Cushing, 1986a, note 45, for more details.) This general question of the determination of a potential from the experimental phase shifts (which can be gotten rather directly from the observed scattering cross sections) was an important one for the theoretical nuclear physics of the time.

Jost himself characterizes the relevance of the criticism of Heisenberg's S-matrix program to theoretical physics as follows.[29]

> Summing up this special line of development it seems fair to state that the *criticism* of Heisenberg's S-matrix program has stimulated a qualitatively new kind of theoretical research, in which non relativistic wave mechanics serves, not as an analytic tool to calculate some experimental cross-section, but rather as an experimental playground for the discovery of general relationships, which might also be useful elsewhere.

This 'experimental playground' would remain an important tool for research in the S-matrix program, even well into the 1960s.

2.5 The S matrix and nuclear theory

Contact with a long-established line of work in theoretical nuclear physics was made by Hu (1948a). As background, let us say a few words about the relevant nuclear theory. (We return to the compound nucleus model in Chapter 9.) Breit and Wigner (1936, p. 519) had developed (in analogy with the resonance scattering of light; cf. Eqs. (A.86)–(A.96)) a theory of scattering near a (complex) resonance energy E_c:

> ... [T]heories of ... cross sections ... for the capture of slow neutrons ... explain the anomalously large capture cross sections as a sort of resonance of the s states of the incident particles.

Breit (1940a) summarized this as follows:

> Many nuclear reactions show excitation curves with pronounced peaks which suggest that there is a resonance of the nuclear system to certain energies. (p. 506).

> ... [T]he optical case has been cleared up by Weisskopf and Wigner [1930]. .... The band of photons emitted in such a jump is found to have an intensity distribution of the resonance type
>
> $$\frac{\text{const.}}{(E - E_{ji})^2 + \Gamma^2/4}$$
>
> with a peak at the emission center E_{ji}. (p. 507)

Breit (1940b) used this approximation to calculate Wheeler's S matrix for the decay of unstable states as

$$S_{\alpha\beta} \approx \delta_{\alpha\beta} + \frac{c_{\alpha\beta}}{E - E_c} \tag{2.39}$$

where the $c_{\alpha\beta}$ are constants. This particular form was based on simplicity (Breit, 1940b, p. 1070):

> The simplest possibility for [poles of the scattering matrix a_{lj}] will be considered
>
> $$a_{lj} \approx \frac{c_{lj}}{E - E_c} + d_{lj}.$$
>
> Here the c_{lj}, d_{lj} are constants and the equation is meant to be only an approximation in the neighbourhood of the complex eigenvalue $[E_c]$.

Very interestingly, Siegert (1939) seems to have been the first to consider the S matrix as an analytic function of k to derive a Breit–Wigner type formula:

> We investigate the singularities of the cross section which occur at certain complex values of the energy. Those singularities which lie near enough to the real axis, cause a sharp resonance maximum on the real axis and we can replace the cross section there by its singular part added to a smooth function of the energy. (p. 750)

Of course, one can 'see' these results today in Siegert's paper, but his motivations and notations were quite different.

An important insight into the origin and generality of the Breit–Wigner resonance formula was obtained by Wigner and his co-workers. We outline below the *formal* derivation of a Breit–Wigner type of formula. Later, we shall discuss a physical or more intuitive explanation of such resonance formulas.

In a series of papers, Wigner (1946a, 1946b) and Wigner and Eisenbud (1947) introduced the derivative matrix $R(E)$ and based a multichannel nuclear scattering theory upon it.[30] If a is the range of the potential, then for a simple single-channel case we have (cf. Eq. (A.37))

$$R(E) = [r\psi_E(r)/d(r\psi_E)/dr]_{r=a} = \frac{\tan(ka+\delta)}{k} \quad (2.40)$$

where $\psi_E(r)$ is the wave function for the problem. If we set $\psi_E(r) = e^{i\delta}\cos(ka+\delta)\phi_E(r)$, then *outside* the nuclear surface $(r=a)$ we can write $\phi_E(r)$ as

$$\phi_E(r) = \frac{1}{r}\left\{\frac{1}{k}\sin[k(r-a)] + R\cos[k(r-a)]\right\} \quad (2.41)$$

with $R(E) = [r\phi_E]|_{r=a}$. By using the complete set of real basis states $\phi_\lambda(r)$ defined by

$$H\phi_\lambda = E_\lambda \phi_\lambda, \quad (2.42a)$$

$$\left.\frac{d(r\phi_\lambda)}{dr}\right|_{r=a} = 0, \quad (2.42b)$$

$$\int_{\substack{\text{nuclear}\\\text{surface}}} \phi_\lambda^2 \, dV = 1, \quad (2.42c)$$

one can prove (Wigner and Eisenbud, 1947) that $R(E)$ has the *exact* expansion

$$R(E) = \sum_\lambda \frac{\gamma_\lambda \gamma_\lambda}{(E_\lambda - E)} \quad (2.43)$$

where the γ_λ are independent of E. For example, in the case of the square well (Eq. (A.38)) one can show directly that

$$R(E) \equiv \frac{\tan(Ka)}{K} = \sum_\lambda \frac{2}{a}\frac{1}{(E_\lambda - E)}. \quad (2.44)$$

In the multichannel case Eq. (2.41) becomes

$$\phi_{\alpha\beta}(r_\beta) = \frac{1}{r_\beta}\left\{\frac{\delta_{\alpha\beta}}{k_\alpha}\sin\left[k_\alpha(r_\alpha - a_\alpha)\right]\right.$$

$$\left. + R_{\alpha\beta}\cos\left[k_\beta(r_\beta - a_\beta)\right]\right\} \quad (2.45)$$

and Eq. (2.43) generalizes to

$$R_{\alpha\beta}(E) = \sum_\lambda \frac{\gamma_{\lambda_\alpha}\gamma_{\lambda_\beta}}{(E_\lambda - E)} \quad (2.46)$$

where the γ_{λ_α} are again independent of E. Notice that Eq. (2.46) implies that

$$\frac{dR_{\alpha\alpha}}{dE} > 0, \quad (2.47)$$

whenever this quantity is defined. The S matrix and this R matrix are related as (cf. Eq. (A.39a))[31] (Wigner and Eisenbud, 1947)

$$S_{\alpha\beta} = e^{-ik_\alpha a_\alpha}\left[\frac{1 + i\sqrt{k_\alpha}\,R_{\alpha\beta}\sqrt{k_\beta}}{1 - i\sqrt{k_\alpha}\,R_{\alpha\beta}\sqrt{k_\beta}}\right]e^{-ik_\beta a_\beta}. \quad (2.48)$$

In the neighborhood of a (real) resonance energy E_λ, $R_{\alpha\beta}$ may be approximated by one term in the sum of Eq. (2.46). Equation (2.48) then gives the Breit–Wigner one-level formula

$$S_{\alpha\beta} \approx e^{-ik_\alpha a_\alpha}\left[\delta_{\alpha\beta} - \frac{2ik_\alpha^{\frac{1}{2}}\gamma_{\lambda_\alpha}k_\beta^{\frac{1}{2}}\gamma_{\lambda_\beta}}{(E - E_\lambda) + i\sum_\nu k_\nu\gamma_\nu^2}\right]e^{-ik_\beta a_\beta}, \quad (2.49)$$

which has the same general *form* as Eq. (2.39) for $\alpha \neq \beta$.

One should not let the formalism obscure the central structure of the argument. It is essentially the completeness of the eigenfunctions ϕ_λ (Eqs. (2.42)) that allows one to obtain the 'resonance' form (Eqs. (2.44) and (2.46)) for $R(E)$ and that, in turn, leads to the Breit–Wigner form (Eqs. (2.48) and (2.49)). The interior of the nucleus remains effectively an unknown black box, since $R(E)$ is defined in terms of quantities at the nuclear surface $(r = a)$ *only*.

Motivated by Heisenberg's S-matrix program, Hu (1948a) used directly the analytic properties of S to derive (in the context of potential theory) the Breit–Wigner form in the neighborhood of a complex pole of the S matrix near the positive energy axis. He also clarified the common distinction that had been made in the literature between the 'potential scattering' that produced the smooth background against which the 'resonance scattering' of Eq. (2.39) stood out (cf. Eqs. (A.112)–(A.114) and the accompanying discussion). Hu showed that the 'potential' and 'resonance' terms both had the same source, the difference in their behavior depending upon the distance of the complex pole from the positive energy axis. He also made the useful observation (which would become the convention eventually adopted) that, since

$$S(k)S^*(k^*) = 1, \tag{2.50}$$

a value of k_n on the negative imaginary axis such that $S(k_n) = 0$ corresponds to a value $k_n^* = -k_n$ such that $S^*(k_n^*) = \infty$, or $S(-k_n) = \infty$. That is, the *poles* of the S matrix on the positive imaginary axis locate the bound-state energies. From now on we follow this convention illustrated in Figure 2.1. The upper half k-plane maps onto the entire $E = k^2$ physical sheet and the lower half k-plane onto the second (Riemann) sheet in the energy plane.

Hu (1948b) and Eden (1948, 1949a, 1949b) began to study the complicated problem of the analytic continuation of the S matrix when there were production thresholds present. As an aside, let us stress that it is important to appreciate, both here and in subsequent developments discussed throughout the text, that we use the term 'analytic' to include functions that have poles, branch points, and even essential singularities. The key property of such analytic functions is that once they are known exactly in any region however small, then they are (in principle) determined everywhere in their domain of analyticity. We have already

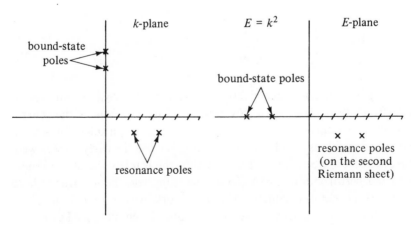

Figure 2.1   The k- and E-planes for the S matrix.

seen that Ning Hu was working with Heitler in Dublin when Pauli informed them of Heisenberg's S-matrix program. When Heisenberg visited Cambridge for two months early in 1948, Eden worked with him and again for a month in Göttingen later that year. Eden also recalls[32] that Heisenberg had visited Cambridge briefly a year earlier while Dyson was there.

We can easily see how discontinuities in the S or T matrices come about from Eq. (2.20). Let us take $T_{ii}(E)$ to be a function of the complex energy variable E and appreciate that the symbolic sum in Eq. (2.20) implies conservation of energy and momentum. Then, if E is real and below the two-particle scattering threshold (i.e., E is less than the energy $2m_0c^2$ which follows from the Einstein mass–energy equivalence relation $E = Mc^2$), it follows that Im $T_{ii} = 0$. This simply states that a reaction cannot occur if the total energy available is less than the rest mass of the interacting particles. That is, the optical theorem relates the total cross section σ_t to Im T_{ii}. Below threshold, $\sigma_t = 0$ and, hence, Im $T_{ii} = 0$ there. However, once we cross that threshold (by an increase in the total energy E), we have Im $T_{ii} \neq 0$. Such behavior (of the change in Im T_{ii}) is readily illustrated with the function $F(z)$ of the complex variable $z = |z|\,e^{i\phi}$

$$F(z) = \sqrt{-z} = (e^{-i\pi}z)^{\frac{1}{2}}. \tag{2.51}$$

For $z = x$(real), $F(x)$ is real for $x < 0 (\phi = \pi)$ and Im $F(x) = 0$ there. For $x > 0$, $(\phi = 0, 2\pi)$ Im $F(x) \neq 0$ and Im $F(x)$ undergoes a discontinuity across the positive real axis. In fact, we find

$$\text{Im } F(x) \begin{cases} \underset{z \to x^+}{\longrightarrow} -\sqrt{x} & \tag{2.52a} \\ \underset{z \to x^-}{\longrightarrow} \sqrt{x} \end{cases} x > 0 \qquad \tag{2.52b}$$

where $z \to x^\pm$ means, respectively, approaching the positive real axis (or cut) from above $(\phi = 0)$ or below $(\phi = 2\pi)$. Investigations of the general analytic structure of multiparticle scattering amplitudes (of several complex variables) is a horrendous problem and the early papers were inconclusive. Later, Eden (1952), within the framework of the renormalized perturbation expansion of quantum field theory, was able to reach some definite conclusions. His work was a precursor of an important area of activity several years later. Eden played a key role in that subsequent development and we return to it in a later chapter.

2.6   Causality and dispersion relations

However, the outgrowth of the Heisenberg *S*-matrix program most fruitful for the eventual use of dispersion relations in elementary particle theory followed from a suggestion made by Kronig (1946). Whereas Kramers (Heisenberg, 1944) had suggested that analyticity be incorporated into the *S*-matrix program of Heisenberg, Kronig now recalled that causality had been a principle useful in deriving dispersion relations for the index of refraction for the propagation of light waves (Kronig, 1926; Kramers, 1927, 1929) (cf. Eqs. (A.80)–(A.85) and the accompanying discussion). In his paper, Kronig (1946) states:

> As is well known, the scattering of monochromatic light by atoms is governed by a relation between the real and imaginary parts of the scattering amplitudes, leading to a familiar connection between the index of refraction and the coefficient of absorption of matter in bulk. ... This relation is a direct consequence of the natural requirement that an electromagnetic field, vanishing at the place of the atom for all times $t < 0$ and beginning to act only thereafter, cannot cause the emission of scattered waves before the time $t = 0$. In analogy one would expect that a centre of force, influencing the waves associated with material particles in a small region around it, will not give rise to scattered waves before $t = 0$ if the primary wave field at the centre is chosen to be equal to zero until this instant.
>
> It hence would seem reasonable to postulate for the scattering of particles a connection between the real and imaginary parts of the scattering amplitudes of the same type as in optics. On the side of the theory it remains to be discussed if this demand is correlated with Heisenberg's condition for *S* given above [the unitarity relation, $SS^\dagger = 1$] and what is a suitable extension for the case that new particles are created in the scattering process.

Put very simply, and at a minimum, here the principle of causality[33] means that an effect (or information) cannot instantaneously be communicated across a finite distance in zero time. For example, a light wave traveling at a speed c cannot arrive at an observer located a distance l from a source before a time l/c has elapsed once the light has been turned on. It is enough to indicate here how causality can plausibly be connected with analyticity. (For more details and references, see Eqs. (A.63) − (A.70), and the accompanying discussion, in the Appendix.) Suppose we have a ('reasonable') function $f(t)$ that remains identically zero until $t = 0$ (i.e., $f(t) \equiv 0, t \leqslant 0$). This could, for example, represent the response of a system to a light wave that reaches it at $t = 0$. We define the

Fourier transform $f(\omega)$ of $f(t)$ as

$$f(\omega) = \frac{1}{\sqrt{2\pi}} \int_{-\infty}^{\infty} f(t) e^{i\omega t} \, dt.$$

Since $f(t)$ vanishes for $t \leqslant 0$, this becomes

$$f(\omega) = \frac{1}{\sqrt{2\pi}} \int_{0}^{\infty} f(t) e^{i\omega t} \, dt.$$

In the upper half complex ω-plane (Im $\omega > 0$) of Figure 2.2 (where C is the integration contour), the integrand above has an exponential damping factor $e^{-\text{Im}\,\omega t}$. With some additional technical assumptions, this allows one to prove (and we have *not* done this here), that $f(\omega)$ is an analytic function of ω in the upper half complex ω plane. The point for our present purposes is that causality (in the sense that $f(t)$ vanishes identically prior to some time) implies analyticity for $f(\omega)$. We shall encounter variations in this type of argument again.

Kronig's was a qualitative suggestion about the use of causality. The question that now occupied several theorists was just *how* this requirement of causality was to be implemented in the S-matrix program and precisely what the connection between causality and analyticity would be. We do not mean to imply that *all* of their efforts were directly connected with Kronig's insight.

Jost, Luttinger and Slotnik (1950) outlined how to calculate higher-order quantum electrodynamics (QED) corrections to the perturbation expansion of the S matrix from lower-order terms by using the unitarity

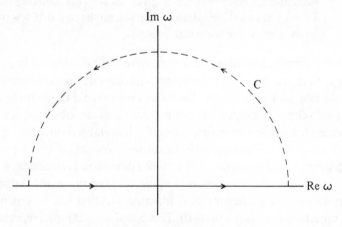

Figure 2.2   The complex ω plane for f(ω).

condition and by also exploiting the analytic properties of these matrix elements to write a dispersion relation for them. In this paper, Wheeler and Toll are thanked for that suggestion. In fact, these authors (Jost *et al.*) had, through Robert Karplus, access to a preliminary version of Toll's thesis (1952) on dispersion relations. Jost 'was ignorant of Kronig's (1946) remark ... and was under the impression that we [Jost *et al.*, 1950] were the first ones to apply dispersion techniques to *S*-matrix elements.'[34] We sketch here this interative procedure for obtaining higher-order corrections from lower-order ones because similar arguments will be important later in the history of the *S*-matrix program. Let us suppress all indices and treat $S(k)$ as a simple function (although the actual problem is more complicated). Then the basic concept is easily illustrated since the S matrix is expanded in a power series in the electric charge as

$$S(k) = 1 + eS_1(k) + e^2 S_2(k) + \dots \tag{2.53}$$

The unitarity condition, $S^*S = 1$, becomes

$$1 + e(S_1 + S_1^*) + e^2(S_1 S_1^* + S_2^* + S_2) + \dots = 1. \tag{2.54}$$

Since the coefficients of each power of e^n must vanish, we find that

$$2 \operatorname{Re} S_2(k) = -|S_1(k)|^2 \tag{2.55}$$

which gives the real part of $S_2(k)$ once $S_1(k)$ has been calculated. But if $S_2(k)$ is an analytic function of k, we can write a dispersion relation (cf. Eqs. (A.67)) as

$$\operatorname{Im} S_2(k) = -\frac{1}{\pi} \mathrm{P} \int \frac{\operatorname{Re} S_2(k') \, dk'}{(k' - k)}. \tag{2.56}$$

(Here the P symbol on the integral indicates that a particular limiting process – the Cauchy principal value – must be taken to make this singular integral well-defined mathematically.) Rohrlich and Gluckstern (1952) calculated forward Delbrück scattering (essentially the scattering of light by the Coulomb field of a nucleus) directly, using the Feynman–Dyson rules, and then using the dispersion-relation approach of Jost *et al.* (1950). The results agreed. Rohrlich in fact was able to use the dispersion relations (at the suggestion of Jost) to locate some errors made in the direct, brute-force approach. Toll, who had done the calculation only one way, at first obtained a result that disagreed with Rohrlich's.[35]

Actually, Rohrlich and Gluckstern's (1952) calculation was not just a perturbation-based use of analyticity as suggested in Eqs. (2.53)–(2.56).

In terms of the (complex) elastic scattering amplitude $f(k, \theta)$, the differential scattering cross section is given as

$$\frac{d\sigma}{d\Omega} = |f(k, \theta)|^2. \tag{2.57}$$

As we have already seen, the optical theorem (Feenberg, 1932; Lax, 1950) relates the total cross section to the imaginary part of the *forward* (i.e., $\theta = 0$) elastic amplitude as

$$\sigma_t(k) = \frac{4\pi}{k} \operatorname{Im} f(k, \theta = 0). \tag{2.58}$$

In essence, Rohrlich and Gluckstern (1952) were able to use the results of a (relatively simple) approximate calculation of $\sigma_t(k)$, obtain $\operatorname{Im} f(k, \theta = 0)$ and then use the general dispersion relation for $f(k, \theta = 0)$,

$$\operatorname{Re} f(k, \theta = 0) = \frac{2k}{\pi} \operatorname{\mathbf{P}} \int \frac{\operatorname{Im} f(k', \theta = 0) \, dk'}{(k'^2 - k^2)} \tag{2.59}$$

to find the complete forward scattering amplitude $f(k, \theta = 0)$. From $f(k, \theta = 0)$, Eq. (2.57) allows the differential cross section in the *forward* direction to be calculated. Bethe and Rohrlich (1952) extended these methods to an approximate calculation of nonforward Delbrück scattering. These were early and important uses of analyticity to calculate experimentally observable quantities.

Schützer and Tiomno (1951) appear to have been the first to attempt to exploit Kronig's (1946) suggestion that the principle of causality be imposed on Heisenberg's S matrix. They (1951, p. 249) credit Wigner with the surmise that the relation (2.46) for the R matrix might have its basis in that principle. Both Wigner and Bargmann at Princeton are thanked for discussions on this problem. Schützer and Tiomno (1951) considered the s-wave scattering produced by a potential of finite range a and implemented causality by the requirement that there be no scattering of an incident wave until that wave reaches the edge of the scatterer (of radius a). They (1951, p. 251) were able to construct a wave packet that is *rigorously* zero until $t = 0$ only because they included negative energies in their Fourier decomposition of $\Psi(r, t)$. As van Kampen points out:[36]

> The absence of negative energies is responsible for most of the complications in my [van Kampen, 1953a] treatment, but also makes it possible for the poles on the imaginary axis that correspond to bound states to exist.

With this assumption of a sharp wave front, Schützer and Tiomno were able to show that the poles of $S(k)$ could only lie on the positive imaginary axis or in the lower half k-plane. They also found that a two-sheeted Riemann surface in the energy plane is required for $S(E)$. They were not able to deduce all of the properties of the R function previously found by Wigner and Eisenbud (1947).

Van Kampen, while working in Denmark, had written a paper (1951) on the analytic continuation of the S matrix, pointing out an equivalence between the bound-state energies E_n and the phase $\eta(k)$. He subsequently went to Princeton where he wrote the paper on the implications of causality for the S matrix. He acknowledges (1953b, p. 1276) Wigner and Bargmann for discussions on this later work. Van Kampen (1953a) applied the causality principle to the Maxwell equations for the scattering of light by a spherically symmetric center of finite size (a). As a step toward establishing the mathematical properties of Heisenberg's S matrix on general grounds, van Kampen was able to show that the S matrix (in this case) was an analytic function of k and that it satisfied a dispersion relation, as well as to derive the analytic properties of Wigner's R function. One undesirable feature of the results was that they depended explicitly upon the size a of the scattering center. Van Kampen (1953b) then considered the implications of causality for the S matrix in nonrelativistic Schrödinger theory, the same problem studied by Schützer and Tiomno. One basic difficulty in a nonrelativistic theory is that there is no maximum velocity of wave propagation. So van Kampen replaced their assumption of a wave packet with a sharp front by a restriction on the probability of finding the particle outside the scattering center. His motivation was the following (1953b, p. 1267):

> The causality condition [for a light wave traveling at the velocity c] was formulated as follows: If at a large distance r_1 from the center of the sphere [of radius a] the ingoing wave packet is zero for all $t < t_1$, then the outgoing packet shall be zero at r_1 for all $t < t_1 + 2(r_1 - a)/c$.
>
> Obviously, for nonrelativistic particles a modification is necessary, since no maximum velocity exists, and one is inclined to postulate: If at any distance r_1 the ingoing wave packet is zero for all $t < t_1$, then the outgoing wave packet must also be zero for $t < t_1$ [as assumed by Schützer and Tiomno (1951)]. ... However, ... there is a serious objection.
>
> The difficulty is that *there are no ingoing or outgoing wave packets that are rigorously zero up to a certain time* ...

For a finite-range potential, he again obtained a dispersion relation for

$S(k)$. He also discussed (1953b, pp. 1275–6) the practical physical indistinguishability between a strictly finite-range potential and a long-range one (such as an exponential potential) and contrasted this with the radical difference in the analytic properties of the S matrix in these two cases. In practice, of course, we consider scattered particles as asymptotically free.

Finally, Toll (1952), who was a student of Wheeler's at Princeton, in a widely-referenced but unpublished Princeton Ph.D. dissertation, conducted an exhaustive study of the connection between causality and dispersion relations for light, including a calculation of Delbrück scattering (Rohrlich and Gluckstern, 1952). Toll and Wheeler (1951) applied dispersion relations to pair production processes. A summary of Toll's thesis was eventually published (Toll, 1956).

Wigner (1955a) used the property (2.47) of Eq. (2.43) for $R(E)$ to derive a famous inequality on the derivative of the phase shift,

$$\frac{d\delta}{dk} > -a - \frac{1}{2k}, \tag{2.60}$$

where a is the range of the potential.[37] He (1955b) also gave a beautiful and elementary summary discussion of the implications causality had for the relation (2.46) satisfied by the R matrix. As Wigner (1955a, p. 145) pointed out, a *lower* bound on $d\delta/dk$ is required by causality:

> It may be useful ... to derive certain general rules about the energy dependence of phase shifts .... The relations to be derived here are based, fundamentally, on what has come to be called 'the principle of causality'. It states that the scattered wave cannot leave the scatterer before the incident wave has reached it.

To see this qualitatively, consider just the s-wave *outgoing* spherical wave (cf. Eq. (A.33)) and from a wave packet with a momentum distribution $g(k)$ sharply peaked about some momentum k_0 as

$$\Psi_{sc}(r,t) = \frac{-i}{2r} \int \frac{g(k)}{k} e^{i[kr + 2\delta(k) - k^2 t]} \, dk. \tag{2.61}$$

As $t \to +\infty$ the rapid phase oscillations of the exponent in the integrand will produce cancellations that make the integral vanish. (This is essentially the Riemann–Lebesgue lemma familiar from standard functional analysis.) This can be avoided if we simultaneously choose r and t so that the phase factor remains constant

$$\frac{d}{dk}[kr + 2\delta(k) - k^2 t] = 0$$

or

$$t = \frac{r}{2k} + \frac{1}{k}\frac{d\delta}{dk}. \tag{2.62}$$

The second term on the right of Eq. (2.62) is often referred to as the delay time. If there were no interaction, the wave would arrive at r at the time $r/2k$. Since the interaction potential $V(r)$ may trap the wave for an arbitrarily long time, there is no fixed upper (positive) bound on $d\delta/dk$. However, if the physical force or the wave speed is to remain bounded, the potential $V(r)$ cannot cause the wave to arrive at r arbitrarily *earlier* than it would have in the absence of $V(r)$. This implies that $d\delta/dk$ must be bounded from *below* by some negative number. So, the causality requirement gives us a qualitative understanding of Wigner's result (2.60). What Wigner (1955a) actually did can be represented essentially as follows. From the inequality of Eq. (2.47) (for the single-channel case) and Eq. (A.38) giving $R(E)$ in terms of $\delta(k)$, one shows by direct calculation of dR/dE that

$$a + \frac{d\delta}{dk} > \frac{1}{2k}\sin[2(ka + \delta)] \geq -\frac{1}{2k},$$

which is Wigner's inequality, Eq. (2.60). An important implication of this bound is that, at a sharp phase-shift resonance where $\delta(k)$ must pass rapidly through (an odd multiple of) $\pi/2$ as k increases (cf. Eq. (A.122)), $\delta(k)$ must *rise* through $\pi/2$ because $d\delta/dk$ can only be large *positive* (but never large negative). Wigner (1955a, p. 146) also pointed out that an expression for $R(E)$ of the form of Eq. (2.43) had previously been shown, by Schützer and Tiomno (1951) and by van Kampen (1953a, 1953b), to be a result of the causality condition. The connection between causality and resonance formulas had become accepted.

2.7   A problem background for the 1950s

In this chapter we have attempted to establish that the extensive interest in dispersion relations and their connection with causality by the early

1950s can be traced directly back to the Heisenberg S-matrix program. Essentially no use has been made here of correspondence which Heisenberg himself had with others about his S-matrix program. Such is not really relevant for the central thesis of this section: a *prima facie* case, based on the published physics literature and on recollections of physicists active during the time span covered here, that work on questions prompted by Heisenberg's S-matrix program generated the background out of which grew the dispersion-theory program of the 1950s and 1960s. (For some correspondence on the S-matrix program, see Grythe (1982) and for the interpersonal relations and sociological background to Heisenberg's S-matrix program, see Rechenberg (1989).) There had been a large concentration of effort by a sizeable fraction of the community of theoretical physicists in the period 1945–54 on problems growing out of suggestions made in Heisenberg's papers. It will become even more apparent in the following chapters that Heisenberg's S-matrix papers raised questions leading to a series of important developments which eventually culminated in the use of dispersion relations in modern high-energy physics. The S-matrix and dispersion-theory programs of the 1960s had a major impact upon the theoretical high-energy physics of that decade. Several of the key contributors to that program (e.g., Chew, Gell-Mann and Goldberger) have stated in their recollections that Heisenberg's original S-matrix program of the early 1940s provided no direct motivation for or influence upon their own work in the late 1940s and early 1950s. While some of these founders of the later dispersion-theory and S-matrix theory program may not have been at the time aware of and involved in the early research on Heisenberg's S-matrix program so that they may not have been conscious of any influence of it on their own original work, this network of problems having roots in the Heisenberg program did, nevertheless, provide an essential background for a crucial advance in the field of relativistic dispersion relations, to which we turn in the next chapter.

But, first let us state here a thesis which can be drawn from this chapter. If Heisenberg's S-matrix program is remembered at all today, it is usually recalled as a program which encountered difficulties quite early and then quickly died out. From this, one could readily form an opinion that the original Heisenberg program was of little importance for subsequent theoretical developments. While it is true that Heisenberg soon abandoned his S-matrix program in favor of his nonlinear field theory of fundamental interactions, Heisenberg's ideas gave rise to a set of questions, such as how the causality requirement, suggested in the mid 1940s by Kronig as a constraint on the S matrix, was to be implemented in the S-matrix theory and how the interaction potential of Schrödinger theory could in principle be determined by scattering data. Work on such problems by many theorists produced a series of papers eventually leading to relativistic dispersion relations, a program usually associated with Goldberger and Gell-Mann. The background necessary for Chew's own early investigations on reaction amplitudes was quite independent of any connection with Heisenberg's S-matrix theory. However, it was the *confluence* of Chew's and Goldberger's programs that led to the S-matrix theory of the 1960s. It is not at all clear that the latter S-matrix program could have become a viable candidate for a theory of strong interactions if it had not been for Heisenberg's seminal papers on an earlier program, one quite different from, yet in the same spirit as, its successor. But, we get a bit ahead of our story here.

2.8 Summary

So, we have seen that although Wheeler had already proposed an S matrix as a useful calculational tool for theoretical nuclear physics in 1937, Heisenberg in 1943 independently introduced the S matrix as the basis for a conceptual framework which might provide an alternative to Hamiltonian quantum field theory. One of his primary motivations for this move was the divergence problems of QFT in the late 1930s, which gave rise to his belief that there might be a fundamental length in nature. Initially, Heisenberg saw the S-matrix theory as an interim program based only on observable quantities. Kramers made the suggestion that the S matrix be considered an analytic function of its energy variable so that scattering data could (in principle) constrain the values of the bound-state energies. Immediately after the Second World War, Heisenberg's S-matrix program looked promising as a general framework within which to construct a coherent theory of interactions among fundamental particles. However, with the great success of the renormalization program in QED, field theory became the main interest among particle theorists. Through the renormalized QED program, the S matrix did become a well-known calculational tool. Heisenberg turned to a nonlinear quantum field theory, but neither this nor his other forays into basic theory proved successful.

Nevertheless, criticism of Heisenberg's S-matrix program, largely within the framework of nonrelativistic Schrödinger theory as a testing ground, did stimulate interest in the inversion problem (of determining the potential $V(r)$ from the scattering data or phase shifts, δ_l) and in a theoretical justification for the Breit–Wigner resonance formula used to fit nuclear scattering data. In this latter undertaking, the analyticity of the S matrix suggested by Kramers was of central importance. Kronig in 1946 recalled that causality (essentially in the form of a first-signal principle) had been related to the analyticity properties of the index of refraction in optics to obtain dispersion relations. He suggested that the causality principle might also play a central role for Heisenberg's S matrix. Some progress was made in deriving the Breit–Wigner formula from the causality requirement and dispersion relations were used as an aid in quantum field theory calculations of scattering amplitudes. In 1955 Wigner showed quite explicitly how the causality requirement implied observable consequences for the behavior of a scattering phase shift near a resonance. Interest in problems such as these provided a background out of which emerged in the mid 1950s the dispersion-theory program for the strong interactions.

3

Dispersion relations

There exist some excellent technical reviews of relativistic dispersion relations (Goldberger, 1960, 1961; Jackson, 1961), as well as Goldberger's (1970) own informal recollections of the period from about 1954–69. In addition, Cini (1980) and Pickering (1989a) have written about some of the sociological influences on that program. We shall comment on these later in this chapter and in the concluding chapter. However, let us begin with a few general observations about the mood of theoretical physics in the United States just after the Second World War, at least in one Physics Department, namely the University of Chicago. This is relevant for what follows, since that Department became a center of activity for the dispersion theory program. Wentzel and Fermi were on the faculty then and Goldberger and Chew were graduate students there. One frequent attendee[1] at the theory seminars at Chicago during those years recalls that at that time (around 1948) the general spirit of many of the younger, exceptionally gifted theorists was that physics was something to *do* (never mind studying the works of the great masters) and that nothing significant had been done (in their areas of interest) prior to this. In fact, many other people had worked on these problems (e.g., fixed-source field theory) before, but there was little sense of history among the younger generation. Wentzel's comments (say, from the audience at a talk) on the previous literature were typically received with impatience. None of this, of course, detracts from the significant and original contributions made by this talented and numerically large generation of younger physicists. While Goldberger and Chew were excited by Heisenberg's ideas, which they learned about from Wentzel's lectures and Møller's articles, they did not really pursue this. The S matrix simply was not a hot topic of discussion then.[2] Furthermore, in the work that led to their 1954 paper (with Thirring) on dispersion relations, Goldberger and Gell-Mann were interested in *applications* and in obtaining relations among finite, experimentally measurable quantities. As quantum field theory went downhill with its inability to handle the strong interactions, there were two major reactions. One was the use of dispersion relations to avoid unreliable perturbation calculations and the other was axiomatic quantum field theory. We begin with the dispersion theory program, about which Goldberger (1970, p. 689) recalled:

> ... Wheeler and Wigner [at Princeton], through their students Toll, Schützer and Tiomno, had gotten the whole subject [dispersion theory] started.

One of the first problems that had to be faced was that of finding some justification for using dispersion relations for processes in which particles can be created and destroyed. That is, dispersion relations and analyticity properties had previously been established (either classically or quantum mechanically) only for interactions in which the number of particles remains fixed during a reaction (as is the case for the nonrelativistic Schrödinger equation). A foundation was needed to extend those relations to situations in which massive particles could be created and annihilated. After World War II there was an explosion in the amount of accelerator scattering data available on high-energy, strongly interacting particles. Quantum field theory was of little help in providing guidance to the experimentalists or in correlating these data. Dispersion relations were promising, if they could plausibly be believed for the data at hand. There was pressure to organize the data and give some coherence to the expanding field of high-energy physics. It was not at first a case of 'high' theory dictating to experiment, but the enterprise was a pragmatic and coupled one.

The key to a justification of dispersion relations was again provided by a causality condition, but this time in the form of a microcausality. The basic motivation is simple enough to explain in general terms. In ordinary quantum mechanics, observables that cannot be measured simultaneously (i.e., incompatible observables) are represented in the theory by operators that do not commute. A typical example is the pair q (position) and p (momentum) for which the commutator does not vanish since $[q, p] = i$. If two observables are compatible, then their commutator does vanish. In relativistic quantum field theory, the basic mathematical entities out of which all observables are finally built are quantized field operators $\phi(x)$. These operators are labeled with space-time variables x (i.e., x stands collectively for $\mathbf{x}$ and t). If two field operators commute, then observables built from (or corresponding to) them should be compatible or *independently* observable. If two space-time points are spacelike separated (i.e., too far apart for a light signal to be able to propagate between them in the elapsed time between the occurrence of the two events: $t_2 - t_1 < |x_2 - x_1|/c$), then an event at one point should be independent of an event at the other point. The first-signal principle of special relativity would lead us to expect that. The events would be causally independent. So, a natural way to implement causality would be to require that $\phi(x)$ and $\phi(y)$ commute if x and y are spacelike separated: $[\phi(x), \phi(y)] = 0$. This is termed *microcausality* because, as stated, it is to hold for *all* spacelike separated x and y, even on a microscopic scale (of, say, subnuclear dimensions). We shall discuss how this causality constraint was employed to argue for relativistic dispersion relations.

This initial success did lead to attempts to establish a general and coherent theoretical framework for high-energy phenomena. Much of this later work was purely theoretical and not motivated directly by experiment.

3.1 Goldberger, Gell-Mann, Thirring and microcausality

In 1953–54, Goldberger and Gell-Mann became interested in the dispersion-theory problem of the type discussed by Schützer and Tiomno and by van Kampen. Around December 1952 to January 1953, they saw the paper by Rohrlich and Gluckstern (1952) on forward Delbrück scattering and the extension by Bethe and Rohrlich (1952) on the use of dispersion relations applied to nonforward Delbrück scattering. Goldberger and Gell-Mann had not been aware of the Kramers–Kronig relations and do not recall having been motivated by the work of Wheeler and Toll (although Toll was in Chicago with his thesis around May, 1953).[2] However, Richard Eden does recall spending the spring of 1954 in Princeton and doing some work there with Goldberger on dispersion relations. In Eden's opinion, Goldberger was aware of Eden's (1952) paper on the analytic structure of the S matrix in quantum field theory, a paper that had its origins in Eden's own earlier interest in Heisenberg's S matrix and that explicitly refers to Heisenberg's program.[3] This does indicate a continuity (by way of problem background, at the very least) with Heisenberg's older program, a point also made by other physicists who were active at that time.[4] Goldberger and Gell-Mann were interested in applications and in finding a calculational tool that involved only finite, measurable quantities. In their attempt to understand, on the basis of first principles, these dispersion relations in situations where particle creation and annihilation are possible, a crucial remark was made to Goldberger (1970, p. 687) by Thirring at Princeton. Thirring observed that the causality condition (Cushing, 1986c) in field theory might be taken care of by the requirement that the commutator of the vector potential $A_\mu(x)$ vanish if x and y are spacelike separated,

$$[A_\mu(x), A_\nu(y)] = 0. \tag{3.1}$$

Gell-Mann, Goldberger and Thirring (GGT) (1954) studied the problem in perturbation theory and derived the Kramers–Kronig dispersion relation for forward photon scattering. There they quite explicitly reference the work of Kronig (1926, 1946), Kramers (1927), Schützer and Tiomno (1951), Toll (1952) and van Kampen (1953a, 1953b). Let us outline the argument used in that paper (GGT) because it is the seminal paper for strong-interaction dispersion relations.

3.2   'Proofs' of dispersion relations

For the interested reader, a few technical details are first given. The general logic of the argument is then discussed.

For a *noninteracting* scalar field of mass m, the (Heisenberg) field operators satisfy the equation of motion[5]

$$(\Box^2 + m^2)\phi(x) = 0. \tag{3.2}$$

Equation (3.2) is essentially a generalization of the wave equation, but now for massive 'particles' (unlike the massless photon that goes with the electromagnetic potential $A_\mu(x)$). This is again an example of a 'reinterpretation' of classical equations (i.e., Maxwell's equations). The canonical commutation relations are

$$[\phi(x), \phi(y)] = i\Delta(x - y). \tag{3.3}$$

For our purposes it is sufficient to note that $\Delta(x)$ vanishes whenever x is spacelike; that is, when

$$x^2 = t^2 - r^2 < 0, \tag{3.4}$$

which means a light signal could not travel the distance r in a time t. So, microcausality is satisfied here. If this field $\phi(x)$ is

Fourier decomposed as

$$\phi(x)=\frac{1}{\sqrt{2k_0}} \sum_k [a_k(x_0)e^{-ikx}+a_k^\dagger(x_0)e^{ikx}], \qquad (3.5)$$

then the creation operators $a_k^\dagger$ and the destruction operators a_k satisfy

$$[a_k,a_k^\dagger]=\delta_{kk'}. \qquad (3.6)$$

One-particle states $|k\rangle$ are created from the vacuum state $|0\rangle$ as

$$|k> =a_k^\dagger|0>, \qquad (3.7a)$$

and a_k annihilates the vacuum as

$$a_k|0> =0. \qquad (3.7b)$$

These operators create and destroy quanta ('particles') of mass m and momentum k.

For an interacting system, the Heisenberg field operators $\phi(x)$ now have as their equation of motion

$$(\Box^2+m^2)\phi(x)=j(x) \qquad (3.8)$$

where $j(x)$ is the interaction current. The commutation relations (3.3) and (3.6) are modified and their form cannot be written down until the $\phi(x)$ themselves have been found. However, the *assumption* of (micro-) causality requires that local field observables be independent at spacelike separated points. Since such observables are built up from expectation values of the field operators, we might *expect* that the microcausality condition can be implemented through the requirement that $\phi(x)$ and $\phi(y)$ commute[6] at spacelike distances,

$$[\phi(x),\phi(y)]=0, \quad (x-y)^2<0. \qquad (3.9)$$

Finally, we define the function $\eta(x-y)$ by the condition

$$\eta(x-y)=\begin{cases}1, & x_0>y_0 \\ 0, & x_0<y_0\end{cases}. \qquad (3.10)$$

A *compelling* argument for the connection between causality and the commutator condition of Eq. (3.9) does not seem to exist (Goldberger, 1961, p. 199):

> Has anyone ever made any effort to correlate the commutator condition with any measurable properties?

The example that GGT (1954) began with was the scattering of scalar particles (represented by $\phi(x)$) from a fixed force center. Since Eq. (3.9) is an operator identity that holds for all spacelike separations, it follows that the expectation value

$$G_{ret}(x, y) \equiv i < 0 | [\phi(x), \phi(y)] | 0 > \eta(x - y) \qquad (3.11)$$

is a function that vanishes whenever $(x - y)^2 < 0$ and/or $x_0 < y_0$. We shall indicate a bit more fully below how $G_{ret}(x, y)$ is related to the Fourier transform of the forward scattering amplitude. However, we already can see that this *might* be reasonable since $\phi(x)$ creates one-particle states from the vacuum so that Eq. (3.11) has something (although it is not exactly clear yet just what) to do with the overlap of such states.

For the case of photons, whose field operator is $A_\mu(x)$, interacting with a quantized matter field, the analogue of (3.9) which GGT (1954) began with was

$$[A_\mu(x), A_\nu(y)] = 0, \quad (x - y)^2 < 0 \qquad (3.12)$$

and from this they formed

$$G_{\mu\nu}(x, y) \equiv i \langle 0, f | [A_\mu(x), A_\nu(y)] | i, 0 \rangle \eta(x - y). \quad (3.13)$$

Here $|i, 0\rangle$ is the initial state (i) of the matter field with no photons (0) present and $|f, 0\rangle$ the final state (f) of the matter field with no photons (0) present. What GGT now did was to compute the $A_\mu(x)$ in a perturbation expansion to order e^2 in the electric charge (e) and then by brute force manipulations to relate the result to the Fourier transform of the forward scattering amplitude for the scattering of light.

The logic of the argument then is that this forward scattering amplitude is the Fourier transform of a function that vanishes identically outside some region. As the discussion following Eq. (A.65) or that given in Section 2.6 indicates, such an amplitude is an analytic function satisfying a Kramers–Kronig type dispersion relation. Although we shall discuss some of the details below, the basic point to be made here is that the microcausality condition, Eq. (3.12), and a perturbation expansion in e together implied the Kramers–Kronig relation for the forward scattering amplitude of light by a matter field. A short time later, Goldberger (1955a) was able to obtain the Kramers–Kronig relation without having to resort to a perturbation expansion for $A_\mu(x)$.

He discovered an application of what would later be known as the reduction technique that allowed him to express the forward scattering amplitude in terms of an expectation value of the *retarded* (because of the presence of the function $\eta(x-y)$) commutator of the current operators $j_\mu(x)$

$$\eta(x-y)[j_\mu(x),j_\nu(y)]\tag{3.14}$$

where these $j_\mu(x)$ are the currents determining the $A_\mu(x)$ from the equation of motion

$$\Box^2 A_\mu(x)=j_\mu(x).\tag{3.15}$$

Notice that from Eqs. (3.12) and (3.15) it follows that

$$[j_\mu(x),j_\nu(y)]=0,\qquad (x-y)^2<0.\tag{3.16}$$

That is, the microcausality condition has been transferred from the fields $A_\mu(x)$ (cf. Eq. (3.12)) to the currents $j_\mu(x)$. Goldberger based his derivation on formal solutions for the $A_\mu(x)$ obtained previously by Yang and Feldman (1950) and by Källen (1950, 1952). We now outline the reduction technique he used. We do not mean to imply that our outline represents the way Goldberger (1955a) obtained his results. However, here the reader is given some idea of what goes on in that important paper. It is also true that Low (1954) had used a reduction result similar to Goldberger's and later Low (1955) indicated how to obtain this result, but by beginning with the formal field-theory perturbation expansion for the S matrix (cf. Eq. (1.23)) in terms of the interaction Hamiltonian density $H_I(x)$ (Mandl, 1959, pp. 83–4)[7]

$$S=\sum_{n=0}^{\infty}\frac{(-i)^n}{n!}\int d^4x_1\ldots\int d^4x_n P\{H_I(x_1),\ldots H_I(x_n)\}\tag{3.17}$$

and verifying directly, for example, the identity $\langle p'k'|S|pk\rangle=\langle p'k'\,\text{out}|pk\,\text{in}\rangle$, which will be defined below. In a sense, though, one must know the form of the result in order to verify it in this fashion.

The basic idea (Jackson, 1961, pp. 11–21) behind the reduction technique as Goldberger developed it for the Heisenberg field operator $\phi(x)$ is that in the remote past $(x_0\to-\infty)$ they approach operators $\phi^{\text{in}}(x)$ representing the incoming fields before the interaction begins. The advantage is that we know more about these asymptotically free operators (cf. Eq. (3.3)) than we do about the $\phi(x)$ themselves. That is, for arbitrary states $|i\rangle$ and $|j\rangle$ we have

$$\lim_{x_0 \to \pm\infty} \langle i|a_k|j\rangle = \langle i|a_k^{\overset{\text{out}}{\text{in}}}|j\rangle. \tag{3.18}$$

For example, the state $a_{k_1}^{\text{in}\dagger}a_{k_2}^{\text{in}\dagger}|0\rangle$ represents a state of incoming particles of momenta k_1 and k_2, that can collide and produce outgoing particles (or waves) (cf. discussion accompanying Eqs. (2.10)–(2.15) and Note 2.10). Similarly, we have $\phi(x)\xrightarrow[x_0 \to +\infty]{} \phi^{\text{out}}(x)$. These $\phi^{\text{in}}(x)$ and $\phi^{\text{out}}(x)$ satisfy the free-field equation of motion, Eq. (3.2), the free-field commutation relations, Eq. (3.3), and have an expansion like Eq. (3.5) in terms of a_k^{in}, a_k^{out} (and their adjoints) with Eq. (3.6) holding for the in and out creation and destruction operators. Because the $\Delta(x-y)$ of Eq. (3.3) satisfies

$$(\Box^2 + m^2)\Delta(x-y) = \delta(x-y), \tag{3.19}$$

we can write a formal solution of Eq. (3.8) as (cf. Eqs. (A.8)–(A.10) for the type of reasoning used)

$$\phi(x) = \phi^{\text{in}}(x) + \int d^4x' \Delta_R(x-x')j(x') \tag{3.20}$$

where $\Delta_R(x) = \eta(x)\Delta(x)$[8]. (That is, one can verify directly that Eq. (3.20) satisfies Eq. (3.8).) This solution is such that $\phi(x)\xrightarrow[x_0 \to -\infty]{} \phi^{\text{in}}(x)$. From the expansion (3.5) and use of the total energy-momentum operator P (actually P_μ) as the translation operator according to

$$\phi(x+a) = e^{iPa}\phi(x)e^{-iPa}, \tag{3.21}$$

it follows that, for single-particle states $|k, \text{in}\rangle$, $|k, \text{out}\rangle$,

$$\langle 0|\phi(x)|k, \text{in}\rangle = \langle 0|\phi(x)|k, \text{out}\rangle. \tag{3.22}$$

The basic idea there is that one can use Eq. (3.21) to 'translate' $\phi(x)$ back into the distant past, where it becomes $\phi^{\text{in}}(x)$. Equation (3.18) implies that

$$\int_{-\infty}^{\infty} dx_0 \frac{\partial}{\partial x_0}\langle j|a_k(x)|i\rangle = \langle j|a_k^{\text{out}}|i\rangle - \langle j|a_k^{\text{in}}|i\rangle. \tag{3.23}$$

These formal results are essential for the reduction technique, which we now illustrate for the scattering of two particles,

$$p + k \to p' + k'. \tag{3.24}$$

The Heisenberg S-matrix element is defined as (cf. Note 2.10)

$$S_{\beta\alpha} = \langle \beta \text{ out}|\alpha \text{ in}\rangle = \langle p'k' \text{ out}|pk \text{ in}\rangle. \tag{3.25}$$

Here we have used α to signify collectively the particles in the initial state and β those in the final state. Use of

$$a_{k'}^{out\dagger}|0\rangle = |k'\ out\rangle, \tag{3.26}$$

Eq. (3.23), Eq. (3.5), the equations of motion and integration by parts finally yields, after considerable manipulation (cf. Jackson, 1961, p. 16),

$$S_{\beta\alpha} = \langle p'|a_{k'}^{out}|pk\ in\rangle$$
$$= \delta_{\beta\alpha} + \int d^4x\ e^{ik'x}\langle p'|j(x)|pk\ in\rangle. \tag{3.27}$$

A similar reduction of $|k\ in\rangle$, plus use of Eq. (3.21), leads to an expression for the T-matrix (cf. Eq. (2.17)) of the form[9]

$$T_{\beta\alpha} = i\int d^4x\ e^{ik'x}\langle p|\eta(x)[j(x),j(0)]|p'\rangle. \tag{3.28}$$

with $k' = p + k - p'$.

All the mathematical details aside, the importance of this reduction technique result is that it gives an *exact* expression for the scattering amplitude $T_{\beta\alpha}$ in terms of the retarded current commutator in the integrand. The vanishing of that commutator in a certain region of integration leads to analyticity properties of the $T_{\beta\alpha}$ (just as occurred for the Fourier transform discussed in Chapter 2).

If we now restrict ourselves to the forward direction ($k' = k, p' = p$), go to the laboratory frame in which $\mathbf{p} = 0$ and consider massless particles for which $k = (\omega, \mathbf{k})$ and $\omega = |\mathbf{k}|$, Eq. (3.28) becomes

$$T(\omega) = i\int dt\ d\mathbf{r}\ e^{i\omega(t - \hat{\mathbf{k}}\cdot\mathbf{r})}\langle p|\eta(x)[j(x),j(0)]|p\rangle. \tag{3.29}$$

Since $[j(x),j(0)]$ vanishes unless $x^2 = t^2 - r^2 > 0$ and since $\eta(x)$ vanishes unless $t > 0$, we see that the t integral in Eq. (3.29) runs only over values of t such that $t > r$ so that $(t - \hat{\mathbf{k}}\cdot\mathbf{r}) > 0$ in the integrand. By the type of argument we have discussed more than once by now, this implies that $T(\omega)$ is analytic in the upper half ω plane and yields a dispersion relation for $T(\omega)$. From Eq. (3.29) we also see immediately that

$$T^*(\omega) = T(-\omega), \tag{3.30}$$

which is the analogue of Eq. (A.68). This is a special case of an important general property known as crossing (Gell-Mann and Goldberger, 1954) and we shall discuss it at length in the next chapter. Although we have found this property from the general expression

(3.29), GGT (1954) obtained this result from their perturbation calculation of $T(\omega)$ and Gell-Mann and Goldberger (1954) argued for crossing on the basis a general property satisfied by the Feynman diagrams representing the perturbation expansion.

Of course, we have glossed over several important points in this sketch. Goldberger (1955a) was dealing not with a massless scalar field $\phi(x)$ but with the four-vector field $A_\mu(x)$ of quantum electrodynamics. Nevertheless, in spite of considerable technical complications, similar manipulations go through. Also, we have argued for the analyticity of $T(\omega)$ for $\text{Im}\,\omega > 0$, but we have overlooked the behavior of $T(\omega)$ as $\omega \to \pm \infty$ along the real axis. Furthermore, there is, in fact, the possibility that $T(\omega) \xrightarrow[\omega \to \infty]{} \omega^n$, where n is a finite power. We shall return to these problems in a later section. In spite of all these caveats, we must not lose sight of the deep and beautiful connection established between causality and analyticity. The essential ingredient in the proof is the microcausality statement, Eqs. (3.9) or (3.12). Of course, this is strictly an assumption appended to the field-theory framework since there are no known rigorous solutions to any realistic (four-dimensional) quantum field theory satisfying such a causality condition.

At this point (say, after Goldberger's (1955a) perturbation-theory independent proof of the forward dispersion relation for photon scattering), only the scattering of massless particles had been handled with any degree of rigor. Here, as throughout the dispersion-theory and S-matrix theory programs, applications of assumed dispersion relations often far outstripped what could be proven (even in a loose sense of that term). This gap between what could be *proven* and what was pragmatically *assumed* to get on with things would become a recurrent characteristic of the dispersion-theory and S-matrix theory programs. In their original paper, GGT (1954) implied that their dispersion-relation proof was valid for massive particles as well. They were criticized for this by van Kampen (GGT, 1954, p. 1613; Goldberger, 1970, p. 687). The basic difficulty when $\omega = \sqrt{k^2 + m^2}$, or $k = \sqrt{\omega^2 - m^2}$, can be seen from the factor

$$e^{ikx} = e^{i(\omega t - \mathbf{k} \cdot \mathbf{r})} = e^{i[\omega t - \sqrt{\omega^2 - m^2}\,\hat{\mathbf{k}} \cdot \mathbf{r}]} \tag{3.31}$$

which does *not* have oscillatory behavior for $0 < \omega < m$. Goldberger (1955b) indicated how this difficulty might be overcome, but a true proof had not yet been constructed. Goldberger (1970, p. 688) himself states about his (1955a) proof: 'Of course, my derivation was not really correct; since the result was correct, I was sure someone would prove it eventually.'

From the analytic properties they established using causality, Gell-Mann, Goldberger and Thirring (1954) did derive the dispersion relation for the electromagnetic forward scattering amplitude $f(\omega)$ (cf. Eqs. (A.66)–(A.70) for details)

$$\operatorname{Re} f(\omega) - \operatorname{Re} f(0) = \frac{2\omega^2}{\pi} \mathbf{P} \int_0^\infty \frac{d\omega' \operatorname{Im} f(\omega')}{\omega'(\omega'^2 - \omega^2)} \tag{3.32a}$$

$$= \frac{\omega^2}{2\pi^2} \mathbf{P} \int_0^\infty \frac{d\omega' \sigma_t(\omega')}{(\omega'^2 - \omega^2)}. \tag{3.32b}$$

These are just Eqs. (A.70c) and (A.70d) which are familiar from classical electrodynamics. Here a previously expected result had been justified within the framework of a wider theory. The limit of Eq. (3.32b) as $\omega \to \infty$ is

$$\operatorname{Re} f(\infty) - \operatorname{Re} f(0) = -\frac{1}{2\pi^2} \int_0^\infty \sigma_t(\omega') \, d\omega'. \tag{3.33}$$

For scattering of light from a bound electron (Rayleigh scattering), we have $f(0) = 0$, while for scattering from a free electron (Thomson scattering) $f(0) = -e^2$ [10] (Kroll and Ruderman, 1954; Low, 1954; Gell-Mann and Goldberger, 1954). If for the quatum electrodynamics (QED) case, the high-energy limit for scattering from a bound electron were $f(\infty) = -e^2$ (as it is classically; cf. Eq. (A.83)), then GGT (1954) would also have obtained the old Thomas-Kuhn sum rule, Eq. (A.85). Such sum rules are important direct relations between experimentally measurable quantities (here e^2 and σ_t). Similar sum rules would prove important in later dispersion-theory and S-matrix theory work. However, the QED value for $f(\infty)$ was unclear then.[11] Nevertheless, they did indicate how to apply the result (3.33) to the scattering of photons by a nucleus and by protons even though no convincing experimental checks of their predictions were yet possible.

3.3   Phenomenological use of dispersion relations

Karplus and Ruderman (1955) assumed that the GGT (1954) dispersion relations were valid for the scattering of massive particles and showed how these could be used to analyze experimental data. They pointed out that Eqs. (3.32) would, in general, have to be supplemented by pole terms (representing bound states) and that, if m is the mass of the lightest particle scattered, then for incident energy $E = \sqrt{k^2 + m^2}$,

$$\text{Im} f = 0, \qquad 0 < E < m^2. \tag{3.34}$$

This requirement can be related to the optical theorem, Eq. (2.9) (since physical scattering cannot take place below $E = m^2$) or to unitarity, Eq. (2.20) (cf. comment preceding Eq. (2.51)). That is, if the energy available is not even enough to include the mass-energy (mc^2) of the scattered particle, then the reaction cannot go and the cross section must vanish. Karplus and Ruderman (1955) restricted themselves to the case of no bound states (and, hence, no pole terms in the scattering amplitude) and observed that Eq. (3.32a) could (in principle) be compared with experiment through the relation

$$\frac{d\sigma}{d\Omega}(\theta = 0) = |f(k)|^2 = [\text{Re } f(k)]^2 + \frac{k^2}{16\pi^2}[\sigma_t(k)]^2. \tag{3.35}$$

That is, from the observed differential scattering cross section ($d\sigma/d\Omega$) and total scattering cross section, one can find $\text{Re } f(k)$ from (3.35) and then compare it with $\text{Re } f(k)$ from Eq. (3.32b). They also suggested using dispersion relations as a guide to doing a phase shift analysis, as follows. Since Eq. (3.35) shows that $d\sigma/d\Omega$ only determines $|f(k)|$, there is an ambiguity in the sign of $f(k)$ and this is reflected in there being more than one solution for the phase shift $\delta(k)$ of Eq. (A.23a). If we consider only s-waves so that the partial-wave expansion for $f(k)$ becomes approximately

$$f(k) \approx \frac{e^{i\delta} \sin \delta}{k} \tag{3.36}$$

and for simplicity of illustration drop $f(0)$ in Eq. (3.32b), we have

$$\text{Re } f(k) \approx \frac{\sin[2\delta(k)]}{2k} = \frac{k^2}{2\pi^2} \int_0^\infty \frac{dk' \, \sigma_t(k')}{(k'^2 - k^2)}. \tag{3.37}$$

Karplus and Ruderman (1955) assumed isospin conservation (a form of charge independence for the nuclear forces) and applied this to 'data'[12] on $\pi^0 p$ scattering and found numerical agreement with the results obtained previously by determining the phase of $f(k)$ through interference (for $\pi^\pm p$ data) with the Coulomb scattering amplitude at small angles. That is, these charged particles (pions and protons) interact through the electromagnetic force (because they carry electric charge) and through the strong (or nuclear) force. Therefore, the full scattering amplitude can be expressed as the sum of a Coulomb (or purely electromagnetic) amplitude and another term representing the strong interactions. This sum of two amplitudes can produce interference effects (just as for optical and acoustic phenomena). This interference dip in the observed cross section can be seen near the forward (i.e., $\theta = 0$) direction. Karplus and Ruderman's basic point was that causality (through dispersion relations) could generally resolve the standard phase shift ambiguity, just as Coulomb inference effects had done in the past.

However, Karplus and Ruderman (1955, p. 772) had used the crossing relation in the form

$$f(-E) = f^*(E) \tag{3.38}$$

which, as Goldberger pointed out, was correct only for uncharged particles. Goldberger, Miyazawa and Oehme (GMO) (1955) provided the correct details for the general pion–nucleon case. They showed that a suitable linear combination of the physical $\pi^+ p$ and $\pi^- p$ elastic amplitudes satisfied the dispersion relations previously written down by Goldberger (1955b) for massive particles. These proper linear combinations do satisfy simple crossing relations like Eq. (3.38), even though the individual amplitudes themselves do not. Goldberger, Miyazawa and Oehme (1955) gave the correct generalizations of Eqs. (3.32) and (3.33) for $\pi^\pm p$ scattering. If the subscript $+$ is used to denote quantities for $\pi^+ p$ scattering and $-$ those for the $\pi^- p$ reactions, then the GMO dispersion relations for the real part of the scattering amplitudes are

$$D_+(k) - \frac{1}{2}\left(1 + \frac{\omega}{\mu}\right)D_+(0) - \frac{1}{2}\left(1 - \frac{\omega}{\mu}\right)D_-(0)$$
$$= \frac{2f^2}{\mu^2}\frac{k^2}{\omega^2 - \mu^2/2M} + \frac{k^2}{4\pi^2}\int_\mu^\infty \frac{d\omega'}{k'}\left[\frac{\sigma_+(\omega')}{\omega' - \omega} + \frac{\sigma_-(\omega')}{\omega' + \omega}\right], \tag{3.39a}$$

$$D_-(k) - \frac{1}{2}\left(1 + \frac{\omega}{\mu}\right)D_-(0) - \frac{1}{2}\left(1 - \frac{\omega}{\mu}\right)D_+(0)$$
$$= -\frac{2f^2}{\mu^2}\frac{k^2}{\omega^2 + \mu^2/2M} + \frac{k^2}{4\pi^2}\int_\mu^\infty \frac{d\omega'}{k'}\left[\frac{\sigma_-(\omega')}{\omega' - \omega} + \frac{\sigma_+(\omega')}{\omega' + \omega}\right]. \tag{3.39b}$$

Here $\omega = \sqrt{k^2 + \mu^2}$, μ is the pion mass, M the proton mass, f the pion–nucleon coupling constant (about which more will be said later), and principal-value integrals are understood. The great value of these justly famous dispersion relations was that they were directly applicable to realistic situations represented by the experiments that could actually be performed at the time. Anderson, Davidon and Kruse (1955) applied these relations to the phase shift analysis of πp data between 0 and

1.9 Bev, and Davidon and Goldberger (1956) did further work on $\pi^{\pm}N$ data. The parameter f had to be adjusted to obtain good agreement between Eqs. (3.39) and the experimental data. They used a value of f found by other means. The determination of f, by several independent means, will prove to be an important point in subsequent sections. Goldberger, Nambu and Oehme (1957) eventually published the details of the dispersion relations for the much more complicated case of nucleon–nucleon scattering.

3.4   Pragmatic attitude of many practitioners

It should be evident that much of this work on dispersion relations was characterized by an attitude that placed a premium on getting practical things done, at time in an almost cavalier fashion. For example, in his (1955b) 'proof' of dispersion relations for massive boson fields, Goldberger essentially assumed without any proof or argument part of the answer he wanted (cf. Goldberger, 1955b, note added in proof, p. 985). 'We made up rules as we went along, proceeded on the basis of hope and conjecture, which drove the purists mad'.[13] As Goldberger (1961, p. 196) himself observed: 'It is perhaps of historical interest to relate that almost all of the important applications (Chew, Goldberger, Low and Nambu, 1957a, 1957b) of the nonforward dispersion relations were carried out before even the forward scattering relations were proved rigorously.' Pickering (1989a) has characterized as 'pragmatic' this attitude in the applications of dispersion relations. Cini (1980) has made a similar observation. On the other hand, those interested in a more rigorous approach were not particularly impressed by this style. As Jost states[14]:

> If I remember correctly (and if you allow me a judgment) very few of the so called 'purists' paid much attention or took much interest in Goldberger's conjectures ... . [T]hey were absorbed in their own work.

This contrast between the pragmatism of the Americans, who were much concerned with experimental results, and the rigor of the Europeans, who were distant from experiment, is consistent with the view presented by Cini (1980). He also sees this concern with phenomenology as the key to understanding the renewed interest in dispersion relations, an old idea in physics. The locus of high-energy experimental physics was largely in America during this period.

Still, Goldberger usually knew the right answer even if he couldn't always prove it. The nature of the problem was now rather well defined and this provided the basis for a viable research program: to 'prove' dispersion relations for a wider class of cases (e.g., for massive particles; in the nonforward direction; eventually, in two or more variables; etc.) on the basis of general field-theory axioms and to apply these relations to many experimentally accessible situations. For example, when John C. Polkinghorne (who would later become a leader of the 'Cambridge school' of S-matrix theorists) went to the California Institute of Technology as a postdoctoral fellow in 1955 to work with Gell-Mann, he learned from Gell-Mann that dispersion relations might provide an on-mass-shell, alternative formulation of quantum field theory.[15] Polkinghorne (1956) gave a heuristic proof for single-variable dispersion relations for mesons scattered from a nucleon, allowing for the possibility of meson creation or annihilation. He specifically referred to the Heisenberg program (Polkinghorne, 1956, p. 217) and suggested the possibility that it might provide an alternative formulation to quantum field theory. Once again we see a continuity going back to the early motivations for the S-matrix program.

If rigorous proofs were not available, one assumed the dispersion relations needed for a particular application (subject, of course, to their being no recognizable contradictions). It was only in 1957 that Symanzik (1957) provided a rigorous proof for the forward case. Dispersion relations in the energy variable, but now for fixed, nonzero values of the scattering angle, were heuristically derived, for example, by Salam (1956), by Salam and Gilbert (1956) and by Capps and Takeda (1956). More rigorous proofs of nonforward relations were given by Bogoliubov, Medvedev and Polivanov (1958) and then by Bremermann, Oehme and Taylor (1958) using the theory of several complex variables. No proof good for all angles was ever constructed.

3.5   More general proofs

There were several attempts to derive general results that would remain valid independently of the details of any particular quantum field theory or model. Thus, Lehmann, Symanzik and Zimmermann (LSZ) (1955) began by assuming the existence of Heisenberg field operators and then deriving as much as possible about the S-matrix elements. Nambu (1955) derived dispersion representations of quantum field Green's functions using the causality requirement. However, the attempt probably most in the spirit of Heisenberg's old S-matrix program (which LSZ (1955) explicitly cited) was that of Lehmann, Symanzik and Zimmermann (1957) who started with a pure S-matrix formalism. They began by assuming the existence of a scattering operator that connected the causal in and out operators and then sought to prove from this the existence of causal field operators that interpolated between the asymptotic in and out operators. Theirs was one attempt to provide a general framework based on first principles. They were careful (1957) to stress that it remained unknown whether a causal S matrix actually exists:

> We begin with a pure S-matrix formalism [Heisenberg, 1943a, 1943b], (p. 319)

> As is customary we call a field operator $A(x)$ causal – without discussing here the physical interpretation – if $[A(x), A(y)] = 0$ for $(x-y)^2 > 0$. (p. 323)

> It is an entirely open question whether any scattering matrix exists which is causal in the sense of this definition or whether this is too stringent a demand. (p. 324)

This program was in a sense the converse of their earlier work (LSZ, 1955). That is, this new approach reversed the emphasis from fields being primary to the S matrix being the entity of central interest. Many impressive mathematical results were obtained within and as an outgrowth of the LSZ program. Lehmann (1958) used the LSZ (1957) formalism (which relies heavily on reduction techniques similar to those discussed earlier in this chapter) and causality to establish the analyticity of the two-particle scattering amplitude in the momentum-transfer variable (i.e., roughly $\cos \theta$ where θ is the center of momentum scattering angle) at *fixed* energy. This included the region of momentum transfer for which single dispersion relations in the energy variable had been established. Lehmann's proof employs Dyson's (1958) integral representation of expectation values of causal commutators of the field operators. This was a remarkable and important paper, since Lehmann was the first to show explicitly how causality implied analyticity in the momentum transfer variable, even when that variable was outside the physical region (i.e., outside the values of that variable accessible to experiment). Much subsequent work, which was largely, but not exclusively, a European undertaking, attempted to make these discussions more mathematically rigorous and more general. Res Jost was a central figure in much of this work, both by his own technical contributions and especially by the influence he had upon theorists associated with the ETH (Eidgenössische Technische Hochschule) in Zürich where Jost was a Professor. Steinmann (1960), Ruelle (1961) and Hepp (1965) related the LSZ formalism to Wightman's (1956) mathematically rigorous axiomatic formulation of quantum field theory[16], although the LSZ program does not follow from Wightman's axioms alone. Bros, Epstein and Glaser (1965) established crossing (which we discuss in the next section) for the four-particle (i.e., two in and two out) amplitude within the LSZ framework and they (1972) proved some analyticity properties for the n-particle amplitude. Still, it was not possible to underpin completely the type of dispersion relations and analyticity properties of general scattering amplitudes that were being used in applications.[17] We do not pursue this axiomatic QFT program further here, mainly because it is an involved and highly technical story requiring a separate study of its own and because it is not really central to our purposes in this study. Nevertheless, the motivation for the axiomatic study of scattering amplitudes was nicely summarized by Lehmann (1959, pp. 153–4):

> It is hoped that by obtaining explicit properties of the scattering matrix we may:
> (a) Correlate and thereby partially understand experimental results.
> (b) Test the correctness of the axioms by confronting them with empirical facts.
> Another motivation is of course the lack of anything better. That is, despite many attempts, nobody has succeeded in formulating *specific* interactions ... in such a way that reliable quantitative calculations can be made, in the case of strongly interacting relativistic particles.

This last paragraph gives some indication of the value placed on getting on with a program in the sense of handling the enormous amount of experimental data then available.

For reference later, let us give here a list (Table 3.1) of the prominent workers in the dispersion-theory program (Goldberger, 1961, p. 181), just to indicate both the international character of the undertaking and the fact that the enterprise was dominated by Americans.

3.6 Other applications of dispersion relations

In this section, we indicate that use of dispersion relations was not restricted to scattering amplitudes alone (Chew, Karplus, Gasiorowicz

Table 3.1. *Prominent workers in the dispersion-theory program*

Americans	Europeans	Russians
R. Blankenbecler	R. Eden	N. Bogoliubov
G. Chew	S. Fubini	E. Fainberg
R. Cutkosky	R. Jost	A. Lagunov
S. Drell	H. Lehmann	L. Landau
F. Dyson	R. Omnès	I. Pomeranchuk
W. Frazer	J. Polkinghorne	D. Shirkov
M. Gell-Mann	K. Symanzik	
M. Goldberger	J. C. Taylor	
F. Low	J. G. Taylor	
S. MacDowell	W. Thirring	
S. Mandelstam	W. Zimermann	
Y. Nambu		
R. Oehme		
J. Toll		
S. Treiman		
J. Wheeler		
E. Wigner		
D. Wong		
F. Zachariasen		

and Zachariasen, 1958). This also allows us to introduce some concepts that will be useful in subsequent chapters. The basic problem here is to discuss the electromagnetic structure of strongly interacting particles (i.e., of those particles, such as the proton or the pion, that undergo both strong and electromagnetic interactions). Let us begin with the relatively simple example of electron(e)–pion(π) scattering, say

$$e^+ e^- \to \pi^+ \pi^-, \tag{3.40}$$

illustrated in Figure 3.1(a). Figure 3.1(b) shows the approximation to order e^2 for the scattering amplitude of the process in (3.40). This amplitude is found by computing the S-matrix element between the initial and final states of Eq. (3.40). It turns out to be essentially the product of the matrix element for the process $e^+ e^- \to \gamma$, the photon propagator (i.e., the function representing the propagation of the photon) and the matrix element for $\gamma \to \pi^+ \pi^-$. This last vertex is given as

$$\langle q'|j_\mu(0)|q\rangle = e(q' + q)_\mu F_\pi(t) \tag{3.41}$$

where $F_\pi(t)$ is the electromagnetic form factor of the pion and t, the four momentum transfer, is defined as

$$t = (q' - q)^2. \tag{3.42}$$

(Equation (3.41) is another example of the great usefulness of the currents j_μ that also entered so crucially into the derivation of dispersion relations as in Eq. (3.16).) The function $F_\pi(t)$ incorporates all of the (unknown and in practice uncalculable) strong-interaction effects of the $\gamma \to \pi\pi$ vertex. In the nonrelativistic limit (which will not be the case generally of interest to us here), $F_\pi(t)$ is simply the Fourier transform of the charge distribution $\rho(\mathbf{r})$ of the electric charge of the pion,

$$F_\pi(t) \overset{\text{N.R}}{=} \int \rho(\mathbf{r}) e^{i(\mathbf{q} - \mathbf{q}') \cdot \mathbf{r}} d^3 r, \tag{3.43}$$

where, by definition,

$$F_\pi(0) = 1. \tag{3.44}$$

Reduction techniques similar to those outlined earlier in this chapter allow one to obtain a dispersion relation for $F_\pi(t)$ as

$$F_\pi(t) = 1 + \frac{t}{\pi} \int_{4m^2}^{\infty} \frac{dt' \operatorname{Im} F_\pi(t')}{t'(t'-t)}. \tag{3.45}$$

Either formal field theory manipulations or the unitarity condition of Eq. (2.20) show that the imaginary (or absorptive) part of $F_\pi(t)$ is related to all the intermediate processes that can connect the γ state to the $\pi\pi$ state. Since the two-pion state is the lowest-mass intermediate state allowed by energy-momentum conservation and the discrete conservation laws (such as charge conservation), the lowest order

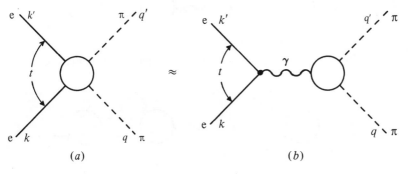

(a)   (b)

Figure 3.1   The electromagnetic form factor of the pion.

approximation is (see Figure 3.2)

$$\text{Im } F_\pi \approx F_\pi^* \, e^{i\delta} \sin \delta \tag{3.46}$$

where δ is the p-wave $\pi\pi$ phase shift for $\pi\pi \to \pi\pi$ scattering. Here $e^{i\delta} \sin \delta$ is just the $\pi\pi$ elastic scattering amplitude in the $l = 1$ angular momentum state. Figure 3.2 illustrates the important principle that any reaction proceeds *via* (or 'communicates' with) all possible intermediate states to which it can be connected (subject to the constraints of the operative conservation laws). An important point is that the integral equation resulting when Eq. (3.46) is substituted into Eq. (3.45) can be solved exactly (Omnès, 1958) so that if the p-wave $\pi\pi$ phase shift were known, $F_\pi(t)$ would be (approximately) known.

The problem of more immediate physical interest is the electromagnetic structure of the nucleon. This was the case first discussed by Federbush, Goldberger and Treiman (1958) in terms of dispersion relations. The electron–proton process

$$ep \to ep \tag{3.47}$$

can, much as in Figure 3.1, be approximated by the one-photon intermediate state in terms of the nucleon form factor[18] for

$$\gamma \to N\bar{N}. \tag{3.48}$$

Again, a dispersion relation can be written for $F_N(t)$ (Drell and Zachariasen, 1961). Now, however, the diagram for Im $F_N(t)$ becomes Figure 3.3. That is, the equation for $F_N(t)$ involves $F_\pi(t)$ (which, as we have just seen, is known in terms of the $\pi\pi$ phase shifts) and the amplitude for $\pi\pi \to N\bar{N}$. The electromagnetic structure of the pion

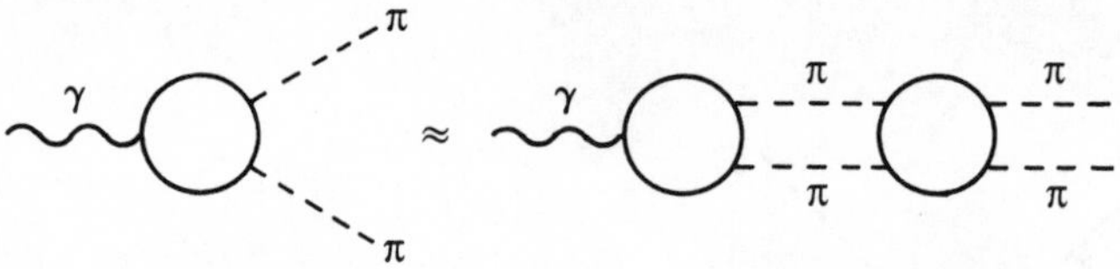

Figure 3.2   The pion form factor and pion–pion scattering.

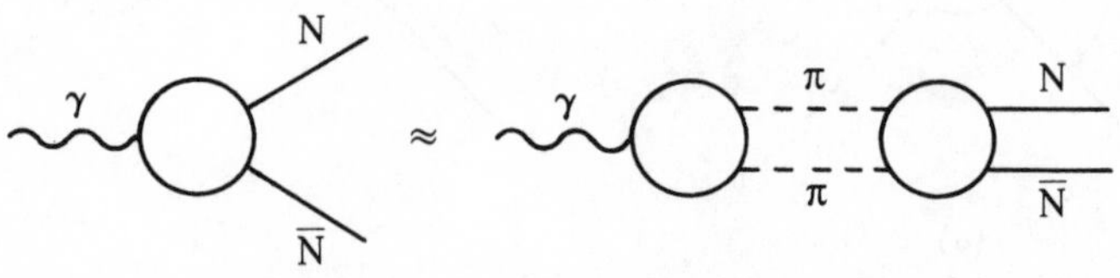

Figure 3.3   The nucleon form factor and pion–nucleon scattering.

plays an essential role in the electromagnic structure of the nucleon. In a later chapter we discuss how the amplitudes for $\pi\pi\to\pi\pi$ and $\pi\pi\to N\bar{N}$ were eventually found and what the experimental consequences of $F_\pi(t)$ and $F_N(t)$ are. A central point to appreciate, though, is that the 'coupling' of various processes illustrated in Figures 3.1–3.3 is essentially a result of unitarity (recall Eqs. (2.19) and (2.20)) in the form

$$\text{Im } T_{ij} = \sum_k T_{ik} T_{jk}^*. \tag{3.49}$$

That is, speaking loosely but graphically, unitarity ties one process to many others. This insight is essential for the bootstrap concept to be discussed later.

3.7 Summary

In the early 1950s, Goldberger and Gell-Mann at the University of Chicago saw Rohrlich and Gluckstern's paper on the use of dispersion relations to calculate scattering amplitudes in quantum field theory. They were interested in finding a calculational scheme that involved only finite, measurable quantities. Dispersion relations appeared to be a likely tool, but Goldberger and Gell-Mann sought some understanding of them in terms of first principles, such as causality. Thirring pointed out that in quantum field theory the causality requirement can be implemented by demanding that the effects of the field operators $\phi(x)$, $\phi(y)$ acting at spacelike separated points x and y be independent (cf. Eq. (3.1) or Eq. (3.9)). In their first paper on dispersion relations, Gell-Mann, Goldberger and Thirring in 1954 used this commutator causality condition and perturbation theory to derive the Kramers–Kronig dispersion relation for forward photon scattering to lowest order in e^2. The key point was that causality had led to this dispersion relation. Soon afterwards, Goldberger obtained the same dispersion relation without having to use a perturbation expansion for the electromagnetic field.

Only the forward scattering of massless particles (here, the photon) had been handled with any degree of rigor. In the case of massive particles, or for scattering in nonforward directions, there remained technical gaps in the argument. Nevertheless, dispersion relations were simply *assumed* to hold in these cases as well and were successfully applied to the analysis of both elelctromagnetic and strong interaction scattering data. This was indicative of a pragmatic attitude of many theorists at the time. Success was used as a justification for this approach. Rigorous proofs lagged far behind these applications and many of these assumed results were never proved on the basis of an underlying theory. A research program had been born.

4

Another route to a theory based on analytic reaction amplitudes

Particularly important sources for this and subsequent chapters are correspondence with Professor Chew[1] and Chew's own recent recollections (Chew, 1989), as well as the *Festschrift* for him (De Tar, Finkelstein and Tan, 1985), especially the interview it contains with Chew (Capra, 1985). Let us give an overview of the next few chapters. Geoffrey Chew is a central figure in this chapter because he was extremely important for the early development of this program and because he provides an example of how one person came to propose a radically new theoretical approach.

As we stated previously in Chapter 2, Chew at Chicago became aware of Heisenberg's S-matrix proposal from some lectures given by Wentzel (cf. Wentzel's (1947) review article). However, Heisenberg's ideas seem to have played no direct role in Chew's work leading to the modern S-matrix program. (Only around 1960 did Chew appreciate that he was working with what was essentially Heisenberg's S-matrix.) His publications from the late 1940s to the 1960s can be grouped into three fairly distinct phases:

a. 1948–1952: analysis of hadron scattering data within the framework of potential theory. Chew became an expert in that business.

b. 1953–1955: field-theory calculations, to various orders or in certain approximations, again for hadron processes.

c. 1956–1960: his work with Low, Mandelstam and Frautschi. This marked a transition to problems of greater scope and foundational importance.

All of this theoretical work received major impetus from the rapidly expanding data base provided by the large and active high-energy experimental programs existing after World War II. There was a need to handle these data and experiment had 'outstripped' theory. Motivations came not from 'high theory' but from the details of applications.

In brief summary, Chew specialized in the applications of potential scattering theory (Schrödinger equation) and was struck by the central role of transition *amplitudes* (as opposed to wave functions). This latter idea was impressed upon Chew by some work of Fermi's done in the mid 1930s on the scattering of neutrons by protons that were bound in a molecular structure. We might characterize this point (again, retrospectively) by the Lippmann–Schwinger equation (to which we return below) in the form

$$T_{\beta\alpha}(s) = V_{\beta\alpha} - \sum_\gamma \frac{V_{\beta\gamma} T_{\gamma\alpha}(s)}{E_\gamma - s} \tag{4.1}$$

where the potentials $V_{\beta\alpha}$ drive the equations that produce the physical reaction amplitudes $T_{\beta\alpha}(s)$. It is these amplitudes whose squares provide the observable scattering cross sections. Chew felt early on (around 1950) that there should be no *arbitrariness* in these potentials and amplitudes. He then became involved in quantum field theory calculations (Illinois; Low) and this eventually led to the Chew–Low theory, which would prove an essential ingredient for the *S*-matrix counterpart of force (Chew, 1989). The Chew–Low model was a specific and useful one that led to much theoretically based phenomenology in organizing data. This work impressed upon him the importance of analyticity. Crossing had been learned from Gell-Mann and Goldberger (1954). Dispersion-theory work was done by Chew, Goldberger, Low and Nambu (CGLN 1957a, 1957b). Chew (1958b) recognized and stressed the importance of the particle–pole correspondence as a general principle and this, *via* extrapolation, allowed further experimental checks. Mandelstam (1958) provided the necessary generalization of the 'potential' to the relativistic case. Regge's paper (1959) (via Mandelstam) indicated how to supply the necessary asymptotic conditions on the scattering amplitude. By 1959–60 the, what might now be termed, 'classical' *S*-matrix theory had been born. Once again (perhaps similar to the definition of a problem area as provided by Goldberger's formulation of relativistic dispersion relations), a general approach to a large class of important problems had been outlined and defined. Fairly simple analytical and calculational tools, based on general principles (e.g., unitarity, crossing and analyticity), allowed much fruitful contact with a large body of data. A fair representation of this early development may be to see Chew in the pivotal 'sociological' role of organizing a work force by putting together (his own and others') ideas in order to implement one central concept: *no arbitrariness*. (There is an interesting analogy in this concept with the importance for Einstein of general, overarching principles in theory construction: cf. Rosenthal–Schneider (1980), p. 14.) Chew has stressed the importance for him of close collaboration with others in his work[2], a point also recently discussed by Frazer (1985). Low (1985) recalls that Chew's 'work has had an extraordinary unity' and Goldberger (1985) that 'he had, at any given time, a program.' Chew's personality was a significant factor in his being able to establish a 'school' of theoretical high-energy physics. The central characteristic of this program was the extraction of general properties from specific models and their direct application to the experiments of the day. There was a fruitful and ongoing mutual interplay between theory and experiment.

4.1 Some technical preliminaries

This chapter focuses on Chew's formulation of the impulse approximation, his work on the field-theory model of meson scattering by static nucleons, the Chew–Low model, the pole–particle correspondence, and the Chew–Low extrapolation technique. However, we summarize here a few technical matters that are necessary as background for the understanding of those developments.

First, let us indicate where the Lippmann–Schwinger equation (Eq. (4.1)) comes from. In Eq. (A.11b) the scattering amplitude is expressed in terms of the matrix element $\langle \phi_f | V | \psi_i^+ \rangle$ where ϕ_f is a plane-wave state for the final momentum $\mathbf{k}_f$ and ψ_i^+ the scattering state for the incident particle of momentum $\mathbf{k}_i$. The square of this matrix element gives the probability that, if the system begins in a state ψ_i^+, then, under the influence of the interaction V, it will end up in the state ϕ_f. Let us now use the index β collectively for all the final state variables and α for the initial state ones. If we write the Hamiltonian as

$$H = H_0 + V, \tag{4.2}$$

then the time-independent Schrödinger equation for ψ_α^+ is

$$H\psi_\alpha^+ = E_\alpha \psi_\alpha^+ \tag{4.3}$$

and the free-particle ('plane wave') states satisfy

$$H_0 \phi_\alpha = E_\alpha \phi_\alpha. \tag{4.4}$$

The formal solution for ψ_α^+, having the proper incident 'plane wave' and outgoing scattered-wave boundary conditions, is

$$\psi_\alpha^+ = \phi_\alpha - (H_0 - E_\alpha - i\varepsilon)^{-1} V \psi_\alpha^+, \qquad (4.5)$$

where the limit $\varepsilon \to 0^+$ is understood. This is the (nonrelativistic) Lippmann–Schwinger (1950) equation. (In the $\mathbf{r}$ representation, $\psi_\alpha^+(\mathbf{r}) = \langle \mathbf{r} | \psi_\alpha^+ \rangle$ and Eq. (4.5) becomes Eq. (A.10a).) From it, we obtain an integral equation for the scattering amplitude $T_{\beta\alpha}(s)$ with $s = E_\alpha + i\varepsilon$,

$$
\begin{aligned}
T_{\beta\alpha}(s) &\equiv \langle \phi_\beta | V | \psi_\alpha^+ \rangle \\
&= \langle \phi_\beta | V | \phi_\alpha \rangle - \langle \phi_\beta | V (H_0 - s)^{-1} V \psi_\alpha^+ \rangle \\
&= V_{\beta\alpha} - \sum_\gamma \frac{V_{\beta\gamma} T_{\gamma\alpha}(s)}{E_\gamma - s}.
\end{aligned}
\qquad (4.6)
$$

This is Eq. (4.1), the Lippmann–Schwinger equation for the transition amplitude $T_{\beta\alpha}(s)$.[3] This equation shows (as we would expect) that the interaction V directly determines the transition amplitude. It is equivalent to the Schrödinger equation.

Next, if the unitarity condition of Eq. (2.20) is written out in detail for elastic scattering, it becomes

$$\operatorname{Im} T(\hat{\mathbf{k}}_\beta \cdot \hat{\mathbf{k}}_\alpha; k) = \frac{\pi k \sqrt{s}}{2} \int d\Omega_k T^*(\hat{\mathbf{k}}_\gamma \cdot \hat{\mathbf{k}}_\beta; k) T(\hat{\mathbf{k}}_\gamma \cdot \hat{\mathbf{k}}_\alpha; k)$$

$$(4.7)$$

where $k\sqrt{s}/2 \to km$ in the nonrelativistic limit. If $T(\hat{\mathbf{k}}_\beta \cdot \hat{\mathbf{k}}_\alpha; k)$ is expanded in partial waves (cf. Eq. (A.23a)), then the partial-wave amplitudes $T_l(k)$ satisfy the unitarity condition

$$\operatorname{Im} T_l(k) = 2\pi^2 k \sqrt{s} |T_l(k)|^2. \qquad (4.8)$$

If $\delta_l(k)$ is the phase of $T_l(k)$, then the form of Eq. (4.8) requires that

$$T_l(k) = \frac{1}{2\pi^2 k \sqrt{s}} e^{i\delta_l(k)} \sin \delta_l(k). \qquad (4.9)$$

The point for our purposes later is that the form (4.9) for $T_l(k)$ automatically satisfies unitarity, where $\delta_l(k)$ is the phase shift. That is, if an amplitude T is subject to unitarity, then that amplitude *must* have the form of Eq. (4.9) (and vice versa).

So far we have avoided the complications that are due to the spin and charge of the particles involved. We shall continue to do this as much as possible, since our main concern is with the basic concepts involved. Nevertheless, the spin of the nucleon (N) and the charge states of the pion (π) and nucleon will be essential below. By a nucleon N we mean either a proton p or neutron n and by a pion π any of its three charge states π^+, π^-, or π^0. The isotopic spin (or isospin) formalism for the strong interactions treats the proton and neutron as members of the nucleon isospin doublet ($\tau = \frac{1}{2}$) with $\tau_3 = +\frac{1}{2}$ for p and $\tau_3 = -\frac{1}{2}$ for n. This is very much like the quantum-mechanical treatment of an electron with spin $s = \frac{1}{2}$ and projections $s_3 = \pm\frac{1}{2}$. (The technical reason for this is that rotations in real three-dimensional space and in isospin space are both governed by the same mathematical group, often denoted by SU(2).) Similarly, the pion is assigned to an isotriplet ($t = 1$) with $t_3 = +1$ for π^+, and $t_3 = 0$ for π^0, and $t_3 = -1$ for π^-. The assumed 'charge independence' (somewhat of a misnomer) of the strong interactions is implemented formally by requiring the pion–nucleon interaction to be invariant under rotations in isospin space. Loosely and a bit misleadingly speaking, charge independence means that the strong interaction should be indifferent to the charge state (e.g., p versus n or $\tau_3 = +\frac{1}{2}$ versus $\tau_3 = -\frac{1}{2}$) of the strongly interacting particles. Rotations in isospace can, among other things, 'rotate' a proton into a neutron (or 'interchange' them). (This isospin space is a convenient mathematical construct, not a physical space like the familiar space-time one.) That is, just as with the usual addition rules for quantum-mechanical spins, the total isospin vector, $\mathbf{T} = \mathbf{t} + \boldsymbol{\tau}$, can have the values $\frac{3}{2}$ and $\frac{1}{2}$. There are just *two* independent isospin amplitudes in terms of which all the $\pi N \rightarrow \pi N$ reactions can be expressed. For example, we find that

$$\langle \pi^+ p | T | \pi^+ p \rangle \equiv T_{\pi^+ p} = T^{\frac{3}{2}}, \tag{4.10a}$$

$$\langle \pi^- p | T | \pi^- p \rangle \equiv T_{\pi^- p} = \tfrac{1}{3} T^{\frac{3}{2}} + \tfrac{2}{3} T^{\frac{1}{2}}, \tag{4.10b}$$

$$\langle \pi^0 n | T | \pi^- p \rangle \equiv T_{\text{ch·ex.}} = \tfrac{\sqrt{2}}{3}(T^{\frac{3}{2}} - T^{\frac{1}{2}}), \tag{4.10c}$$

where $T^{\frac{3}{2}}$ and $T^{\frac{1}{2}}$ are the isospin $\frac{3}{2}$ and $\frac{1}{2}$ amplitudes, respectively. The three physical amplitudes are then linearly related as

$$T_{\pi^+ p} = T_{\pi^- p} + \sqrt{2}\, T_{\text{ch·ex.}}. \tag{4.11}$$

Here, as typically, the invariance group has reduced the number of independent amplitudes.

Now a pion in an orbital angular momentum state $l = 1$ can couple with a nucleon of spin $s = \frac{1}{2}$ to form total angular momentum states $J = \frac{3}{2}$

and $J=\frac{1}{2}$. So, the phase shifts $\delta(k)$ for pion–nucleon scattering, rather than being labeled simply $\delta_l(k)$, will depend upon both T and J, as say $\delta_{TJ}(k)$. The convention that has actually been adopted historically is to use $2J$ and $2T$ as subscripts. Thus, $\delta_{33}(k)$ is the phase shift for the $T=\frac{3}{2}$, $J=\frac{3}{2}$ state.

4.2   Fermi and the impulse approximation

For the historical development of this chapter we can begin with Fermi's (1936) paper 'On the motion of neutrons in hydrogenous substances.' In Section 10 ('Scattering of neutrons with bound hydrogen atoms') Fermi developed an approximation under the assumption that the de Broglie wavelength (λ) of the neutron is much greater than the range (ρ) of the n–p force and than the n–p scattering length (a_0). Here the de Broglie wavelength λ is a measure of the effective size of the wave packet associated with the neutron of momentum p (where $\lambda = h/p$). The range ρ is the distance beyond which the nuclear force is essentially zero and the scattering length a_0 is (*very* crudely) the distance over which the n–p wavefunction differs appreciably from zero. Under these conditions, (i.e., $\lambda \gg \rho$ and $\lambda \gg a_0$), Fermi devised (1962, p. 1007) an approximation to the Schrödinger equation when the bound particle (here the proton in the hydrogen atom) is confined by the intermolecular potential while the projectile particle (a neutron) interacts strongly but only for a very short time with the bound particle. During this short but intense interaction, all the other variables remain essentially constant and the neutron simply scatters from the (fixed) bound proton. This is not the same as treating the neutron and proton as free particles. Fermi was able to express this scattering of a neutron from a (molecularly) bound hydrogen atom directly in terms of the parameters for free n–p scattering and a molecular binding parameter. Since this problem was the scattering of slow neutrons from, say, paraffin, the approximations made were for low energies. In 1947 Fermi suggested to Chew and Goldberger the possibility of extending these ideas to a very different regime, that of high-energy scattering (Goldberger and Watson, 1964, p. 684; Capra, 1985, p. 253). Chew (1948) worked on proton–deuteron scattering for his Ph.D. thesis problem (Chamberlain, 1985, p. 12). While Chew was still a predoctoral fellow, he and Goldberger (1948) did some Schrödinger equation calculations (including spin and isospin) to fit n–p scattering data and then they (Chew and Goldberger, 1949a, 1949b) made some extensions of the effective-range formula (an approximation for low-energy scattering, cf. Eq. (A.22)) for n–p and p–p scattering. Chew and Goldberger (1950) wrote a paper on 'pick up', or sudden rearrangement, processes in high energy nuclear reactions in which, say, a bombarding neutron picks up a proton from a target nucleus to produce a deuteron as a product. (A rigorous treatment of such processes was later given by Gell-Mann and Goldberger (1953)). However, Chew continued to concentrate on p–d scattering as an outgrowth of his thesis work (1948). In a series of papers (Chew, 1950a, 1950b, 1951a, 1951b; Chew and Wick, 1951, 1952; Chew, Goldberger, Steinberger and Yang, 1951; Chew and Lewis, 1951; Chew and Goldberger, 1952), Chew and his collaborators developed the impulse approximation, the name given to his extension of Fermi's older approximation. Chew (1950a, p. 219) stressed the use of measurable physical quantities only:

> A theoretical approach has been developed to relate n–d inelastic scattering to free n–p and n–n scattering cross sections at the same energy, *without reference* to the nature of nuclear forces.

Again, Chew and Lewis (1951, p. 779) make a similar point:

> 'A phenomenological treatment of charged and neutral photomeson production from deuterons is presented which is independent of the details of the meson theory.'

The central idea here is that the scattering between the bound and free particle is essentially given by the amplitude these particles would have if neither were bound, except that the bound particle is given the momentum distribution (or 'energy') corresponding to its actual bound state. Chew and Wick (1952, p. 637) summarize the approximation as follows.

> There are many circumstances ... under which the many-body character of a problem [of nuclear collisions involving more than two nucleons] is only a secondary feature, and the collision may be decomposed, as it were, into a superposition of simple two-body collisions .... [However,] the crude additivity rule for cross sections ... is seldom applicable. The impulse approximation is another expression of this same thought.

> Consider as a typical case a simple particle (say a nucleon) striking a complex system (deuteron, light nucleus). The assumptions [of the impulse approximation] are then the following: (I) The incident particle never interacts strongly with two constituents of the system at the same time; (II) the amplitude of the incident wave falling on each constituent (nucleon) is nearly the same as if that constituent were alone; and (III)

the binding forces between the constituents of the system are negligible during the decisive phase of the collision, when the incident particle interacts strongly with the system.

The theme of these approximations is to express the reaction amplitudes for complex scattering processes in terms of *two-body* amplitudes (which can be related to measurable quantities or reasonable extrapolations from them). This method was successfully applied to the analysis of a wide range of (then-current) experimental situations: elastic and inelastic n–d scattering, elastic p–d scattering (in terms of the free n–p and p–p amplitudes), the reaction $\pi^+ + d \rightarrow p + p$, and pion photoproduction from deuterons. Chew's early work was closely associated with analysis of experimental data (e.g., Chew and Steinberger, 1950a, 1950b; Chew and Moyer, 1950, 1951; Moyer and Chew, 1951; Taylor and Chew, 1950; Lax, Feshbach, Chew and Lewis, 1951). The general framework was a Schrödinger theory (say something like Eqs. (4.2)–(4.6)), but the goal was to extract results that related (measurable) amplitudes in a model-independent way. Chew was struck by the possibility of expressing everything in terms of *reaction amplitudes* (driven by potentials) rather than in terms of *wave functions*. (See Eq. (4.1) for our attempt to represent this idea). At this stage in his career, Chew saw no particular point in philosophy (Capra, 1985, p. 266). Recently, Chew has recalled of this period:[4]

> It is true that the published record of the 1948–52 period shows no indication of my uneasiness with Hamiltonian (Lagrangian) theory based on potentials, since I used the Lippmann–Schwinger equation. Part of the reason is that Goldberger and Wick did not share my uneasiness, and part is that I could at that time find no language to express my feelings. The root of my feelings was Fermi's rule for the scattering of slow neutrons by molecules, which became generalized into the impulse approximation. Here the scattering amplitude for a 'complex' particle was expressed directly through scattering amplitudes for 'constituent' particles, with no reference to a potential. It was somehow evident to me, even then, that such a representation should have a generalization.

> My dissatisfaction with QFT stemmed from my dissatisfaction with arbitrary potentials in a *Lagrangian*.

Notice the importance Chew attaches to the absence of any 'language' (or framework) within which to express his views. In Section 9.4 we shall return to the role that the language provided by a model plays in establishing a conceptual framework. There our example will be the compound nucleus model of nuclear reactions.

4.3　The Chew model and phenomenology

Once at the University of Illinois and able to work with Francis Low, who was an expert in quantum field theory, Chew began a study of the fixed-source theory applied to the scattering of pions by the nucleon. To get some idea what the static model is like, let us consider the (fictitious) interaction of neutral particles (say 'pions'), represented by the field $\phi(\mathbf{r})$, with a static (i.e., fixed or at rest) source ('nucleons') distributed according to the density function $\rho(\mathbf{r})$.[5] The interaction Hamiltonian is

$$H_{\mathrm{I}} = \gamma \int \rho(\mathbf{r})\phi(\mathbf{r})\,\mathrm{d}\mathbf{r}. \tag{4.12}$$

That is, this interaction in Hamilton's form of the field equation produces a dynamical equation for the field $\phi(\mathbf{r})$ in which $\rho(\mathbf{r})$ appears as the source term (much as in, say, Poisson's equation, $\nabla^2\phi = \rho$, where ρ is the source term (say, electric charges) and ϕ is the electrostatic potential). If we take two static point nucleons located at $\mathbf{r}_1$ and $\mathbf{r}_2$ and evaluate only the one-meson exchange graphs of Figure 4.1, we find the energy perturbation or 'potential' $V(r)$ between them to be

$$\Delta E = V(r) = \frac{\gamma^2 \mathrm{e}^{-mr}}{4\pi r}. \tag{4.13}$$

This is the well-known Yukawa potential where $r = |\mathbf{r}_1 - \mathbf{r}_2|$, m is the meson mass and γ is the $NN\pi$ coupling constant. We state this result only to indicate that such static models can give qualitatively reasonable results for low-energy phenomena (in which the recoil of the nucleus is quite small).

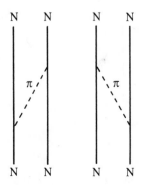

Figure 4.1　The one-pion nucleon exchange force.

The static model Chew actually considered had the interaction Hamiltonian

$$H_I = \sqrt{4\pi} \frac{f}{\mu} \sum_{\lambda=1}^{3} \tau_\lambda \int d\mathbf{r} \rho(\mathbf{r}) \boldsymbol{\sigma} \cdot \nabla \phi_\lambda(\mathbf{r}). \qquad (4.14)$$

Here τ and $\boldsymbol{\sigma}$ are, respectively, the isospin and spin operators for the nucleon, $\phi_\lambda(\mathbf{r})$ the pion field, μ the pion mass, and λ is the isospin index (or label). This model was not new (Pauli (1946) had reviewed it in his lectures at MIT in 1944), but Chew pressed the model pretty far in analyzing the available πN data. He (1953) first observed that a static model (i.e., a fixed nucleon N) with a point source for the interaction with pseudoscalar (π) mesons could not reproduce in lowest-order perturbation in the coupling constant the observed phase shifts for the πN data. As Figure 4.2 indicates, there is a πN resonance (or 'bump') around 200 Mev (pion laboratory kinetic energy) and the shape of the angular distribution (or differential scattering cross section) at that energy requires that this resonance be a p-wave ($l = 1$) one. The basic idea is that at a resonance just *one* partial wave (i.e., value of l) dominates the expansion of Eq. (A.23) so that $f(k,\theta)$ becomes approximately $f_l(k)P_l(\cos\theta)$. This implies that the differential cross section $d\sigma/d\Omega$ behaves, as a function of θ, as $|P_l(\cos\theta)|^2$ (cf. Eq. (A.14)). For $l = 1$, $P_l(\cos\theta) = \cos\theta$ so that $d\sigma/d\Omega \sim \cos^2\theta$. As Figure 4.2 (*b*) indicates, there is a $\cos^2\theta$ variation. (Actually, the $1 + 3\cos^2\theta$ form is

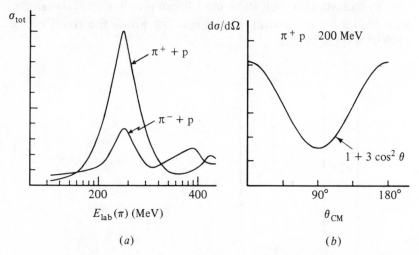

Figure 4.2   The (3,3) resonance in pion–nucleon scattering (after Jackson (1958), pp. 13 and 15).

produced by the nucleon spin of $\frac{1}{2}$ coupling to the pion angular momentum $l=1$.) The essential point to appreciate, though, is that the specific *shape* of the angular distribution at resonance is the *signature* of a particular angular momentum (or 'spin') for the resonance state. Here, this is the δ_{33}, or (3,3), resonance.[6] The important point is that of the four $l=1$ phase shifts (δ_{11}, δ_{13}, δ_{31}, and δ_{33}) only δ_{33} is large and positive near this resonance at 200 Mev. Although the point-source model does not allow this (3,3) resonance, Chew (1953) showed that, for an extended source model, the lowest (that is, second) order diagrams of Figure 4.3 did allow δ_{33} to be positive while δ_{11}, δ_{13} and δ_{31} were all negative. Unfortunately, δ_{33} could not be made large while the others remained small. Chew (1954a, 1954b) next studied the renormalization theory of this model and developed a new approximation method for doing perturbation calculations in terms of renormalized quantities. With this expansion he (1954c) then made a fourth-order calculation of these phase shifts and found that δ_{33} could resonate while the others remained small and negative. This allowed a fit to the data. Chew (1954d) compared the predictions of this model with the experimental data on πN scattering, on the photoproduction of pions ($\gamma + N \rightarrow \pi + N$), and on the anomalous magnetic moments of the nucleons. (The *anomalous* magnetic moment refers to the departure of the actual magnetic moment of a particle from its basic Bohr magneton value of $eh/2mc$.) The agreement was quite good and, from this data, Chew suggested a value of $f^2 = 0.058$ for the πN coupling constant from a fit to the data. Similar to what he had previously done within the context of potential theory, this program, now in a quantum field theory framework, had once again obtained relations among observable quantities.

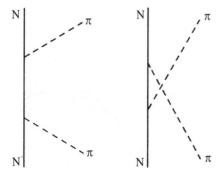

Figure 4.3   Pion scattering from a static nucleon.

Although the Chew model, as it became known, was phenomenologically successful, its theoretical basis remained uncertain since the results had been obtained from a few low-order terms in a perturbation expansion, whose convergence properties were unknown. Low (1954), in a paper we discussed in Chapter 3, began with Dyson's definition of the S matrix (Eq. (3.17)) and obtained an *exact* set of nonlinear integral equations involving only renormalized quantities. These were equations for (essentially) the πN scattering amplitude. In the limit of static nucleons and the approximation of just one-meson intermediate states, this Low equation reduced to a set of three coupled, nonlinear integral equations for the δ_{11}, $\delta_{13} = \delta_{31}$ and δ_{33} phase shifts. Low (1954) also pointed out that a linear integral equation could be obtained for the pion photoproduction amplitude. The early Chew–Low work really revitalized strong-interaction dynamics since it allowed considerable headway to be made in organizing empirical data using fairly elementary analytical techniques.

4.4   A digression: crossing symmetry

We must now digress to define *crossing symmetry*.[7] To begin, let us illustrate this with some simple examples. Consider the (somewhat artificial) case of charged pions scattering *only via* their electromagnetic interactions, say

$$\pi^+\pi^+ \to \pi^+\pi^+. \tag{4.15}$$

The lowest-order, one-photon exchange diagrams are shown in Figure 4.4. The Feynman rules for these two diagrams

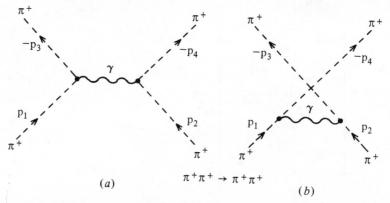

$$\pi^+\pi^+ \to \pi^+\pi^+$$

(a)   (b)

Figure 4.4   One-photon contributions to $\pi^+\pi^+ \to \pi^+\pi^+$ scattering.

yield, respectively, for the corresponding amplitudes

$$e^2 \frac{s-u}{t}, \tag{4.16a}$$

$$e^2 \frac{s-t}{u}, \tag{4.16b}$$

where we have defined the kinematic variables

$$s = (p_1 + p_2)^2 = (p_3 + p_4)^2, \tag{4.17a}$$

$$t = (p_1 + p_3)^2 = (p_2 + p_4)^2, \tag{4.17b}$$

$$u = (p_1 + p_4)^2 = (p_2 + p_3)^2, \tag{4.17c}$$

with

$$s + t + u = 4m^2. \tag{4.18}$$

We have neglected multiplicative factors (such as 4π, etc.) in Eqs. (4.16). It will also become clear why we have chosen to label the outgoing momenta in Figure 4.4 as $-p_3$ and $-p_4$. First, notice that if we were to exchange the 'legs' 3 and 4 (actually, the four momenta p_3 and p_4 as $p_3 \Leftrightarrow p_4$), then Figures 4.4 (a) and 4.4 (b) would interchange. In terms of the variables of Eqs. (4.17), this is equivalent to the interchange $t \Leftrightarrow u$. But, from Eqs. (4.16), we see that this interchange or 'substitution' (i.e., $t \Leftrightarrow u$) transforms (4.16a) into (4.16b) and vice versa. This is an example of the substitution law known for some time in quantum electrodynamics. Since the full scattering amplitude for $\pi^+\pi^+ \to \pi^+\pi^+$ is then given (in our one-photon approximation) as the sum of the terms of Eqs. (4.16),

$$T_{\pi^+\pi^+}(s,t,u) = e^2 \left[\frac{s-u}{t} + \frac{s-t}{u} \right], \tag{4.19}$$

we see that this amplitude is invariant under the interchange $t \Leftrightarrow u$. This is an example of crossing symmetry and it is a result of the fact that for each diagram in the Feynman series the crossed diagram also occurs. This result remains valid when *all* higher-order Feynman diagrams are included.

For strong-interaction physics this crossing property was discovered and first stated by Goldberger and Gell-Mann. Thus, we find in Goldberger (1954, p. 29):

> One can divide all Feynman diagrams into two classes in one of which the meson lines are uncrossed and in the second of which the meson

lines are crossed. Since the total matrix element is just the sum of these two it must satisfy the requirement, in the forward direction, that $M_{ji}(q) = M_{ij}(-q)$ where q is the four-momentum of the meson.

Similarly, Gell-Mann and Goldberger (1954, p. 1436) observed that the scattering amplitude for photon-nucleon scattering $(\gamma + N \rightarrow \gamma + N)$ satisfies the symmetry relation

$$T_{\nu\mu}(q', q) = T_{\mu\nu}(-q, -q').$$

This symmetry condition depends only on the fact that for every 'crossed' diagram there is a corresponding 'uncrossed' one and vice versa.

Since we have introduced the concept of crossing here, let us discuss it a bit more, for reference later. Take the reaction

$$\pi^+ \pi^- \rightarrow \pi^+ \pi^-, \tag{4.20}$$

illustrated in the one-photon approximation in Figure 4.5. In terms of the variables

$$s' = (p'_1 + p'_2)^2 = (p'_3 + p'_4)^2, \tag{4.21a}$$

$$t' = (p'_1 + p'_3) = (p'_2 + p'_4)^2, \tag{4.21b}$$

$$u' = (p'_1 + p'_4)^2 = (p'_2 + p'_3)^2, \tag{4.21c}$$

$$s' + t' + u' = 4m^2, \tag{4.21d}$$

the scattering amplitude for (4.20) is

$$T_{\pi^+\pi^-}(s', t', u') = -e^2 \left[\frac{t' - u'}{s'} + \frac{s' - u'}{t'} \right]. \tag{4.22}$$

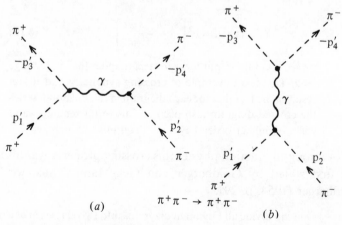

Figure 4.5   One-photon contributions to $\pi^+\pi^- \rightarrow \pi^+\pi^-$ scattering.

This amplitude satisfies the crossing relation of symmetry under the interchange $p'_2 \Leftrightarrow p'_3$ or $s' \Leftrightarrow t'$. Now, let us make a very important observation.[8] Suppose we make the interchanges $s' \Leftrightarrow u$, $u' \Leftrightarrow s$, $t' \Leftrightarrow t$. This produces the replacement $T_{\pi^+\pi^+} \Leftrightarrow T_{\pi^+\pi^-}$. That is, the amplitudes for the reactions (4.15) and (4.20) are given by the *same* amplitude, once the values, or roles, of the variables are extended, or interchanged. In fact, Figure 4.6 can represent both of these reactions if we agree to reverse the sign on the four momentum and change a particle to an antiparticle (here $\pi^+ \Leftrightarrow \pi^-$) when we read the diagram against the direction of the arrow on a leg. Thus, the diagram of Figure 4.6 has three 'channels' (or ways of being read):

$$s: 1+2 \to \bar{3}+\bar{4} \tag{4.23a}$$
$$\pi^+\pi^- \to \pi^+\pi^-$$

$$t: 1+3 \to \bar{2}+\bar{4} \tag{4.23b}$$
$$\pi^+\pi^- \to \pi^+\pi^-$$

$$u: 1+4 \to \bar{3}+\bar{2} \tag{4.23c}$$
$$\pi^+\pi^+ \to \pi^+\pi^+$$

So much for technical background.

Chew recalls[9] having learned about crossing and analyticity from Gell-Mann and Goldberger. Gell-Mann (1987, p. 484) has given his own representation of this encounter.

> Goldberger and I [around 1956] tried to teach these notions [the use of crossing, analyticity and unitarity to generate scattering amplitudes] to Geoffrey Chew. It was very difficult, because he resisted

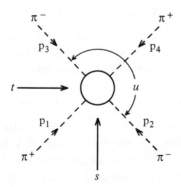

Figure 4.6   The pion–pion elastic scattering amplitude.

furiously. Among other things, he disliked the idea that there were mysterious boundary [or asymptotic] conditions (and perhaps other conditions) that would distinguish one theory from another and complete the information necessary to give the whole S-matrix.

4.5   The Chew–Low-model

Let us now continue with Chew and Low's work. Wick (1955, p. 339), in a review article on some applications of quantum field theory to strong interaction physics, characterized the importance of Chew's use of field theory models as follows:

> Thanks to Chew's efforts, the Yukawa theory has at last achieved some contact with the quantitative aspects of meson physics. ... Chew's work also contained another useful lesson. It showed ... that it was possible to apply to this theory ideas and methods derived from modern field theoretic work, without getting involved into too abstruse formalisms.

A few years later, this mix of some field theory results with phenomenology to yield a useful analysis and a correlation of experimental data on strong interactions was also exemplified by the use of the one-pion exchange force from field theory (much as in Eqs. (4.12) and (4.13)) to provide the long-range part of the nucleon–nucleon interaction. This known long-range contribution was then combined with a phenomonological representation for a phase-shift analysis of nucleon–nucleon scattering (Cziffra, MacGregor, Moravcsik and Stapp, 1959; MacGregor, Moravcsik and Stapp, 1959). Michael Moravcsik[10], originator of this 'modified phase shift analysis' method, has emphasized that such a mix of theory and phenomenology for the strong interactions was seen as quite unconventional at the time.

> I vividly recall the morning at the Berkeley Rad Lab (I believe it was in Peter Cziffra's office) when he, Chew, and I happened to be gathered, and I suggested this scheme. Geoff's first reaction was that one cannot do that, it won't work, you cannot mix theory and phenomenology in such a way. It took only an hour, however, to convince him.

In their original paper on this new method, we find acknowledgement of the importance of Chew's work on the dominance of a scattering amplitude by the nearest poles (i.e., those singularities closest to the region of the variable accessible to experiment). Such dominance gave reason to believe that only a few remaining low partial waves would have to be fit phenomenologically (Cziffra, MacGregor, Moravcsik and Stapp, 1959, p. 880).

> It has been suggested by one of us (M. J. M.) that the conventional phase shift analysis of nucleon–nucleon scattering experiments be replaced by a modified scheme in which the contribution due to the exchange of one pion is explicitly included in the scattering amplitude. This approach was motivated by some conjectures of Chew [1958b] on the behavior of the scattering amplitudes in the nonphysical region of the complex $\cos \theta$ plane.

This semiphenomenological approach in which some firm theoretical concepts are smoothly interfaced with a phenomenological component was a truly quantitative verification of the Yukawa idea that forces are generated by particle exchange (as in Figure 4.1 and the discussion leading to Eq. (4.13)). Today, nearly thirty years later, modifications of this technique are still used to study the nucleon–nucleon interaction. One may even wonder whether much real progress has been made beyond the 'one-meson' level (Moravcsik, 1985). (The 'Paris' and 'Bonn' nuclear potentials are modern examples of this, as we shall see in Chapter 8.)

In 1956 Chew and Low (1956a, p. 1570) effectively combined the Low equation and the Chew model.

> Recently, it has been shown [Chew, 1954d] that a crude static model of the pion–nucleon interaction, based on the Yukawa idea, is quite powerful in correlating certain experiments; however, the relation of the model to a true theory has been obscure. One of the main purposes of this paper ... is to show that the most important predictions of the model are actually independent of its details and thus may also be predictions of a 'true' theory.

This paper would prove to be another example of a skill Chew developed of abstracting general results from model theories, something that is reminiscent of the old Heisenberg program and that would be essential for the later S-matrix theory program.

Chew and Low (1956a) gave a relatively simple derivation of their model by a method suggested by Wick (1955). Their static model was characterized by the Hamiltonian

$$H = H_0 + H_I, \tag{4.24a}$$

$$H_0 = \sum_k a_k^\dagger a_k \omega_k, \tag{4.24b}$$

$$H_I = \sum_k [V_k a_k + V_k a_k^\dagger]. \tag{4.24c}$$

(The meson mass μ has been set equal to unity.) The scattering amplitude matrix elements for n meson states Ψ_n are given as (Wick, 1955)

$$T_q(n) = \langle \Psi_n^{(-)} | V_q \Psi_0 \rangle \qquad (4.25)$$

where Ψ_0 is the single-nucleon state. With the Lippmann–Schwinger (1950) formalism, Chew and Low (1956a) derived a nonlinear equation for $T_q(p)$. If we define[11]

$$t_{qp}(z) = -\sum_n \left[\frac{T_p^\dagger(n)T_q(n)}{E_n - z} + \frac{T_q^\dagger(n)T_p(n)}{E_n + z} \right], \qquad (4.26)$$

then

$$T_q(p) = \lim_{z \to \omega_p + i\varepsilon} t_{qp}(z). \qquad (4.27)$$

Here E_n is the energy of the n-meson state. Equation (4.26) is an *exact* equation for the (*one*-meson) elastic πN scattering amplitude. The function $t_{qp}(z)$ embodies all of the following properties:
   (i) satisfies unitarity
   (ii) satisfies crossing as

$$t_{qp}(z) = t_{pq}(-z) \qquad (4.28)$$

   (iii) has a simple pole at $z = 0$
   (iv) behaves as $1/z$ as $z \to \infty$
   (v) has branch cuts on the real axis,

$$-\infty < x < -1, \qquad 1 < x < +\infty.$$

Part of the importance of the Chew–Low model was that it provided a specific instantiation of these generally hoped-for properties. A simple dispersion integral can be written down for $t_{qp}(z)$ satisfying all the properties (i)–(v) (Chew and Low, 1956a, p. 1573). If the one-meson approximation is now made, then these $t_{qp}(z)$ can be expressed in terms of three new functions each satisfying a unitarity relation like Eq. (4.8). Therefore, (cf. Eq. (4.9)) these new functions depend only upon the three πN phase shifts $\delta_{11}, \delta_{13} = \delta_{13}$ and δ_{33} (as well as, of course, upon $v(p)$, the Fourier transform of the nucleon source function $\rho(r)$) and have simple dispersion integral representations. A numerical solution to the Chew–Low equation was obtained by Salzman and Salzman (1957).

We can illustrate the method of solution of the Chew–Low equation (in the one-meson approximation) with a simple example[12], although the actual Chew–Low problem is much more involved. If there were only one spatial dimension and hence just one function $F(z)$ (this is the analogue of $t_{pq}(z)$ above) to be determined, then $F(z)$ would have the following properties:

(i) $F^*(z^*) = F(z)$; $F(x)$ – real for $0 < x < 4$ (4.29a)

(ii) $\operatorname{Im} F(x) = \begin{cases} 0, & 0 < x < 4 \\ |F(x)|^2/2\sqrt{x(x-4)}, & x > 4 \end{cases}$,

       – unitarity   (4.29b)

(iii) $F(z) = F(4 - z)$ – crossing   (4.29c)

(iv) $F(z) \xrightarrow[z \to 1]{} 1/(z - 1)$ – bound-state pole   (4.29d)

(v) $F(z) \xrightarrow[|z| \to \infty]{} 1/(z - 1) - 1/(z - 3) \equiv f(z) \to -\dfrac{2}{z^2}$

       – asymptotic condition   (4.29e)

Statement (i) follows from analyticity once $F(x)$ is required to be *real* (i.e., $F^*(x) = F(x)$, or $\operatorname{Im} F(x) \equiv 0$) below the scattering threshold at $x = (1 + 1)^2 = 4$. This latter requirement is connected with the optical theorem, as we have discussed previously (cf., Eq. (3.34)). The unitarity condition (ii) is essentially Eq. (2.20) or Eq. (4.8), where the factor $[x(x-4)]^{-\frac{1}{2}}$ arises from the two-dimensional structure of this particular model. Similarly, in this model there are just *two* channels so that $s + u = 4$ and $s \Leftrightarrow u$, or crossing, interchange amounts to $s \Leftrightarrow 4 - s$ (or $z \Leftrightarrow 4 - z$). This is property (iii). The assumption of a particle of mass $m = 1$ requires a (bound-state) pole at $z = 1$, which is (iv). Finally, (v) is a statement about how $F(z)$ behaves at infinity. It is a technical condition that allows a (well-defined) dispersion integral to be written down. (Other asymptotic forms are possible, but this has been assumed to simplify the example here.)

If we define a new function

$$G(z) = \frac{f(z)}{F(z)} \xrightarrow[|z| \to \infty]{} 1, \tag{4.30}$$

then unitarity (ii) on $F(x)$ implies

$$\operatorname{Im} G(x) = -\frac{f(x)}{2\sqrt{x(x-4)}}, \qquad x > 4. \tag{4.31}$$

Now a dispersion relation can be written for $G(z)$ as

$$G(z) = 1 - \frac{1}{2\pi} \int_4^\infty \frac{dx' f(x')}{\sqrt{x'(x'-4)}} \left[\frac{1}{x'-z} + \frac{1}{x'+z-4} \right].$$

(4.32)

This integral can be carried out and the result expressed in terms of elementary functions. Of course, this solution is not unique since we must add to the right hand side of Eq. (4.32) any pole terms in $G(z)$ caused by the zeros of $F(z)$ (cf. Eq. (4.30)). This nonuniqueness of solutions is a general characteristic of solutions to the static Low equation in the one-meson approximation. Castillejo, Dalitz and Dyson (1956) demonstrated that such solutions depend upon an infinite number of adjustable parameters. It was hoped that an exact treatment of the full problem would reduce this arbitrariness.

Among many remarkable properties obtained by Chew and Low (1956a) is the prediction that the quantity

$$\frac{k^3}{\omega} \cos \delta(\omega)$$

(4.33)

should be a straight line for small ω and should extrapolate to $1/f^2$ as $\omega \to 0$.[13] This is a generalization of the effective range formula of Eq. (A.22) and comes from the expansion of an integral like (4.32) for Re $G(x)$.[14] Experimental values for Eq. (4.33) for δ_{33} fell on a straight line and gave $f^2 = 0.08$. Thus, a value for f^2 had been arrived at by two independent methods (recall Chew, 1954d, and the discussion in Section 4.3 above) and the results agreed. The Chew–Low model also gave indications that δ_{33} would resonate, but did not require it to do so. Chew and Low (1956b) then applied their model to the photoproduction of pions $(\gamma + N \to \pi + N$, one of the reactions considered in Chew, 1954d) and expressed the low-energy amplitude for this process in terms of the πN scattering phase shifts. Their result satisfied the Kroll–Ruderman (1954) prediction that this low-energy amplitude should allow another determination of f^2. The photomeson production data were consistent with the $f^2 = 0.08$ obtained from πN scattering, a totally different process. Unitarity, crossing and analyticity were all essential ingredients in the extremely successful Chew–Low model and were expected to transcend this particular model (Chew and Low, 1956a, p. 1578). There was, in fact, a large industry at Illinois on the static-model Yukawa theory (Chew, 1956).

The collaboration of Chew, Goldberger, Low and Nambu (1957a, 1957b) brought together the concepts of the Chew–Low theory and fully relativistic dispersion relations. This was a natural collaboration, resulting from visits between Chicago and Urbana, of Goldberger and Nambu, who had worked on single-variable dispersion relations, and of Chew and Low who had developed their model from fixed source theory.[15] The validity of the necessary dispersion relations was *assumed* there (Chew, Goldberger, Low and Nambu, 1957b, p. 1345).

> Relativistic dispersion relations for photomeson production analogous to the pion–nucleon scattering dispersion relations, are formulated without proof. The assumption that the 33 resonance dominates the dispersion integrals then leads to detailed predictions about the photomeson amplitude.

The papers showed that the dominance of the (3,3) (i.e., $2J = 2T = 3$) resonance reduced the exact problem essentially to the Chew–Low model. Here the (3,3) resonance was assumed to 'exhaust' or saturate (i.e. to make the overwhelmingly most important numerical contribution to) the dispersion integrals. Once the (3,3) resonance had been fed in, the s ($l = 0$), d ($l = 2$) and the remaining p-wave ($l = 1$) phase shifts were calculated. It turned out that relativity and spin had caused no essentially new complications.

4.6   Chew's particle-pole conjecture

Chew (1958b), now back at Berkeley, proposed a method to determine the πN coupling constant from nucleon–nucleon data. The conjecture was based on several plausibility arguments using Feynman diagrams, crossing relations, consistency of his method with what had already been found for nonforward dispersion relations, and reference to a suggestion (which we discuss in the next chapter) made by Mandelstam (1958) for dispersion relations for πN scattering. Let us illustrate Chew's central idea with the diagrams of Figure 4.7 for πN scattering. Neglecting kinematic and spin factors, the amplitude for these two diagrams is[16]

$$\frac{g^2}{s - M^2} + \frac{g^2}{u - M^2} \qquad (4.34)$$

where M is the nucleon mass. There is a pole in the s-channel, corresponding to Figure 4.7(*a*), and one in the u-channel for Figure

4.7(b). This amplitude (or the corresponding sum of diagrams) has symmetry under $s \Leftrightarrow u$ and poles at $s = M^2$ and at $u = M^2$. The residue is essentially the square of the pion–nucleon coupling constant, g^2. (A similar behavior can be seen for the diagrams of Figures 4.4 and 4.5 and the corresponding amplitudes (4.19) and (4.22), except there the poles are at s (or t or u) $= 0$ and the residue is e^2 (the electromagnetic 'coupling constant' squared).) Finally, Figure 4.8 is the one-pion exchange diagram for nucleon–nucleon scattering whose amplitude would be

$$\frac{g^2}{t - \mu^2}. \tag{4.35}$$

In all cases, an amplitude has a pole in a given channel at the value of

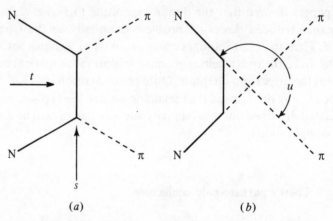

(a)   (b)

Figure 4.7   One-nucleon exchange graphs for pion-nucleon scattering.

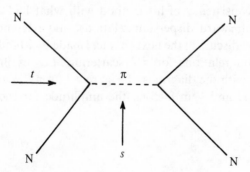

Figure 4.8   One-pion exchange graph for nucleon-nucleon scattering.

that variable corresponding to the exchange of a stable particle (here a pion or a nucleon). Chew observed that the higher-order Feynman diagrams contributing to these amplitudes have no poles at these locations. Therefore, if for, say, N–N scattering we multiply by $(t - \mu^2)$ and extrapolate to $t = \mu^2$, we should obtain g^2. The problem, of course, is that for the N–N scattering of Figure 4.8 the variables s and t have the values in the center of momentum (c.m.) frame

$$s = 4(M^2 + k^2) \geqslant 4M^2, \tag{4.36a}$$

$$t = -2k^2(1 - \cos\theta) \leqslant 0. \tag{4.36b}$$

Here k is the center of momentum three momentum and $0 \leqslant \theta \leqslant \pi$ is the scattering angle. Clearly, $t = \mu^2 > 0$ lies outside the physically accessible region. Chew's hope was that the amplitude (once multiplied by $(t - \mu^2)$) would be smooth enough to allow the necessary extrapolation. If that should be the case, then N–N scattering data would yield a value for g^2 to be compared with that found from πN. Furthermore, since g^2 and f^2 are related as[17]

$$f^2 = \frac{1}{4\pi}\left(\frac{\mu g}{2M}\right)^2 \tag{4.37}$$

there is yet another check on these values. *Any* process with an NNπ vertex could, in principle, be used to find the value of g^2. Thus, Figure 4.9 would lead us to expect the low energy photomeson production cross section to be related to $g^2 e^2$ or to $f^2 e^2$, as it was shown to be in the Chew–Low model. Again, we have a nice example of consilience of inductions.

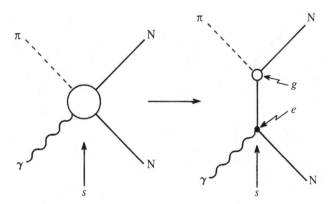

Figure 4.9   Pion photoproduction from a nucleon.

This pole–particle correspondence was an important insight for the S-matrix program (Chew, 1989). But, as Chew (1958b, p. 1382) admitted, 'This whole procedure of course amounts to nothing more than an optimistic conjecture.' It may seem to the reader that this pole–particle correspondence is obvious and so not very startling. However, it is self-evident only within the framework of perturbative quantum field theory. It is by no means evident that strong-interaction quantum field theory has a convergent expansion since the coupling constant g(cf. Eq. (4.37)) is of the order of 10 (i.e., *not* small compared to unity). Chew postulated this correspondence and extrapolation procedure for the *strong* interactions. So, even though part of his motivation for the conjecture came from perturbative quantum field theory, the conjecture itself could stand on its own, subject to direct experimental confirmation or refutation. In fact, the extrapolation of data to obtain the pion–nucleon coupling constant proved quite useful and reliable. The skepticism about the applicability to the real world of the Feynman-diagram pole–particle correspondence is illustrated by Chew's (1989, p. 605) own recollection:

> In 1958 I wrote a paper *conjecturing* that the nucleon-nucleon scattering pole of Figure [4.8] could be verified by extrapolation of scattering data, but I remember finding it difficult to believe that such would actually work. (It did.) Slightly later, Low and I made a corresponding conjecture about the pole of Figure [4.10]. The fact that Francis expected this latter conjecture to be verified was comforting to me; I had enormous respect for Francis's judgement.

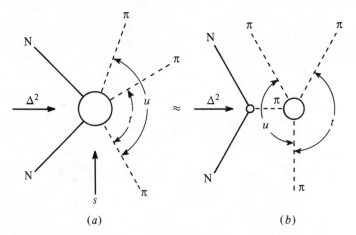

Figure 4.10   Chew–Low extrapolation to the pion pole.

This is again an example of abstracting a (hoped for) general property from a specific model or theory.

Chew and Low (1959) used this pole structure of the scattering amplitude to propose a means of extracting experimentally unmeasurable amplitudes from known production amplitudes. Consider the example

$$\pi + N \rightarrow \pi + \pi + N, \tag{4.38}$$

illustrated in Figure 4.10(a). If the differential cross section could be extrapolated to the pion pole at $\Delta^2 = \mu^2$, then its residue should be essentially the $\pi\pi \rightarrow \pi\pi$ scattering cross section. The $\pi\pi$ elastic amplitude could not be directly measured for the practical reason that there are no stable pion targets and facilities for colliding pion beams did not exist. For the reaction $\pi N \rightarrow \pi\pi N$ of Figure 4.10, this Chew–Low extrapolation gives[18]

$$\frac{\partial^3 \sigma}{\partial s \partial t \partial (\Delta^2)} \sim \frac{\Delta^2 f(\Delta^2)}{(\Delta^2 - \mu^2)^2} |T(s, t; \Delta^2)|^2. \tag{4.39}$$

Here $f^2(\Delta^2) \xrightarrow[\Delta^2 \rightarrow \mu^2]{} f^2$, the πN coupling constant, and $T(s, t; \Delta^2) \xrightarrow[\Delta^2 \rightarrow \mu^2]{} T(s, t)$, the physical $\pi\pi \rightarrow \pi\pi$ scattering amplitude. Since the physical $\pi\pi \rightarrow \pi\pi$ differential cross section is just

$$\frac{\partial \sigma_{\pi\pi}}{\partial t} \sim |T(s, t; \Delta^2 = \mu^2)|^2, \tag{4.40}$$

we see that $\partial \sigma_{\pi\pi}/\partial t$ can be found by the extrapolation (again, up to known factors)

$$(\Delta^2 - \mu^2)^2 \frac{\partial^3 \sigma}{\partial s \partial t \partial (\Delta^2)} \xrightarrow[\Delta^2 \rightarrow \mu^2]{} \frac{\partial \sigma_{\pi\pi}}{\partial t}. \tag{4.41}$$

This means of obtaining $\sigma_{\pi\pi}$ will be relevant for our discussion of the $\pi\pi$ system in the next chapter.

Once again, we see the importance of the data and of the physics practice as the origin of the ideas that emerged. There was a premium in *doing* something that gave directly testable results.

4.7 Summary

Early on in his career, Chew was impressed with Fermi's approximate calculation (for scattering from a bound state) in which only two-body ('measurable') neutron–proton *amplitudes* appeared. This formulation, not in terms of potentials and wave functions, but rather in terms of observable quantities, was to form the basis of a recurrent theme in Chew's own work. As an outgrowth of his calculations for deuteron–proton scattering using a modification of Fermi's method, Chew developed the impulse approximation, which proved very successful in the analysis of strong-interaction scattering data in the early to mid-1950s. He came to believe that these potentials, Lagrangians or amplitudes should have *no arbitrariness* in them.

His phenomenological applications of fixed-source quantum field theory led to a collaboration with Low on the highly influential Chew–Low (static nucleon) model for the strong-interaction scattering amplitude. This model incorporated the crossing symmetry he had learned of from Goldberger and Gell-Mann and exhibited important features of analyticity (cf. Eq. (4.28)), especially a pole in the energy variable. Their model found wide applicability and its general features appeared to transcend the limited domain of its derivation. Fully relativistic dispersion relations for pion–nucleon and nucleon–nucleon scattering soon followed with a collaboration of Chew, Goldberger, Low and others. In the late 1950s Chew proposed an extrapolation scheme, based on pole–particle correspondence, to determine (from scattering data) the pion–nucleon coupling constant. This in turn led to the Chew–Low method of extrapolating unknown scattering cross sections. A characteristic common to several of these developments is the origin of central concepts in highly *technical* problems, rather than in general, overarching philosophical principles.

5
The analytic S matrix

The lines of research outlined in Chapters 3 and 4, while never completely independent, merged into a common subject of study at the end of the 1950s. Since quantum field theory had come to a calculational impasse in strong-interaction physics and since people could not prove on the basis of general field theory axioms all of the analytic properties of scattering amplitudes needed to calculate relations among physically measurable quantities, it became the fashion to postulate or guess (with various degrees of justification) general properties and to make up the rules of the game as one went along. Goldberger was one of the most prolific practioners of this art. He recalls[1] as a particularly appealing aspect of that project:

> I think the fact that one could deal with physical matrix elements in a largely model independent way was the most attractive aspect to me of S-matrix theory

Let us indicate how this analytic S-matrix program came into existence.

By 1956 Gell-Mann was of the opinion that quantum field theory had no hope of explaining high-energy phenomena and he sketched an alternative program based on general principles and employing unitarity as a central tool in calculations. He acknowledged Heisenberg's original S-matrix program as the ancestor of this project. A crucial step for the dramatic forward progress of this alternative approach to conventional quantum field theory was taken by Mandelstam when he proposed a specific form for a relativistic (double) dispersion relation in the momentum transfer (essentially the scattering angle) variable as well as in the energy variable. This *Mandelstam representation*, as it became known, not only allowed many new types of empirically testable calculations to be made, but also led to an entire 'industry' of attempting to prove such double dispersion relations on the basis of QFT. The same 'bootstrap' iteration procedure that Gell-Mann had discussed was quite evident when the Mandelstam representation was combined with the unitarity condition. The outlines of an independent dynamical scheme then became apparent. A purely formal and at first apparently unrelated mathematical technical 'trick' developed by Regge offered concrete hope of closing the circle on a complete dynamical theory for the strong interactions. This advance is a nice example of physical intuition being led by the mathematics.

Around this same time, quantitative, numerical calculations by Chew and Mandelstam led to the recognition of a bootstrap, or self-sustaining, solution to the equations for an important strong-interaction process. A particular resonance was shown to generate itself as a self-consistent solution to the dynamical equations governing the strong interactions. This was the birth of bootstrap dynamics, which provided a program independent of quantum field theory, at least at the calculational level. Some, but by no means all, of the practitioners of this new art saw the bootstrap program as anti-field theory. An essential hallmark of this entire sequence of developments is that highly technical advances provided the *source* for new ideas in particle physics.

5.1   Gell-Mann: a new approach to QFT

Already in their 1955 paper, Lehmann, Symanzik and Zimmermann had indicated that, from the equations they derived for (finite) S-matrix elements, the usual perturbation expansion of quantum field theory could be (formally) recovered. At the 1956 Rochester Conference, Gell-Mann (1956) sketched how Lorentz invariance, crossing, analyticity and unitarity, together with suitable boundary conditions in momentum space, were almost enough to specify the field theory. Gell-Mann mentioned as an aside (just for amusement)[2] the Heisenberg S-matrix program of 1943 (Gell-Mann, 1956, p. III-31; 1987, p. 484). He pointed out that, as more dispersion relations were assumed (hopefully to be justified somehow on, say, the basis of microscopic causality), one could generate an expansion in powers of a coupling constant to recover the S matrix of field theory. He (1956) gave what would turn out to be a rather prescient outline of the future S-matrix program. This was not *anti*-quantum field theory, but simply an alternative approach.

> Goldberger and myself and others started some years ago to look at the dispersion relation for forward scattering of γ-rays from ...

protons. The idea was to write down a rule, an exact law, which depends on very simple assumptions ... which would be independent of all the details of field theory. We found that just by using microscopic causality and a few other very simple things, we could get dispersion relations expressing the real part of the photon scattering amplitude in the forward direction in terms of total cross sections for photons producing *anything*. This is ... simply the Kramers-Kronig relation. ... Goldberger and company have generalized these relations. ... The character of the program has changed. ... As one writes down more and more dispersion relations (... characteristic of amplitudes in field theory), ... one begins to come dangerously close to finding all of the results of field theory ... [W]e seem to have come very close to having prescriptions which ... seem to yield the *S* matrix in field theory. (pp. III-30–III-31)

One can try, for example, to carry out Heisenberg's program set forth in 1943. ... The analog of Heisenberg's program here is to ask whether we can write down a set of formulas relating only amplitudes with particles having their physical masses .... Heisenberg had realized that one cannot eliminate all but the physical amplitudes without paying a price, namely, one has then to extend the *S* matrix to the region of unphysical energies, energies below the rest masses of the particles, for example. Exactly the same sort of thing has to be done in present-day work. ... In this way one tries to carry the program through. (pp. III-31–IV-32)

As a means of generating the *S* matrix in place of the usual field equations, [these principles] seem very hopeful .... [These] calculations ... are presumably better than perturbation calculations .... Maybe this approach will throw some light on how to construct a new theory that is *different*, and which may have some chance of explaining the high energy phenomena which field theory has no chance whatsoever to explain. (p. IV-35–IV-36).

In this program all four momenta are on the mass shell (i.e., they all satisfy the relation $p^2 = E^2 - \mathbf{p}^2 = m^2$), unlike the case in field theory. The price paid is that these mass-shell amplitudes must sometimes be evaluated at unphysical values of the energy (i.e., at values not accessible through physical processes). Gell-Mann (1956) illustrated his idea with the example of pion–nucleon scattering as shown in Figure 5.1. In terms of the pion–nucleon vertex function (or 'coupling constant') of Figure 5.1(*a*), one could use unitarity for the imaginary part of the πN scattering amplitude, Figure 5.1(*b*), to generate the series of Figure 5.1(*c*). (This is similar to Jost, Luttinger and Slotnik's 1950 procedure of Eqs. (2.53)–(2.56) above.) Gell-Mann assumed the existence of nonforward dispersion relations and concluded (1956, p. III-34):

> We now probably have almost enough equations to carry through the Heisenberg program. What one would do then is to calculate the scattering in second order. One would calculate the imaginary part of the amplitude in fourth order by squaring that second order amplitude. One would then use the dispersion relation to calculate the real part of the amplitude in fourth order and so on. In this way, by using dispersion relations and conservation of probability [unitarity] ... one may hope to generate the entire S matrix from more or less fundamental principles.

Chew (1958a, p. 95) in his rapporteur's talk at the 1958 CERN Conference, picked up on this same theme a bit later:

> The fourth and last application of dispersion relations is the most interesting but so far has made the least progress. It is the attempt to use them as fundamental dynamical relations which replace the usual field questions. There has been hope that in this way ... many of the difficulties associated with the Lagrangian approach to field theory could be avoided. Gell-Mann pointed out several years ago that ... the dispersion relations can be expanded in a power series which reproduces the content of conventional perturbation theory. This observation continues to stimulate efforts to treat the relations as integral equations which ... give a complete description of the system.

5.2   The Mandelstam representation

A decisive step in the dispersion-theory program was taken by Mandelstam (1958). Citing recent work in dispersion theory and Gell-Mann's (1956) conjecture that these relations might be able to

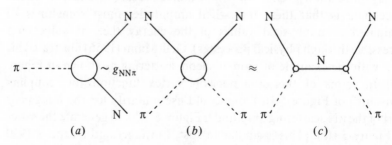

Figure 5.1   The pion–nucleon coupling constant graph.

replace field theory, Mandelstam sought to extend the static-model equations of the Chew–Low theory to allow a calculation of the πN scattering amplitude by an iterative procedure. In conjecturing his double dispersion relations[3], he referred to a previous representation by Nambu (1955, 1957) of certain Green's functions (i.e., vacuum expectation values of products of quantum field operators, a special case of which, Eq. (3.13), yielded forward dispersion relations). Nambu (1955, p. 394) had appreciated that dispersion relations, involving only physically meaningful quantities, could be used to circumvent renormalization problems.

> The causality condition also serves as a means of obtaining renormalized quantities without recourse to the usual subtraction procedure.

Eden recalls[4] that when he met Nambu in Seattle in 1956, Nambu said that he was working on the problem of extending Eden's earlier (1952) work on the analytic properties of reaction amplitudes. This led to Nambu's paper (1957) which was a stimulus for Mandelstam's 1958 paper. Mandelstam has characterized[5] the influence of Nambu's work as follows.

> Although things were really not as simple as Nambu thought, I was intrigued by his suggestion that multi-variable dispersion relations might exist.

Here again we can detect a thread leading back to the old Heisenberg program.

Mandelstam considered the problem of the equivalence of quantum field theory and dispersion relations in the introductions to his seminal papers.

> It is ... tempting to ask the question whether or not the dispersion relations can actually replace the more usual equations of field theory and be used to calculate all observable quantities in terms of a finite number of coupling constants – a suggestion first made by Gell-Mann [1956]. At first sight, this would appear unreasonable, since, although it is necessary to use all the general principles of quantum field theory to derive the dispersion relations, one does not make any assumption about the form of the Hamiltonian other than that it be local and Lorentz-invariant. However, in a perturbation expansion these requirements are sufficient to specify the Hamiltonian to within a small number of coupling constants if one demands that the theory be renormalizable and therefore self-consistent. It is thus very possible that, even without a perturbation expansion, those requirements are sufficient to determine the theory. (1958, p. 1344)

The usual dispersion relations can be proved by examining the restrictions imposed by causality on the four-point Green's function, provided that the momentum transfer is sufficiently small. It is unlikely that a corresponding proof can be carried out [for a double dispersion representation], or indeed that the representation follows from these requirements alone. The general principles of field theory contain much more information, since the causality condition enables one to deduce analytic properties of all the Green's functions, which are related to one another by the unitarity conditions. It is therefore very possible that the representation is a consequence of the general principles of field theory, but it seems at present a matter of considerable difficulty to carry out such a proof.

In the absence of a rigorous treatment making use of all the information available from the general principles of field theory, therefore, it should be useful to examine the analytic structure of transition amplitudes in perturbation theory. (1959a, pp. 1741–2).

Again, there is no sense of an *opposition* between quantum field theory and dispersion relations.

Mandelstam appealed to perturbation theory as a guide here too. Whereas the previous dispersion relations we have seen were in the

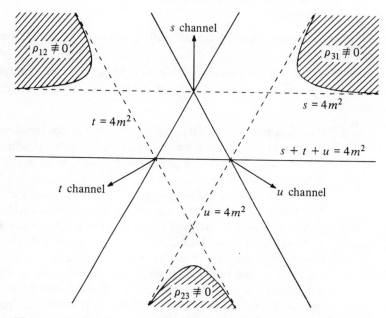

Figure 5.2   Regions of support for the Mandelstam double spectral functions.

energy variable (say s of Eq. (4.17a)), these new dispersion relations were in both independent variables. That there are just two independent variables (for the elastic scattering of two spinless particles) can be understood by realizing that the magnitude of the momentum $k = |\mathbf{k}|$ and the scattering angle θ completely specify the scattering process. For two identical particles (each of mass m) observed in the center of momentum frame (i.e., the particles approach each other with equal but opposite momenta or velocities), the relativistically invariant variables s and t take on particularly simple and easily interpreted forms. We find that $s = 4(k^2 + m^2) = 4E_{cm}^2$, where E_{cm} is the total energy of the particles in the center of momentum (also often termed the center of mass) frame. Similarly, we have $t = -2k^2(1 - \cos\theta) = -|\mathbf{k}_f - \mathbf{k}_i|^2$, where $(\mathbf{k}_f - \mathbf{k}_i)$ is the momentum transfer (i.e., the change in momentum between the incident momentum $\mathbf{k}_i$ of a particle and its final, outgoing momentum $\mathbf{k}_f$). (Reference to Figure 4.6 will provide a particular example of such an elastic scattering process, where we have considered the s-channel, $\pi^+\pi^- \to \pi^+\pi^-$.) Even though Mandelstam considered the πN problem, we illustrate his double dispersion relations here for the (somewhat fictitious) case of equal mass particles as

$$
\begin{aligned}
A(s,t,u) = &\frac{1}{\pi} \int \frac{ds'\sigma_1(s')}{(s'-s)} + \frac{1}{\pi} \int \frac{dt'\sigma_2(t')}{(t'-t)} + \frac{1}{\pi} \int \frac{du'\sigma_3(u')}{(u'-u)} \\
&+ \frac{1}{\pi^2} \int\int \frac{ds'\,dt'\rho_{12}(s',t')}{(s'-s)(t'-t)} + \frac{1}{\pi^2} \int\int \frac{dt'\,du'\rho_{23}(t',u')}{(t'-t)(u'-u)} \\
&+ \frac{1}{\pi^2} \int\int \frac{du'\,ds'\rho_{31}(u',s')}{(u'-u)(s'-s)}.
\end{aligned}
\tag{5.1}
$$

The single spectral functions σ_i and the double spectral functions ρ_{ij} are zero except when their variables are above the allowed intermediate-state thresholds. Figure 5.2 shows the actual regions of support for the double spectral functions.[6] (Any point on that diagram satisfies the constraint $s + t + u = 4m^2$, so that, as stated above, these are just *two* independent variables).

Notice that the representation of Eq. (5.1) places s, t and u each on the same footing and allows one easily to 'cross' from one channel (or reaction) to another just by allowing s (or t or u) to play the role of, say, the 'energy' variable (or of the momentum transfer variable). Each single or double integral represents a different class of contributions (from various possible virtual or intermediate states) to the full scattering amplitude $A(s,t,u)$. Thus, previous studies of the vertex function (cf. Figure 5.1(a)), for example by Karplus, Sommerfield and Wichmann (1958), had shown it to have a structure representable by the single integrals in Eq. (5.1). For instance, a perturbation diagram like that of Figure 5.3(a) would be included in the single integrals and one like 5.3(b) in the double integrals. Mandelstam (1958, p. 1349) also explicitly pointed out that an older fourth-order calculation of Compton scattering (Brown and Feynman, 1952) had just the structure of the double dispersion representation of Eq. (5.1) (see also, Mandelstam, 1959a). The important point for our purposes here is that Eq. (5.1) displays the analytic and crossing properties of the scattering amplitude $A(s, t, u)$.[7] We can also recover the single-variable dispersion relations from it. Mandlestam (1958) performed a tedious mathematical calculation to evaluate the elastic unitarity integral of Eq. (4.7) for the representation of Eq. (5.1). (In our case, unitarity is elastic for $(2m)^2 \leqslant s \leqslant (4m)^2$.) This produces a nonlinear relationship among the quantities in Eq. (5.1), one similar in structure to that of the Chew–Low equations.[8] We discuss this below. The unitarity integral over θ could be carried out only because Eq. (5.1) was a dispersion relation in $t = -2k^2(1 - \cos\theta)$, as well as in $s = 4E_{cm}^2$. Mandelstam subsequently proved (1959a) his conjectured representation in low-order perturbation theory. He also reconstructed (1959b) a perturbation series for the scattering amplitude using the assumptions of unitarity and analyticity. He was able to begin with pole terms, which represent the Born (or high-energy) approximation to the scattering amplitude, use (two-particle) unitarity, and then find a better approximation to the amplitude from the dispersion relation. This had been the nature of what Gell-Mann (1956) had conjectured.

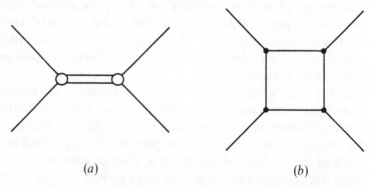

(a)   (b)

Figure 5.3   Contributions to the single and double spectral functions.

5.3   Proofs of double dispersion relations

Rigorous general proofs of double dispersion relations were never found, although a whole industry was spawned of examining the integrals corresponding to an arbitrary Feynman diagram and showing that, to each order in perturbation theory, the Mandelstam representation held (e.g., Eden 1960a, 1960b, 1960c; Polkinghorne and Screaton, 1960a, 1960b; Tarski, 1960; Landshoff, Polkinghorne and Taylor, 1961). Unfortunately for this project, singularities were found (Eden, Landshoff, Polkinghorne, and Taylor, 1961) that caused a breakdown of the Mandelstam representation in some cases.

However, the Mandelstam representation was thoroughly studied within the framework of potential scattering theory. Blankenbecler, Goldberger, Khuri and Treiman (BGKT) (1960) considered the scattering amplitude $f(k^2, \cos \theta)$ for a potential $V(r)$ which is a superposition of Yukawa potentials

$$V(r) = \int_m^\infty d\mu \sigma(\mu) \frac{e^{-\mu r}}{r}, \qquad m > 0, \tag{5.2}$$

where $\sigma(\mu)$ is a given weight function. Using results on Fredholm integral equations, they established that $f(k^2, \cos \theta)$ has the proper cut structure in the (complex) variables k^2 and $\cos \theta$ to satisfy a Mandelstam representation. They also proved that there is no essential singularity at $k^2 = \infty$, but had to *assume* there was none at $\cos \theta = \infty$. Except for this last point, they had established a double dispersion relation for $f(s, t)$ as[9]

$$f(s, t) = f_B(t) + \sum_{i=0}^n \frac{\Gamma_i(t)}{s + s_i} + \frac{1}{\pi^2} \int_0^\infty ds' \int_0^\infty \frac{dt' \rho(s', t')}{(t' - t)(s' - s)} \tag{5.3a}$$

where

$$s = k^2, \tag{5.3b}$$

$$t = -2s(1 - \cos \theta) \tag{5.3c}$$

and where $f_B(t)$ is the Born approximation

$$f_B(t) = -\frac{1}{4\pi} \int d\mathbf{r} \, e^{-i(\mathbf{k}_f - \mathbf{k}_i) \cdot \mathbf{r}} V(r) = -\int_m^\infty \frac{d\mu \sigma(\mu)}{\mu^2 - t}. \tag{5.3d}$$

They also explicitly constructed an iteration procedure using Eq. (5.3a) and unitarity which would generate $\rho(s,t)$ once $\sigma(\mu)$ (or $f_B(t)$) had been given. In other words, they had used a nonlinear equation to solve the Schrödinger equation! Goldberger (1970, p. 690) once commented on the irony of this:

> I often wonder if some student in the year 2000 will produce a brilliant linear differential equation from which all the consequences of nonrelativistic quantum theory can be extracted without a resort to the complicated nonlinear dispersion approach.

The basic reason this iterative scheme works is related to the region of support of the double spectral function shown in Figure 5.4.[10] When Eq. (5.3a) is substituted into the elastic unitarity condition, Eq. (4.7), the integral over $\cos\theta$ carried out, and the discontinuity of $\mathrm{Im}\, f(s,t)$ taken, the resulting equation for $\rho(s,t)$ has the form

$$
\rho(s,t) = \int_m^\infty d\mu_1 \sigma(\mu_1) \int_m^\infty d\mu_2 \sigma(\mu_2) K(s,t;\mu_1^2,\mu_2^2)
$$

$$
- \frac{2}{\pi^2} \int_{4m^2}^\infty ds' \int_{4m^2}^\infty dt' \frac{\rho(s',t')}{(s'-s)} \int_m^\infty d\mu_2 \sigma(\mu_2) K(s,t;t',\mu_2^2)
$$

$$
+ \frac{1}{\pi^4} \int_{4m^2}^\infty ds' \int_{4m^2}^\infty dt' \int_{4m^2}^\infty ds'' \int_{4m^2}^\infty \frac{dt''\rho(s',t')\rho(s'',t'')}{(s'-s)(s''-s)}
$$

$$
\times K(s,t;t',t''). \tag{5.4a}
$$

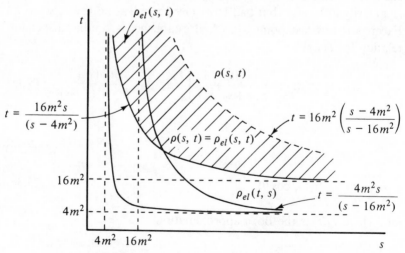

Figure 5.4   Regions of support for the elastic double spectral function.

Since Eq. (5.4a) has a rather complicated appearance, let us rewrite it symbolically as

$$\rho(s,t) = [\sigma, \sigma, K(s,t); m^2, m^2] + [\rho, \sigma, K(s,t); m^2, 4m^2]$$
$$+ [\rho, \rho, K(s,t); 4m^2, 4m^2] \quad (5.4b)$$

in order to bring out a simple and important property. Here the symbol [ ] stands for a functional of the enclosed arguments and vanishes whenever $K \equiv 0$. The kernel $K(s,t; t't'')$ vanishes identically (for any given s) unless t is in a region bounded asymptotically by (see Figure 5.4)

$$t > (\sqrt{t'} + \sqrt{t''})^2. \quad (5.5)$$

(Since $\rho(s,t) = \rho(t,s)$ here, we also have the asymptotes in the variable s.) Therefore, Eq. (5.4) has the following very nice iterative property. For $t < 9m^2$, only the first integral in Eq. (5.4) contributes and $\rho(s,t)$ is determined *exactly* above the boundary curve of Figure 5.4 or that of Eq. (5.5). Since $\sigma(\mu)$ is given, we can find $\rho(s,t)$ here. Next, for $t < 16m^2$, only the first two integrals contribute and $\rho(s',t')$ in the second integral is needed only for values found by the first iteration. We can continue these iterations and find $\rho(s,t)$ everywhere and then, by Eq. (5.3a), $f(s,t)$ itself.[11]

The relativistic case is more complicated for two reasons. The Mandelstam representation has more terms (compare Eq. (5.1) with Eq. (5.3a)) and, more importantly, elastic unitarity does not hold for all values of s because there are inelastic thresholds. It is the second point that presents an essential difficulty. Nevertheless, one can write $\rho(s,t)$ as

$$\rho(s,t) = \rho^{(el)}(s,t) + \rho^{(inel)}(s,t) \quad (5.6)$$

where $\rho^{(el)}(s,t)$ is that part of $\rho(s,t)$ determined by elastic unitarity (something like Eq. (5.4)). In the nonrelativistic case above, the 'potential' $\sigma(t)$ was the discontinuity in t of everything in $A(s,t,u)$ (that is, the $f(s,t)$ of Eq. (5.3a)) not involving $\rho^{(el)}$. If we now analogously define a 'potential' for a given channel, it will have contributions from the cross-channel spectral functions. If these potentials were given, we could again iteratively determine the amplitude[12]. Since these potentials are not given in advance, the self-consistency nature of the problem is evident. In the next chapter we return to this generalized concept of potential with a specific example of a bootstrap calculation. This relativistic generalization of the concept of 'force' which had been missing (cf. Chapter 4 above) became clear to Chew from Mandelstam's double dispersion relation. Forces between hadrons are generated by hadron exchanges in the crossed channels.

5.4   Regge's innovation

The asymptotic behavior of $f(k^2, \cos \theta)$ as $\cos \theta \to \infty$ was settled for potential scattering by Regge (1959). Although a beautiful piece of classical mathematical analysis in its own right, this paper was to have its greatest impact on the S-matrix program for the conjectures to which it would lead. In two previous papers, Regge (1958a, 1958b) had studied the analytic properties of the Jost function (cf. (2.34)) and the question of false zeros of $S(k)$ in potential theory. Problems that had arisen out of Heisenberg's old S-matrix program were not irrelevant for Regge's work. In his 1959 paper Regge cites the papers of Wigner and von Neumann (1954), of Gel'fand and Levitan (1951b), of Jost and Kohn (1953) and thanks Symanzik for advice. What Regge did in his 1959 paper was to consider the partial-wave index l to be a continuous complex variable and to examine the properties of the wavefunctions $\psi_l(r)$ and of the partial-wave amplitudes $f_l(k)$ as functions of this complex l. Again, for the Yukawa potential of Eq. (5.2), Regge was able to show that the partial-wave expansion for $f(k, \cos \theta)$ (cf. Eq. (A.23)) could be transformed into a contour integral in the complex l plane

$$f(k, \cos \theta) = \sum_{l=0}^{\infty} (2l+1) f_l(k) P_l(\cos \theta) \tag{5.7a}$$

$$= \frac{\mathrm{i}}{2} \int_C \frac{(2l+1)\, \mathrm{d}l f(l, k) P_l(-\cos \theta)}{\sin(\pi l)} \tag{5.7b}$$

where the contour C is shown in Figure 5.5. By studying the integral equation satisfied by the $\psi(l, k; r)$, Regge proved that the *only* singularities of $f(l, k)$ in the l-plane are simple poles. From the known properties of the Legendre functions $P_l(-\cos \theta)$ and those he had deduced for $f(l, k)$, he was able to deform C to run (essentially) along the vertical line $\mathrm{Im}\, l = -\frac{1}{2}$ and pick up the contributions from the poles of $f(l, k)$. As $\cos \theta \to \infty$, the vertical line integral behaves as $(\cos \theta)^{-\frac{1}{2}}$ and can be neglected. The sum from the poles is dominated by $(\cos \theta)^N$, where N is the real part of the location of the pole farthest to the right in the l-plane. That is, if $l = \alpha(k)$ is the trajectory (subsequently termed the 'Regge trajectory') of that pole, then $N = \mathrm{Re}\, \alpha(k)$. The remarkable result proved by Regge is that, for all k, there is some *fixed* largest (finite) value of N

such that $\operatorname{Re}\alpha(k) < N$ for any trajectory. The upshot of all this is that $f(k, \cos\theta)$ behaves no worse than $(\cos\theta)^N$ as $|\cos\theta| \to \infty$. That allows a (possibly subtracted) dispersion relation to be written for $f(k, \cos\theta)$ in the variable $(\cos\theta)$. This, plus BGKT (1960), finally established the Mandelstam representation for potential-scattering theory (for Yukawa potentials which, after all, are not unexpected from field theory; cf. Eq. (4.13)). Regge also (1960) considered the connection between bound states and substraction terms in potential scattering theory and showed that the number of subtractions increases with the strength of an attractive potential. Charap and Fubini (1959) also demonstrated how a nuclear potential could be defined in terms of a general field-theory formalism. Their potential, when used in a Schrödinger equation, reproduced the scattering amplitude to a given level of approximation.

5.5 A bootstrap mechanism

On a practical level, assumption of a Mandelstam representation allowed considerable advances to be made in iterative calculations of scattering amplitudes (Chew, 1960b). This was an essential factor in the popularity of the S-matrix program among practicing physicists. The analytic properties of partial-wave amplitudes, defined as (cf. Eq. (4.9))

$$A_l(k^2) \equiv \frac{1}{2} \int_{-1}^{1} A(k^2, \cos\theta) P_l(\cos\theta) \, d(\cos\theta) \tag{5.8a}$$

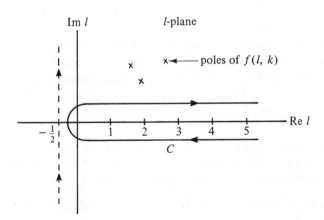

Figure 5.5 The complex l-plane and the Sommerfeld–Watson transformation.

$$=\frac{\omega}{k}e^{i\delta_l}\sin\delta_l, \qquad \omega=\sqrt{k^2+m^2}, \tag{5.8b}$$

could now be deduced with the use of Eq. (5.1) since the $\cos\theta$ projection can be carried out explicitly. For $\pi\pi\to\pi\pi$ scattering, Chew and Mandelstam (1960) studied the analytic properties of these $A_l(v)$, as a function of $v=k^2$. These amplitudes have a right-hand cut beginning at $v=0$ and determined by unitarity in the direct channel. There is also a left-hand cut $-\infty<v<-m^2$ coming from unitarity in the crossed channels. Chew and Mandelstam (1960) were able to develop approximate coupled integral equations for the phase shifts, keeping only s- and p-waves. They had chosen to concentrate on the $\pi\pi$ problem (even though no direct experimental data were available for that process, but only the Chew–Low extrapolation procedure of Eq. (4.41)) because the pion is the least massive strongly-interacting particle, the $\pi\pi$ system is closed under crossing (i.e., each channel is a $\pi\pi$ channel), and the $\pi\pi$ amplitude is needed (as we have seen at the end of Chapter 3) to understand NN→NN and $\pi\pi$→NN̄ processes, the last of which in turn is involved in the nucleon electromagnetic form factor (through the pion's electromagnetic form factor) (Frazer, 1961). Using the Mandelstam representation and the analytic properties of partial-wave amplitudes derived therefrom, Frazer and Fulco were able to approximate (1960a) the $\pi+\pi\to N+\bar{N}$ amplitude and to study quantitatively (1959, 1960b) the nucleon electromagnetic form factor. They found that the large radius of the proton required the pion–pion p-wave isospin-1 (or $J=1$, $T=1$) phase shift to resonate. They gave a rough prediction of the mass of this resonance, which was subsequently found experimentally (Erwin, March, Walker and West, 1961), and is now known as the ρ meson.[13] Chew and Mandelstam (1961) showed that a $J=1$, $T=1$ resonance fed into their $\pi\pi$ equations would sustain (or 'reproduce') itself as output. This was the first example of a 'bootstrap' mechanism and that term appears in the 1961 Chew–Mandelstam paper, although the term was first used (Capra, 1985, p. 258) publicly by Chew in this context in his 1959 Kiev Conference summary talk:

> The attractive force ... could come from the exchange of a P-wave pion-pair. If a resonance were present in the P-state, this force would be strong: a 'bootstrap' mechanism (Chew, 1960a, p. 332).

And, again, Chew and Mandelstam (1961, p. 752) spoke of this same mechanism:

> Self-consistent solutions can be found in which a P-wave resonance is

> sustained by a 'bootstrap' mechanism: that is, a strong attraction force ... results from the exchange of a resonating pair of P-wave pions.

Chew (1989, p. 605) recalls the impression this resonant solution made on him.

> Then, in 1959, just before the Kiev Conference, Mandelstam and I encountered a mind-boggling phenomenon. We found that a spin-1 $\pi\pi$ resonance could be generated by a force due to Yukawa-like 'exchange' of this same resonance.

The bootstrap concept had arisen as a result of calculations. Frazer and Fulco (1960c) used the Mandelstam representation to write partial-wave dispersion relations for πN scattering. They showed that, when all but the singularities nearest threshold are neglected, the approximate solution to the parital-wave dispersion relations is the Chew–Low effective range formula, Eq. (4.33), for δ_{33}. We discuss these bootstrap solutions in the next chapter. Here, though, we have an example of how, once a set of ideas or procedures 'hang together', there is some presumed resonableness (or 'rationality') to the enterprise.

5.6   An independent *S*-matrix program

By the time of Chew's 1958 CERN Conference summary report (Chew, 1958a), an analytic *S*-matrix theory was very much in the air. At the 1959 CERN Conference, Regge's (1959) proof was *mentioned* (i.e., its existence) and an *S*-matrix program was in use in Berkeley. A bootstrap mechanism (Chew, 1960a, p. 332) was cited for the $\pi\pi$ program. At the 1959 Kiev Conference, Landau remarked to Chew that Chew and Low had made a discovery of major significance in the pole–particle relationship[14] and expressed the opinion that Chew's work with Mandelstam on $\pi\pi$ dynamics was concerned too much with approximate calculations (Capra, 1985, p. 255). Gell-Mann (1987, p. 485) also recalls that Landau condemned field theory and welcomed an independent dispersion-theory approach.

> Landau ... went on to enunciate a special dogma, starting from his impression that field theory was no damned good because all the coupling constants were zero. He proposed that while field theory was wrong, the program of using dispersion relations, crossing relations, and unitarity to calculate the *S*-matrix was right and could be used as a substitute for field theory. ... Landau now proposed that it was correct and field theory incorrect. I could never understand that point of view and I still cannot.

In Kiev in 1959, I had terrible arguments with Landau on the subject of condemning field theory and welcoming dispersion theory. As usual, it was impossible, really, to argue. He did not give an inch. Then Geoffrey Chew adopted the same sort of point of view in La Jolla in 1961.

Landau's performance in Kiev also made an impression on Jost[15]:

I remember a vivid exchange between L. Landau and H. Lehmann in one of the sessions, in which Landau admitted not to have followed the recent developments in field theory.

Polkinghorne recalls[16]:

Landau's talk at Kiev (which was given in an informal session since Bogoliubov squeezed him out of more than a few minutes on the formal program) was an extraordinarily exciting and powerful occasion.

This anti-field-theory attitude of Landau's is evidenced by his paper (to which we return in Section 7.1) given at the Kiev Conference in which he argued (as he and Pomeranchuk, 1955, and Pomeranchuk, 1956b, had earlier) that any local field theory has no solutions except that for *noninteracting* particles. That is, he claimed that the only consistent QFT was a trivial or empty one. Previously, he (Landau *et al.*; 1954a, 1954b, 1954c, 1954d; Landau, 1955) had discussed field theory singularities actually present but not seen in perturbation theory. (That is, a power series expansion can fail to converge for large values of the energy, say, because the function itself in singular there.) Such concerns with perturbation theory went back to Dyson (1952; Edwards, 1953). Landau (1959) also formulated a set of rules, based on the diagrams of field theory (but not upon the validity of perturbation theory), which allowed one to find the locations of the singularities of scattering amplitudes. This was really a set of conjectures on Landau's part (Chew, 1989), but they proved very useful. They included the pole–particle correspondence as a special case. Cutkosky (1960) then gave a rule – a generalization of the unitarity condition – that expressed the Mandelstam double spectral functions in terms of Feynman graphs.[17] Polkinghorne and Screaton (1960a, 1960b) obtained the Landau rules from perturbation theory. Directly from unitarity and analyticity, Polkinghorne (1962a, 1962b) derived the Landau and Cutkosky rules as the conditions representing the minimum set of singularities that must be present. This contributed to a sense of progress that ever more detailed results, formerly associated with and obtainable only from QFT, were now flowing from these general principles *alone*. This nice picture was complicated somewhat when, subsequently, a new class of non-Landau singularities was discovered (Fairlie, Landshoff, Nuttall and Polkinghorne, 1962).

Mandelstam is generally credited[18] with recognizing the importance of Regge's (1959) work on complex angular momentum for the relativistic double-dispersion theory problem. What attracted Mandelstam to Regge's paper was that (in potential theory) Regge '... showed how to calculate the subtraction constants [which are determined by the asymptotic behavior of the scattering amplitude] (in fixed-s dispersion relations) and also how the number of subtractions was correlated with the proximity of resonances. It seemed to provide what previously had been missing.'[19] In Chew's own 1961 book, *The S-Matrix Theory of Strong Interactions* (Chew, 1961c), Regge's nonrelativistic proof is *mentioned* but still not used for the relativistic problem. By the time of Chew's 1962 Cargése lectures (Chew, 1963a), Regge poles are *fully* in. (Interestingly enough, Regge himself never backed the cavalier applications of his results (which had been proved only in nonrelativistic Schrödinger theory) to the relativistic problem.) For Chew, the main interest in Regge's work was the bearing it had on the asymptotic behavior (or 'boundary conditions') of the scattering amplitude for the $\pi\pi$problem. There was an intense interest in Chew's program by the early 1960s. He was invited to (and did) give the *same* set of lectures at *two* international summer schools in 1960 (Chew, 1961a). At this time, lectures and papers by Chew were highly sought after. Furthermore, during those years it was difficult to find a field theory course taught in some graduate programs in the United States.

A summary of the mass-shell field-theory program in the early 1960s was given by Mandelstam (1961, pp. 209–10) at the Twelfth Solvay Conference on Physics.

> Calculations in elementary-particle physics performed with the aid of the analytic properties of transition amplitudes, if successful, would be a means of implementing an approach to quantum field theory originally put forward in 1955 by Lehmann, Symanzik and Zimmermann and developed by other physicists. The main departure of this approach from the conventional one is in the suggestion that it is unnecessary to specify the Lagrangian in a local field theory; the other postulates of the theory determine it to within a small number of coupling constants. Though this is in striking contrast to nonrelativistic quantum mechanics, where a knowledge of the Lagrangian is necessary to define the problem, it has in fact been tacitly assumed in

quantum field theory. For instance, in the pion–nucleon system, the only relativistic Lagrangian which leads to a consistent theory is

$$\mathscr{L} = \mathscr{L}_0 + ig\bar{\psi}\gamma_5\psi\tau_i\phi_i + \lambda\phi_i^2\phi_j^2$$

where $\mathscr{L}_0$ is the Lagrangian for non-interacting fields, and g and λ are arbitrary constants. Any other local Lagrangian is unrenormalizable, i.e. it leads to infinities in observable quantities, and is therefore unacceptable. This conclusion has only been proved in any order of perturbation theory and is therefore not rigorous, but it is at least very plausible.

If one accepts the hypothesis that the Lagrangian is determined by the other postulates of quantum field theory, one should be able to reformulate the theory without using the Lagrangian at all. By doing so one would avoid having to mention infinite unrenormalized coupling constants and masses, since the only place where they occur is in the Lagrangian itself. It is then necessary to introduce directly the requirements of unitarity and causality, which were previously introduced by suitably restricting the Lagrangian – more precisely, by demanding that it be Hermitian and local. The unitarity equation is straightforward, though in applying it approximations have to be made since it involves an infinite number of intermediate states. The consequences of the causality condition are less simple to determine. We adopt the condition in the form $[\Phi_1(x_1), \Phi_2(x_2)] = 0$, where Φ_1 and Φ_2 are any two operators and x_1 and x_2 are space-like separated. To take this as the statement of the causality condition is physically reasonable, as the lack of commutativity of two operators is associated with the interference between measurements of the corresponding observables and, if the points are space-like separated, such interference can obviously not occur without signals propagating faster than light.

This makes it especially clear that Mandelstam saw the dispersion-theory program as a mass-shell field theory rather than as a program opposed to quantum field theory. Independence from any specific Lagrangian is not equivalent to a renunciation of the concept of a quantized field. The general belief (or hope) was that the requirement of suitable high-energy behavior for the scattering amplitude would be equivalent to renormalization.

5.7   Summary

By 1956 Gell-Mann had indicated that the requirements of Lorentz invariance, crossing, analyticity, unitarity and suitable asymptotic boundary conditions alone *might*, in principle, be enough to specify (uniquely) the scattering amplitudes of quantum field theory, without having to resort to any specific Lagrangian. He did not consider this to be an *anti*-quantum field theory program, but rather an alternative to the standard QFT approach of using Lagrangians. Even by 1958 Chew expressed the hope that the *S*-matrix equations might provide a complete dynamical description for strong-interaction physics. A series of technical advances provided specific mechanisms that might allow the implementation of such a program.

Mandelstam in 1958 conjectured (partly on the basis of QFT perturbation calculations) that the elastic pion–nucleon scattering amplitude had sufficient analyticity to satisfy a dispersion relation not only in the energy variable (as previously), but also in the momentum transfer (or scattering angle) variable. This representation allowed Chew and Mandelstam to obtain a 'bootstrap' solution for pion–pion scattering, in which a resonance was self sustaining. That is, this resonance fed into the equations reproduced itself as output. Unitarity applied to the Mandelstam representation generated by iteration a set of equations that had a bootstrap structure which might allow one to recover the entire relativistic scattering amplitude from a Born (or 'high-energy') approximation.

Regge, in his 1959 proof of double dispersion relations for nonrelativistic potential scattering, introduced the concept of complex angular momentum, purely as a mathematical technique for solving this problem. He was able to establish a bound on the asymptotic behavior of the scattering amplitude. Others (not Regge) soon conjectured that the form of this bound (established rigorously only for *potential* scattering theory) remains valid for the relativistic case and they also coupled this assumption to crossing symmetry. This provided a plausible mechanism to control the asymptotic behavior of relativistic scattering amplitudes and make double dispersion relations valid, as will be discussed in the next chapter.

This merging of ideas from the Goldberger–Gell-Mann dispersion-theory program with those of Chew and his collaborators produced a theoretical framework that some hoped would provide an alternative to traditional quantum field theory. The positive appeal of this new program was coupled with the problematic status of QFT. Again, it is worth noting that this novel 'philosophical' outlook (an independent and workable SMT) came from the details of highly technical developments, not from some *a priori* commitment to a model of science or of physical theory.

6

The bootstrap and Regge poles

For those who view the S-matrix program as being independent of (and perhaps even opposed to) field theory, we have now arrived at what they might take to be the *beginning* of the S-matrix program. However, considering the field theory roots of nearly all the successful applications of the dispersion-theory and analytic S-matrix program, as well as the failure to date of the more revisionary form of S-matrix theory to attain a coherent abstract formulation and a large base of empirical confirmation, we do not use the term 'S-matrix program' as anything necessarily antithetical to field theory. Both field theory and S-matrix theory have been enormously fruitful, often in different areas, and we can consider them as complementary approaches. We shall see, though, that the developments of this chapter did lead to a proposal for a completely independent S-matrix theory, one that has not been refuted. That autonomous S-matrix program is the subject of the next chapter.

In this chapter we focus on the bootstrap, both its origin and some details of its applications. Pomeranchuk had given various arguments – some based on a simple physical model of scattering processes, others based on apparently reasonable asymptotic properties of an analytic scattering amplitude – that led to certain expectations about the asymptotic behavior of total cross sections and about relations among total cross sections for different processes. In particular, there were what appeared to be reasonably convincing theoretical arguments for the belief that total cross sections should approach *constant* values at high energy (as opposed to vanishing or continuing to increase indefinitely). Some of the first cross sections studied at high energy did have a behavior consistent with such an expectation. This provided a background (or 'prejudice') in terms of which some of the early developments and predictions of the S-matrix program were interpreted.

After he and Mandelstam had been motivated by the result of some numerical calculations to conjecture a bootstrap mechanism that would underlay strong-interaction dynamics, Chew was sufficiently impressed by the early successes and by the promise of an S-matrix program that he 'went public' in a dramatic talk given at the La Jolla Conference in 1961. He emphatically rejected QFT, not so much as being incorrect, but certainly as being empty or useless for the strong interactions. Even at this early date, Chew attempted to justify the assumption of analyticity (or maximum 'smoothness') for the S matrix by appeal to some general philosophical principle. He offered as a candidate his version of the old principle of lack of sufficient reason (i.e., what would 'cause' the S matrix to behave otherwise). It was an argument he would return to in future years. Several intriguing and imaginative new ideas were suggested in the early phase of this S-matrix program. Two that were particularly effective in capturing the imagination of the physics community were nuclear democracy and the bootstrap. Nuclear democracy would place all of the strongly interacting particles on an equal footing with none of them being elementary ('aristocratic' or 'better than' the others). (Not a bad phrase in the egalitarianism and antiestablishmentarianism of the 1960s!) Another was the (somewhat amorphous, but therefore elastic) concept of the bootstrap according to which all strongly interacting particles mutually generated, sustained or 'boostrapped' all others through their interactions with one another. In spite of the generality of this latter idea, it was given specific and useful mathematical instantiations that allowed (successful) numerical predictions to be made. Some of these mathematical techniques remain alive and in use today (in, for example, calculations of nuclear potentials).

The application of Regge's ideas (from potential scattering theory) to the relativistic regime promised a means of providing calculational 'bite' to the bootstrap dynamics program. Early support from experiment came from (linear) Regge trajectories in successfully grouping strongly interacting particles into families of recurrences and for the shrinkage of the forward peak in a certain class of scattering processes. This new approach to strong-interaction physics provided a simple model of high-energy phenomena and, with its relatively elementary analytical machinery, allowed rapid entry into an active field of research for many young workers. By the late 1960s Regge phenomenology became much more complex as new phenomena were encountered in experiments. The theoretical picture also lost much of its simplicity as the general S-matrix principles were pressed for the details of their implications. The program, though, never came into any *in principle* irremediable conflict with experiment and was never refuted. It did lose some of its appeal, however, as relevant calculations became more difficult. It was at this time that supporters of the S-matrix program turned to 'philosophical' arguments about the inherent limitations and inconsistencies of QFT and the great scope and promise of SMT.

6.1   The Pomeranchuk theorems

Two technical developments will provide some background for this chapter. We discuss the first here and the second in the next section. Pomeranchuk (1958) used a (perhaps not wholly convincing) model and dispersion relations to argue that cross sections at high energy approach a constant and that, for example, those for π^+p and π^-p scattering must approach the *same* constant value asymptotically. His simple physical model, based in essence on the finite range of the strong force and on a semiclassical approximation usually taken as valid at high energies, supports the contention that the total cross section should approach a constant at high energy. This model pictured the nucleus as an object which absorbed (captured) a particle hitting it and which left unaffected any particle missing it. The general validity of such a picture was certainly not beyond question. His model also predicted that the elastic scattering amplitude must become purely imaginary (i.e., its real part should vanish) at high energy. Then, taking constant asymptotic values for total cross sections and imaginary asymptotic values for the forward elastic scattering amplitudes as *given*, Pomeranchuk proved that the analytic properties incorporated in the dispersion relations for the scattering amplitudes *required* that certain pairs of total cross sections approach the *same* constant value. There was also an independent, semiquantitative argument, based on the nature of inelastic scattering events, that supported this same conclusion. Data on the total cross section for σ_{π^-p} and σ_{π^+p} at high energy were consistent with Pomeranchuk's predictions. These results became known as the Pomeranchuk theorems. So, by the late 1950s theorists were inclined to accept this Pomeranchuk result of constant asymptotic total cross sections and the equality of total cross sections on a common target when the projectiles in one case were the antiparticles of the projectiles in the other.[1]

We have seen in Eq. (A.23b) that the scattering amplitude can be expanded in terms of $S_l(k)$ as

$$F(s,t) = \frac{4\pi\sqrt{s}}{ik} \sum_{l=0}^{\infty} (2l+1)(S_l-1)P_l(\cos\theta). \qquad (6.1)$$

The optical theorem is (since $f(\theta) = F/8\pi\sqrt{s}$)

$$\sigma_{tot}(s) = \frac{1}{2k\sqrt{s}} \operatorname{Im} F(s,0). \tag{6.2}$$

If we consider the target to be a 'black sphere' of radius a which completely absorbs all partial waves l such that the (classical) trajectory of the projectile would take it through the target

$$l \lesssim l_{max} = ka \tag{6.3}$$

(so that there is *no* outgoing spherical wave for these l) and which does not affect higher partial waves, then we have (cf. Eq. (A.20b))

$$S_l = \begin{cases} 0, & \leq l_{max} \\ 1, & > l_{max} \end{cases}. \tag{6.4}$$

In this model the amplitude $F(s,0)$ approaches for large s

$$F(s,0) \underset{s \to \infty}{\longrightarrow} \frac{\mathrm{i}4\pi\sqrt{s}}{k} \sum_{l=0}^{l_{max}} (2l+1)$$

$$\approx \frac{\mathrm{i}4\pi\sqrt{s}}{k} l_{max}^2 = \mathrm{i}4\pi\sqrt{s}\, ka^2. \tag{6.5}$$

(That is, the arithmetic sum can be carried out explicitly.) This tells us that the forward amplitude becomes purely imaginary at high energy. Furthermore, the optical theorem gives

$$\sigma_{tot}(s) \underset{s \to \infty}{\longrightarrow} 2\pi a^2, \tag{6.6}$$

which is a constant.

The relative sizes of $\operatorname{Re} F(s,0)$ and $\operatorname{Im} F(s,0)$ can be checked experimentally since the relativistically invariant elastic differential cross section $\mathrm{d}\sigma_{el}/\mathrm{d}t$ in the forward direction is just

$$\mathrm{d}\sigma_{el}/\mathrm{d}t|_{t=0} = (\pi/k^2)\,\mathrm{d}\sigma_{el}/\mathrm{d}\Omega|_{t=0}$$

$$= \frac{1}{64\pi k^2 s} |F(s,0)|^2 \underset{s \to \infty}{\longrightarrow} \frac{1}{16\pi s^2}$$

$$\times |[\operatorname{Re} F(s,0)]^2 + [\operatorname{Im} F(s,0)]^2|. \tag{6.7}$$

Therefore, from Eqs. (6.2) and (6.7) we have asymptotically

$$\left[\frac{\operatorname{Re} F(s,0)}{\operatorname{Im} F(s,0)}\right]^2 + 1 = \frac{16\pi[\mathrm{d}\sigma_{el}/\mathrm{d}t]|_{t=0}}{[\sigma_{tot}(s)]^2}. \tag{6.8}$$

Experimental data confirm that $\operatorname{Im} F(s,0) > \operatorname{Re} F(s,0)$ at high energy. That is, the quantity on the right side of Eq. (6.8) can be gotten from experiment. The value of that ratio implies that $\operatorname{Re} F < \operatorname{Im} F$. Data were also consistent with $\sigma_{\text{tot}}(s)$ approaching a constant at high energy. (See Figure 6.1 for a summary of the data available by the mid 1960s (Gasiorowicz, 1966, p. 473). Much later data (cf. Moshe, 1978, Figure 2, p. 258) actually shows a logarithmic *increase* of total cross sections at high energies.)

What Pomeranchuk actually needed for his theorem were the assumptions that $\sigma_{\text{tot}}(s) \to \text{const.}$ and that $F(s,0)$ became predominantly imaginary at high energy. If the dispersion relations of the type of Eqs. (3.39) are used to write a dispersion relation for the amplitude $F^{(-)} \equiv F_{\pi^- p} - F_{\pi^+ p}$ in the forward direction and if we assume $\sigma_{\pi^+ p} \neq \sigma_{\pi^- p}$ asymptotically, then the once-subtracted dispersion relation for $\operatorname{Re} F^{(-)}(s)$ becomes asymptotically

$$\operatorname{Re} F^{(-)}(s) \xrightarrow[s \to \infty]{} \text{const.} \, [\sigma_{\pi^- p}(s) - \sigma_{\pi^+ p}(s)] s \, \ln s \qquad (6.9)$$

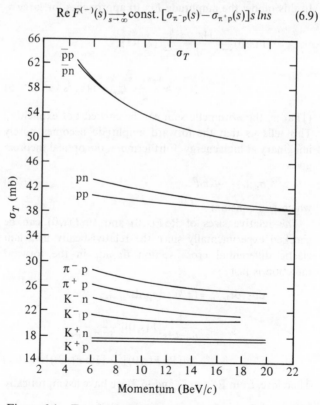

Figure 6.1   Two-body elastic scattering cross sections at high energy (after Gasiorowicz (1966), p. 473).

while the optical theorem of Eq. (6.2) requires

$$\text{Im } F_{\pi^{\pm}p}(s) \xrightarrow[s \to \infty]{} \text{const. } s \tag{6.10a}$$

and

$$\text{Im } F^{(-)}(s) \xrightarrow[s \to \infty]{} \text{const. } [\sigma_{\pi^-p}(\infty) - \sigma_{\pi^+p}(\infty)]s \tag{6.10b}$$

Equations (6.9) and (6.10) together imply that the amplitude is predominantly *real* at high energy. To avoid this we require for the total cross sections the relation

$$\sigma_{\pi^-p}(s) \xrightarrow[s \to \infty]{} \sigma_{\pi^+p}(s) \equiv \text{const.} \tag{6.11}$$

This is the Pomeranchuk result. A previous observation attributed to Okun' and Pomeranchuk (1956; Pomeranchuk, 1956a) supports a similar equality by arguing that as the energy increases more and more inelastic channels open up and must compete for this constant σ_{tot}. Therefore, each inelastic amplitude (and cross section) must vanish at high energy. Take the example of $\pi^-p \to \pi^0n$, a simple charge-exchange process. We see from Eq. (4.11) that

$$F_{\text{ch. ex}} \xrightarrow[s \to \infty]{} 0 \Rightarrow F_{\pi^-p} = F_{\pi^+p}. \tag{6.12}$$

Therefore, $\text{Im } F_{\pi^-p}$ and $\text{Im } F_{\pi^+p}$ become equal and the optical theorem implies

$$\sigma_{\pi^-p} \to \sigma_{\pi^+p},$$

again the statement of Eq. (6.11).

6.2   The Chew–Mandelstam calculation

Chew and Mandelstam (1960) introduced a very useful decomposition of the partial-wave amplitude to facilitate bootstrap calculations. This mathematical technique was an essential tool for implementing the general principles of the S-matrix program. It allowed one to use unitarity, crossing and analyticity to produce specific numerical predictions. And, it was from the results of an application of this method that Chew and Mandelstam were able to recognize a bootstrap mechanism at work. The general bootstrap conjecture was an outcome of a specific, detailed and quite technical mathematical formalism and calculation. Once again, the mathematics had led the physics. Because only some appreciation of the nature of the calculation can give one an understanding of the origin of the bootstrap conjecture, we present below an outline of the Chew–Mandelstam program.

We saw in Chapter 5 that the partial-wave amplitude $A_l(v)$ has both a left and a right cut as illustrated in Figure 6.2 for $\pi\pi$ scattering. Chew and Mandelstam (1960) expressed $A(v)$ (we drop the index l) as[2]

$$A(v) = \frac{N(v)}{D(v)} \tag{6.13}$$

where $N(v)$ (the *numerator* function) has only the left cut and $D(v)$ (the *denominator* function) only the right one. (It can be shown that one always has the freedom to make such a decomposition of A.) For purposes of illustration here we make the (unrealistically simple) assumption that

$$N(v) \xrightarrow[v\to\infty]{} 0, \tag{6.14a}$$

$$D(v) \xrightarrow[v\to\infty]{} 1. \tag{6.14b}$$

That is, we are taking $A(v)$ to vanish asymptotically and Eqs. (6.14) will insure this.

If we set $m = 1$ (i.e., a rescaling of the energy variable) and $v = q^2$, then partial-wave *elastic* unitarity for their $A(v)$ is (recall Eq. (4.8))

$$\operatorname{Im} A(v) = \sqrt{\frac{v}{v+1}} |A(v)|^2, \qquad 0 < v < 3, \tag{6.15}$$

since $v = 3$ now corresponds to the four-pion inelastic threshold. Because of the choice of cuts for N and D, we see

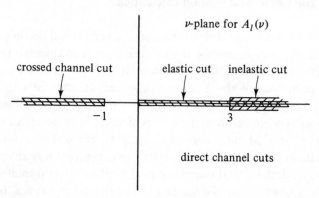

Figure 6.2   The v-plane for the partial-wave amplitudes.

that

$$\text{Im}\, N(v) = D(v)\, \text{Im}\, A(v), \qquad -\infty < v < -1, \quad (6.16a)$$

$$\text{Im}\, N(v) = \text{Im}\, D(v) = 0, \qquad -1 < v < 0, \qquad (6.16b)$$

$$\text{Im}\, D(v) = N(v)\, \text{Im}\left[\frac{1}{A(v)}\right] = -\sqrt{\frac{v}{v+1}}\, N(v),$$

$$0 < v < +\infty. \quad (6.16c)$$

Here we have neglected inelastic effects by using only elastic unitarity in (6.16c). The dispersion relations for N and D are a coupled set of equations,

$$N(v) = \frac{1}{\pi} \int_{-\infty}^{-1} \frac{\mathrm{d}v'\, \text{Im}\, A(v')D(v')}{(v'-v)}, \qquad (6.17a)$$

$$D(v) = 1 - \frac{1}{\pi} \int_{0}^{\infty} \mathrm{d}v' \sqrt{\frac{v'}{v'+1}} \frac{N(v')}{(v'-v)}. \qquad (6.17b)$$

The self-closing nature of the $(\pi\pi)$ problem is now evident. If one knew (or could approximate) Im $A(v)$ on the left cut, then N and D could be calculated and $A(v)$ found from (6.13). Self-consistency would require that the input $A(v)$ agrees with the output one. The difficulty, of course, is getting a reasonable Im $A(v)$ to begin with. In their first paper on the $\pi\pi$ problem, Chew and Mandelstam (1960) found an approximate solution to a modified set of equations with small p-wave phase shifts and later (1961) exhibited the now famous bootstrap solution in which the δ_{33} resonance as input generated itself as output. The problem of asymptotic (or boundary) conditions posed a difficulty in these calculations and this had to be handled by means of a cutoff (in order to truncate the integrals to make them converge and yield finite results).

6.3   Regge poles and asymptotic boundary conditions

In the early 1960s Chew and Frautschi wrote a series of papers in which they attempted to implement a consistent theory of strong interactions based on unitarity, crossing and analyticity, particularly in the form of the Mandelstam representation. They (1960, 1961a) argued, on the basis of diffraction scattering at high energies and the Pomeranchuk rule of constant asymptotic cross sections, that p-wave dominant solutions would be required (as found in the Chew–Mandelstam calculation). That is, there were to be no low-energy bound states or resonances for $J \geqslant 2$. They suggested that this situation allowed the strong interactions to be as strong as was allowed by self-consistency. Chew and Frautschi (1960, p. 583) acknowledged Regge's potential theory result that, for large t, the amplitudes are bounded by $t^{\alpha(s)}$. They *conjectured* that this type of result would hold in a relativistic theory as well. At the Berkeley Conference on Strong Interactions (Dec. 27–29, 1960), Chew (1961b) reviewed this program of dominance by nearby singularities. Regge's papers (1959, 1960) were mentioned, but his results were not used in any essential way even though Mandelstam during discussion (Chew, 1961b, p. 470) did point out the relevance of crossing for the $t^{\alpha(s)}$ behavior. Chew and Frautschi (1961b) proposed a general definition of a potential in the relativistic case. Regge's result plays no role in this paper. Exchanges in the crossed channels provide the 'potentials' that drive the iteration scheme for the double spectral functions $\rho(s, t)$ much as in the nonrelativistic case discussed in the previous section (Eqs. (5.4) and 5.5)). We do not go into this in detail here since the general idea of the bootstrap will be illustrated with some approximate N/D calculations below.

Froissart (1961), using the Mandelstam representation and unitarity, derived a rigorous bound on the scattering amplitude $F(s, t)$ as[3]

$$F(s,t) \underset{s \to \infty}{\lesssim} \text{const.}\, s\, ln^2 s, \qquad t \leqslant 0. \tag{6.18}$$

If we assume a power-law bound, then this becomes

$$F(s,t) \underset{s \to \infty}{\lesssim} \text{const.}\, s^N, \qquad t \leqslant 0, N \leqslant 1. \tag{6.19}$$

Of course, this fit in very nicely with Chew and Frautschi's conjecture about only s- and p-wave resonances being allowed at low energy (or, really, low t for this discussion).

6.4   The La Jolla conference and Chew's rejection of QFT

An extremely important meeting for the S-matrix program and the introduction into it of Regge's work was the June 1961 La Jolla Conference on 'Weak and strong interactions'. Many of the key ideas of the S-matrix program were first put forth there (Gell-Mann, 1987, p. 485; Goldberger, 1970, p. 691; Capra, 1985, p. 260). Unfortunately, no proceedings of that conference were ever published. Gell-Mann (1987, p. 485) recalls suggesting the La Jolla conference to Keith Brueckner, who had just come to San Diego. The actual organization of the meeting was done[4] by William Frazer and David Wong, who were young theorists in the new Department at the University of California at San Diego. The speakers included[4] Chew, Goldberger and Blankenbecler, Regge (on Regge poles), Froissart (on his bound), Amati, Fubini and Stanghellini (on the multiperipheral model), Cutkosky (on his discontinuity rules), Eden (on a class of Landau singularities) and Gell-Mann (on his eightfold way of SU_3). At that conference, Goldberger suggested that families of particles should lie on Regge trajectories (Chew and Frautschi, 1962, p. 43; Blankenbecler and Goldberger, 1962, p. 766). This would prove to be a key insight. Also, it was in preparing a talk for this La Jolla Conference that Chew 'decided to cut the strings from QFT and proceed on the SMT basis'[5] (Capra, 1985, p. 260). Chew's La Jolla talk was a memorable one, even as recalled by Gell-Mann (1987, p. 486).

> But the most dramatic talk was that of Geoffrey Chew, who said, in effect: '... Field theory is no damned good; instead we must use the *S*-matrix theory.' ... [F]or the strong interaction he insisted that we abandon field theory and go over to '*S*-matrix theory'.
>
> [I] have never succeeded in understanding that point of view. While before 1961 one of the great pleasures of working in theoretical physics was the possibility of discussing theories with Geoffrey Chew, after 1961 it became very difficult.

In that talk, Chew (1961d) made an unambiguous break with quantum field theory and enunciated the basic principles of a new program.

> So that there can be no misunderstanding of the position I am espousing, let me say at once that I believe the conventional association of fields with strongly interacting particles to be empty.... I have yet to see any aspect of strong interactions that is clarified by the field concept. Whatever success theory has achieved in this area is based on the unitarity of the analytically continued *S* matrix plus symmetry principles. I do not wish to assert (as does Landau [1960]) that conventional field theory is necessarily wrong, but only that it is sterile with respect to strong interactions and that, like an old soldier, it is destined not to die but just to fade away. (p. 3)
>
> [I]t is appropriate to recall that, although he has now turned his back on the *S*-matrix approach, the original author was Heisenberg in 1943. Heisenberg soon lost interest, I suppose because at that time he lacked the full analytic continuation that is required to give the *S*-matrix dynamical content. (p. 5)

The S matrix is an analytic function of all momentum variables with only those singularities required by unitarity. I have strong faith in the eventual verification of such a conjecture. (p. 6)

My personal inclination here is to resurrect the ancient principle of 'lack of sufficient reason'. ... The fundamental principle ... might be one of maximum smoothness: The S matrix has no singularities except where absolutely necessary to satisfy unitarity. There is no 'reason' for it to have any others. (p. 7)

I am convinced that ... *none* of them [particles corresponding to S-matrix poles] is elementary.

We may be reminded again of the principle of 'lack of sufficient reason.' If one can calculate the S matrix without distinguishing elementary particles, why introduce such a concept? (p. 9)

[W]e may appeal once more to the notion of 'lack of sufficient reason.' ... Steven Frautschi and I have found it natural to postulate that strong interactions are characterized by 'saturation' of the unitarity condition [Chew and Frautschi, 1960, 1961a]; that is, they have the maximum strength consistent with unitarity and analyticity. To us there seems no *reason* for any other strength to occur, and the observed behavior of high-energy cross sections gives strong encouragement to this notion of saturation. (p. 10)

A theory is deemed successful not when it has passed all possible tests, but when it has passed an 'impressive' number and failed none. (p. 13)

If ... a reasonable number of particles now regarded as elementary are successfully 'explained' through the S matrix, then one might be willing to give the theory the benefit of the doubt [even if not *all* particle masses have been computed in terms of just *one* input mass]. (p. 14)

His collaborations with Mandelstam and with Frautschi at Berkeley had been essential for this break. Not only had Chew and Mandelstam (1960, 1961) found some indication of a 'bootstrap' mechanism for the $\pi\pi$ system, but Chew and Frautschi (1960) proposed the 'strip approximation'[6] as a basis for a unified scheme of dynamical bootstrap calculations. About the time of this conference, Chew realized the connection of his ideas with those of Heisenberg's old S-matrix approach. As we have just seen, Chew very forcefully stated his view that, at least for the strong interactions, S-matrix theory must be used and field theory abandoned (Gell-Mann, 1987, p. 485; Capra, 1985, p. 264). There was a good deal of resistance to these views. By no means all of the key contributors to the foundations of what became an autonomous *S*-matrix program agreed with this radical opposition of quantum field theory to *S*-matrix theory. As we have already seen, Gell-Mann[7], Goldberger[8] and Mandelstam[9] considered these two approaches as complementary, while Low[10] saw the *S*-matrix approach as useful but not complete enough to constitute a theory. (Perhaps this is reminiscent of Wentzel's (1947) 'empty frame'.)

Chew recalls that, in this period, 'the most important single empirical discovery was that low mass hadrons, especially the nucleon, lie on Regge trajectories.'[11] This allowed him to break with the long-accepted and nearly universal belief that the proton was elementary. If that worked for the nucleons, then why not for the other hadrons? These conjectures could be tested experimentally by looking at the behavior of forward and backward peaks in two-body scattering data, as we discuss below. This was the beginning of a real philosophical split between Chew, along with some of his collaborators, and several of the major contributors to the dispersion-theory and Regge programs. We use the term 'autonomous *S*-matrix program' to refer to this radical departure. It will be the subject of Chapter 7.

6.5   Early successful predictions

In a post-La Jolla publication, Chew and Frautschi (1961c) proposed a principle of equivalence for all strongly interacting particles[12]. This conjecture placed all of the strongly interacting particles on the same footing, none of them being more elementary than any other. Each helped to bootstrap the others, which in turn contributed to generating it. Gell-Mann coined the expression 'nuclear democracy' (Chew, 1964c, p. 105) to describe this. At Mandelstam's initial suggestion,[13] Chew and Frautschi saw that Regge's work could supply a mechanism for controlling the high-energy behavior of scattering amplitudes. We have seen in Chapter 5 that the large *t* behavior of a nonrelativistic scattering amplitude is $t^{\alpha(s)}$ where $\alpha(s)$ is the Regge trajectory or Regge pole. (This latter term is due to Gell-Mann (Capra, 1985, p. 259).) Chew[14] credits Gell-Mann with recognizing the wide implications of Regge poles:

> Gell-Mann, in particular, was bolder than I initially in grasping the generality of Regge poles.

A qualitative sketch of one of these trajectories in potential scattering is shown in Figure 6.3. For negative energy, $\alpha(s)$ is purely real and if it passes through zero or a positive integer, then there is a bound state in that partial wave. Once s is above the physical scattering threshold $s_0(s_0 = 0$ for the nonrelativistic case), Im $\alpha(s)$ becomes positive so that $\alpha(s)$ cannot produce any more bound-state poles, but only resonances if Im $\alpha(s)$ is small when Re $\alpha(s)$ is a positive integer. If in the t channel we let $s \rightarrow +\infty$ ($s \leqslant 0$ for the *physical* region there), we find $s^{\alpha(t)}$ for fixed t. By crossing, this gives us the asymptotic form of the physical-region scattering amplitude as

$$F(s,t) \xrightarrow[s \to \infty]{} \frac{\gamma(t)[\frac{1}{2}(1 + e^{i\pi\alpha(t)})]}{\sin[\pi\alpha(t)]} s^{\alpha(t)}, \qquad t \leqslant 0. \tag{6.20}$$

Here $\gamma(t)$ is the residue of the Regge pole. From Eqs. (6.2) and (6.7) we then expect

$$\sigma_{\text{tot}}(s) \xrightarrow[s \to \infty]{} \text{const.} \, \gamma(0)s^{\alpha(0)-1} \tag{6.21}$$

and

$$\left. \frac{d\sigma_{el}}{dt} \right|_{t=0} \xrightarrow[s \to \infty]{} \text{const.} \, s^{2[\alpha(0)-1]}. \tag{6.22}$$

There will be a result similar to (6.22) in the backward direction (where $\theta = \pi$, which implies $u = 0$) from an exchange in the u-channel. If the Regge trajectory is approximately linear, then

$$\alpha(t) \approx \alpha(0) + \alpha'(0)t \tag{6.23}$$

and for *small t* we have for $t \leqslant 0$

$$\frac{d\sigma_{el}}{dt} \xrightarrow[s \to \infty]{} g(t)s^{-2[1-\alpha(0)]} e^{-2\alpha'(0)|t|\ln s}, \tag{6.24}$$

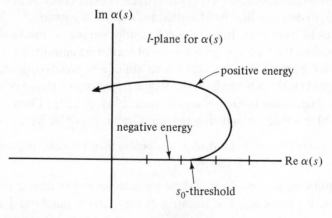

Figure 6.3   A typical Regge trajectory in potential theory.

where $g(t)$ is a lowly-varying (i.e., nearly constant) function. Chew and Frautschi (1961c) state three important predictions of this Regge pole picture (pp. 395–6):

> (a) The width of the peak should decrease logarithmically as [s] increases. (b) The tail of the peak should fall off exponentially with [−t]. . . . A third important prediction is that the same constants $\alpha(0)$, $\alpha'(0)$, etc. should control all forward or backward peaks that relate to the same set of quantum numbers.

We discuss these predictions about the forward peak in detail below. However, it may help the reader if we first show a typical forward peak in the scattering data. Figure 6.4 illustrates the situation for $\pi^- p$ elastic scattering at a particular value of the energy (or momentum). This is a

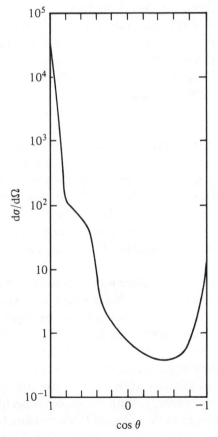

Figure 6.4   Forward and backward peaks in $\pi^- p$ elastic scattering at 4 BeV/c (after Barger and Cline (1969), p. 17).

plot of $d\sigma/d\Omega$ versus $\cos\theta$, where θ is the scattering angle in the center of momentum frame. Notice that the vertical scale is a *logarithmic* one so that the forward peak at $\theta = 0$ (or $\cos\theta = +1$) is more than a factor of 10^4 larger than the cross section at $\theta = 90°$ (or $\cos\theta = 0$). It is truly a sharp, narrow and very high peak. This behavior is characteristic of *elastic* scattering processes. The predictions we discuss below are often given in terms of $d\sigma/dt$, which is a relativistically invariant quantity, rather than in terms of the $d\sigma/d\Omega$ of Figure 6.4. However, since $d\Omega = d(\cos\theta)d\phi$ and $dt = 2k^2 d(\cos\theta)$, these two forms of the differential cross section are related by a simple multiplicative constant (cf. Eq. (6.7), the left-hand side.)

Chew and Frautschi (1961c, p. 396) also pointed out the possibility that if cross sections did approach constants asymptotically, then (cf. Eq. (6.21)) $\alpha(0) = 1$. This special Regge trajectory, which they termed α_{vac} (later known as the Pomeranchuk trajectory or Pomeron, α_P), would control all forward diffraction peaks and have the internal quantum numbers of the vacuum. That trajectory would saturate (i.e., reach) the Froissart limit of Eq. (6.18). With this trajectory, we expect from Eq. (6.24) a logarithmic shrinkage of the diffraction peak as s increases. Furthermore, high-spin field theories were known to be nonrenormalizable and the Froissart bound added credibility to the postulated Regge pole structure for relativistic scattering amplitudes.

Chew and Frautschi (1962) further developed their previously-stated (1960, 1961a) principle that the strong interactions should have the maximum strength consistent with unitarity and crossing. This paper also contained the first example of what would become known as Chew–Frautschi plots, which are (straight-line) graphs of $\mathrm{Re}\,\alpha$ versus s.[15] Figure 6.5 shows their original plot and Figure 6.6 later, more complete versions (Collins and Squires, 1968, pp. 198 and 194, respectively).

Regge poles were now fully 'in' and there was an explosion of activity in the field. Several theorists (mainly at Berkely and at CALTECH) developed the formalism so that it could be put to test by experimental data. Udgaonkar (1962) followed Chew and Frautschi's proposal that the asymptotic behavior of the elastic scattering amplitude is dominated by Regge poles. Using leading Regge terms, he obtained expressions for the differences of total cross sections, such as $[\sigma(\pi^- p) - \sigma(\pi^+ p)]$ and $[\sigma(pp) - \sigma(np)]$, as functions of energy. With the known trajectories he was able to explain deviations from the expected Pomeranchuk limits at then-available energies. A paper by Frautschi, Gell-Mann and Zachariasen (1962; Gell-Mann, 1962a) provided specific and extensive rules for testing the experimental consequences of the Regge-pole hypothesis. Arguing by analogy with resonance theory (cf. Eqs. (2.46) or (2.49)) and demonstrating his result in multichannel potential theory, Gell-Mann (1962b) conjectured that the residue $\gamma(t)$ of a Regge pole in the S matrix factorizes. Figure 6.7 shows the exchange of a Pomeron in the crossed channel of the reaction $a+b \rightarrow a+b$. The factorization conjecture is

$$\gamma(0) \propto g_{aaP} g_{bbP}. \tag{6.25}$$

From Eq. (6.21) we have

$$\sigma_{aa} \propto g_{aaP} g_{aaP}, \tag{6.26a}$$

$$\sigma_{bb} \propto g_{bbP} g_{bbP}, \tag{6.26b}$$

$$\sigma_{ab} \propto g_{aaP} g_{bbP} \tag{6.26c}$$

so that

$$(\sigma_{ab})^2 = \sigma_{aa} \sigma_{bb}. \tag{6.27}$$

A particular (experimentally confirmed) case of this is

$$\sigma_{\pi N}^2 = \sigma_{\pi\pi} \sigma_{NN}. \tag{6.28}$$

Equations (6.26)–(6.28) (along with charge conjugation invariance in

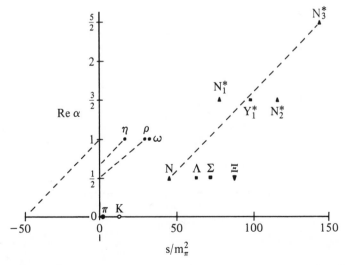

Figure 6.5   An early version of the Chew–Frautschi plot (after Chew and Frautschi (1962), p. 42).

the form $\sigma_{\pi^-\pi^-} = \sigma_{\pi^+\pi^+}$) allow one to recover the Pomeranchuk theorem of Eq. (6.11). Udgaonkar and Gell-Mann (1962) applied the Regge model to scattering from a nucleus. However, some practitioners, such as Zachariasen (1963, 1965, 1966) remained skeptical about the

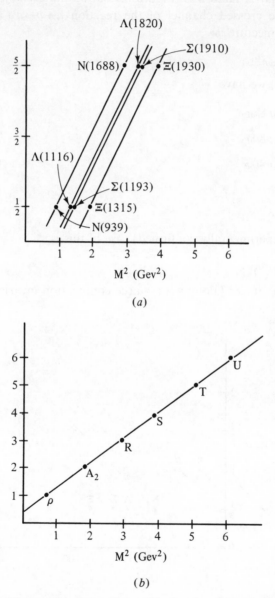

Figure 6.6   Chew–Frautschi plots for the nucleons and the 'vector' meson family (after Collins and Squires (1968), pp. 196 and 198).

bootstrap as a *theory*. At the 1962 Cargése Summer School, he (Zachariasen, 1963, p. X-1) began his lectures with this observation:

> At the present time, the theory of Regge poles, in so far as it can really be called a theory, consists of a set of guesses about the high-energy behavior of scattering amplitudes based on analogies with potential scattering. ... The whole business will become what can honestly be called a theory only when ... it becomes possible to compute the [Regge] trajectories theoretically from first principles. ... [T]he phenomenology of 'Regge-ism' is the only part that has received much attention as yet.

Let us now turn to the experimental support that the Regge pole picture received.[16] We have already indicated that a large number of baryon and meson resonances were found to lie on (nearly) linear Regge trajectories. Such linear trajectories had not been expected on the basis of potential theory, but this correlation of known particle states was considered a definite plus for Regge phenomenology. (Mandelstam (1967, 1969a, 1969b) later showed that constantly rising Regge trajectories could be caused by many-body effects.) One of the most dramatic predictions of the theory was the shrinkage of the diffraction peak in Eq. (6.24). A convenient way to look for this effect in the data is to measure the quantity

$$\chi(s,t) \equiv \text{const.} \frac{d\sigma_{el}/dt}{[\sigma_{tot}(s)]^2} \underset{s \to \infty}{\longrightarrow} \text{const. } e^{-2\alpha'(0)|t|\ln s}. \tag{6.29}$$

We have used Eqs. (6.21) and (6.24) to obtain the asymptotic form on the right. That is, a plot of $\ln[\chi(s,t)]$ versus t (for fixed s) should be a straight line, but the slope of that straight line should change as s increases. Happily for Regge theory, early data on pp scattering (Cocconi, 1962; Diddens *et al.*, 1962a, 1962b; Drell, 1962, p. 898) showed just this expected shrinkage.

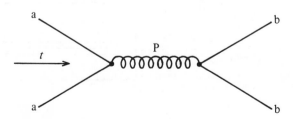

Figure 6.7 Pomeron exchange.

6.6   Degenerating Regge phenomenology

However, later data (Foley *et al.*, 1963) showed no such effect for $\pi^- p$ scattering. (See also Lindenbaum, 1964, 1969.) Figure 6.8 (which is a semilog plot) represents these later data and clearly shows the expected effect for pp scattering and its absence for πp scattering. This was one of many indications that there were special problems about diffractive (elastic) scattering and about the associated Pomeranchuk trajectory α_P. (The term 'diffraction' is used to refer to these large forward peaks for elastic scattering in analogy with a similar peaking effect common in optical phenomena, such as the diffraction of light by an opaque disc or sphere.) We return to the Pomeron below. However, the Regge model had already begun to lose some of its initial simplicity that had made it so appealing. One could still hold onto the model, of course (Collins and Squires, 1968, p. 226):

> The lack of observed shrinkage for many elastic scattering processes, led many people to abandon the Regge pole model. The reasons for this were more emotional than rational since it is already clear . . . that, at present energies, several trajectories are contributing significantly to high energy elastic scattering, and the argument leading to shrinkage only holds in the region where one trajectory is adequate to

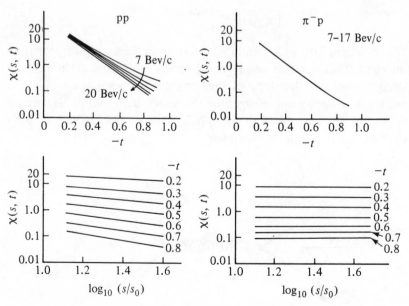

Figure 6.8   Diffraction peak data.

describe the scattering. Thus, although the lack of shrinkage some-what spoils the attractiveness of the model at these energies, it should not lead one to abandon it.

One can reasonably ask about the luck of the sequence of the events. If the diffraction peak in the first observed reaction had not shrunk, would this model have been pursued so vigorously (even though there would have been no conflict of principle)?

A reaction that was to become important for testing the Regge model was πN charge-exchange scattering

$$\pi^- p \to \pi^0 n. \tag{6.30}$$

Since no Pomeron (which must have zero charge) could be exchanged in the crossed channel, the diffractive peak was absent. Figure 6.9 illustrates the exchange of a ρ meson in the t-channel. Although good data on this became available only somewhat later (say around 1966), we discuss it here to indicate a typical problem met by Regge phenomenology. This one-pole model provided a good fit to the data (Stirling *et al.*, 1965) for $d\sigma/dt$ (van Hove, 1967). However, the Regge model predicts zero polarization for this reaction. Polarization is a measure of the amount of nucleon spin flip during the scattering process and is given in terms of the phase difference between the spin flip and the spin nonflip amplitudes. Since the single ρ-exchange Regge model predicts the same phase for both amplitudes (van Hove, 1968), this model requires zero polarization. In fact, though, the data (Bonamy *et al.*, 1966) shows a 10–15% polarization, a clear contradiction to the simple Regge prediction. There were complicating moves that could be, and were, made. The possibility of cuts (associated with branch points, rather than just simple poles in the angular momentum plane) was one and we consider these now. Such theory-saving necessities, however,

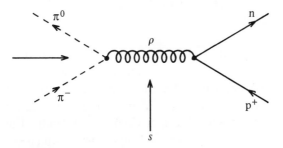

Figure 6.9   Rho trajectory exchange for pion-nucleon charge-exchange scattering.

detracted from the hope that the Regge model might form the basis of a true theory and turned it into curve-fitting phenomenology in terms of many adjustable parameters. This phenomenology continued to be practiced for a long time, but we shall not pursue it further. Zachariasen (1971, p. 50) in a review article on the later Regge program gives a sense of the disappointment that the phenomenology had become so complicated.

> In the form in which it was originally put forward, Regge phenomenology contained a wealth of predications. If Regge poles alone dominated scattering amplitudes at high energies, then the simple form ... dictated very specific relations between scattering amplitudes describing different processes.
>
> As it turned out, however, this admirable vision of high-energy phenomenology has failed. It is simply not true that in general Regge poles alone dominate; other kinds of j-plane singularities, specifically branch points, are also present and are important.

Only a much more restricted set of (model-independent) predictions could now be made. This Regge phenomenology could serve as a nice case study of a degenerating research program in Lakatos' sense of that phrase.

Indications that there might be Regge cuts as well as Regge poles in the relativistic case were not long in coming. Blankenbecler and Goldberger (1962) had mentioned this likelihood and Chew, Frautschi and Mandelstam (1962, p. 1208) gave some oblique indication of trouble from the inconsistency of their equations for $\pi\pi$ scattering. Igi (1962) showed that if there are no singularities other than Regge poles in the J-plane for $t = 0$, then there had to exist the following sum rule for the s-wave πN scattering length a

$$\left(1 + \frac{1}{M}\right) a = -\frac{f^2}{M} \frac{1}{1 - 1/(4M^2)} + \frac{1}{2\pi^2} \int_0^\infty dk[\sigma_{\text{tot}}(k) - \sigma_{\text{tot}}(\infty)].$$

(6.31)

He (Igi, 1963) found that this sum rule was not satisfied and so concluded there must be other singularities present. (This same technique was later applied by Igi and Matsuda (1967) to πN charge exchange scattering to obtain a 'superconvergence' relation in the absence of cuts. We discuss these in Chapter 8.) Amati, Fubini and Stanghellini (1962a, 1962b) examined diagrams in which two Regge poles were exchanged in the crossed channel (Figure 6.10(a)). (Notice that there is no similar phenomenon in potential theory since there is no crossed channel in that case.) They found that this gave rise to a moving cut in the J-plane. Mandelstam (1963a, 1963b) showed that this cut was cancelled by one from other diagrams (Polkinghorne, 1963d). However, he (1963c) also found a more complicated diagram (Figure 6.10(b)) which produced a cut that was not cancelled. The existence of a crossed channel (not present in potential theory) was essential for these cuts to be required. Gribov and Pomeranchuk (1962) had also argued that there would have to be present in the J-plane an essential singularity. The only way to avoid the violation of unitarity which this implied was for there to be a moving cut in J which shielded this essential singularity. The cuts were here to stay and Regge phenomenology was much more complicated now.[17] Still, Chew (1963b) continued to hope for a self-consistent S matrix with Regge asymptotic behavior.

And, in spite of these problems of principle for a Regge pole program, it is important to appreciate that Regge phenomenology continued to be enormously useful in correlating and organizing a huge amount of high-energy scattering data. One only needs, for example, to examine even cursorily Barger and Cline's book on high-energy phenomenology, published in 1969, to see this.

6.7 Applications of the bootstrap

However, the heart of the calculational S-matrix program was the bootstrap hypothesis. Let us see how it fared. Although he did not become a proponent of the autonomous S-matrix program, Gell-Mann played an important role in the bootstrap project. He was quite positive about Chew–Frautschi bootstrap picture of hadrons.[18] An early

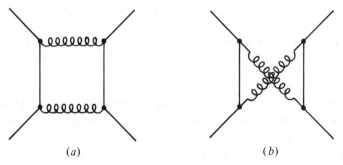

(a)   (b)

Figure 6.10   Regge cut diagrams.

statement of his faith in the bootstrap is (Gell-Mann, 1967, p. 5):

> [T]he q and $\bar{q}$ ... are mathematical.

> If the mesons and baryons are made of mathematical quarks, then the quark model may perfectly well be compatible with the bootstrap hypothesis, that hadrons are made up out of one another.

This is quite interesting since it would drive a wedge between the concept of the bootstrap and nuclear democracy. Gell-Mann is here allowing for the possibility of 'mathematical' elementary entities (quarks), even though the bootstrap would still be operative among the physically observed particles. In a recent retrospective on the early days of the bootstrap conjecture, Gell-Mann (1987) has stressed the importance of the bootstrap idea even for his subsequent work on the quark model.

> [A]bout that [the bootstrap] I was quite enthusiastic.

> In 1963, when I developed the quark proposal, with confined quarks, I realized that the bootstrap idea for hadrons and the quark idea with confined quarks can be compatible with each other, and that both proposals result in 'nuclear democracy', with no observable hadron being any more fundamental than any other. (p. 486)

> [I]t occurred to me that if the bootstrap approach were correct, then any fundamental hadrons would have to be unobservable, incapable of coming out of the baryons and mesons to be seen individually, and that if they were unobservable, they might as well have fractional charge.

> Except for the bootstrap idea, the notion of 'nuclear democracy', I do not know if I would have come so readily to the notion that fundamental hadrons ought to be unobservable and that therefore it was all right for them to have fractional charge. (p. 494)

> As far as I know, something like the bootstrap program might still qualify as a valid description of hadrons, if the right boundary conditions at infinite momenta are imposed on the theory; but it has not proved as useful as the quark description, which led to the development of QCD. (p. 496)

Even today, Gell-Mann believes the bootstrap and feels that Green and Schwarz' (1984c, 1985) recent work on superstrings (to be discussed in Chapter 8) may be a fulfillment of that conjecture.[19]

In fact, Chew credits Gell-Mann's enthusiasm with pushing him over the threshold of the 'complete' bootstrap conjecture.[20] Although the paper by Frautschi, Gell-Mann and Zachariasen (1962) provided specific and extensive rules for testing the experimental consequences of the Regge-pole hypothesis, Gell-Mann felt that, rather than concentrating on just one kind of channel, it was important to concentrate on calculations with an infinite number of intermediate states and many channels.[21] This idea was pursued by Dolen, Horn and Schmid (1967) in their paper on 'duality' and Veneziano (1968) exhibited a specific example of such a function (both of which we discuss in Chapter 8). Gell-Mann also began a series of studies of the conditions under which the 'elementary' particles of conventional field theory might actually lie on Regge trajectories (Gell-Mann and Goldberger, 1962; Blanken-becler, Cook and Goldberger, 1962; Gell-Mann, Goldberger, Low and Zachariasen, 1963; Gell-Mann, Goldberger, Low, Marx, and Zachariasen, 1964; Gell-Mann, Goldberger, Low, Singh, and Zachariasen, 1964) and of the vacuum trajectory (Pomeron) in field theory (Gell-Mann, Goldberger and Low, 1964; Polkinghorne, 1963a, 1964a, 1964b). These papers were definitely not in the spirit of the axiomatic quantum field theory:

> 'In particular, the contribution of axiomatic field theory to calculations has been less than any preassigned positive number, however small.'[22]

As we have seen, they were, nevertheless, critical of *opposing* quantum field theory to *S*-matrix theory (Goldberger, 1969a, p. 86). Not all quantum field theories in fact Reggeize. When some of Mandelstam's papers (which we mentioned above) on the Gribov singularities made the Regge picture seem quite hopeless, Gell-Mann's active participation in the *S*-matrix program waned and he concentrated on other theoretical problems.

But, as we indicated previously, calculational applications of the bootstrap turned on approximate techniques for solving the N/D equation, Eq. (6.17), for the partial-wave amplitudes. Balázs (1962a, 1962b, 1962c, 1963) approximated the left-cut contributions by pole terms and obtained numerical solutions for NN, πN and N$\bar{\text{N}}$ scattering. To illustrate the basic idea of an N/D bootstrap calculation, we discuss the ρ bootstrap of Zachariasen (1961) and Zachariasen and Zemach (1962). It was this elementary application of the bootstrap idea (here to generate the ρ meson from the $\pi\pi$ interaction) that was a clear and, for many, a convincing demonstration of the bootstrap mechanism.[23] Goldberger (1970, p. 690) has stated:

> '[I] first felt comfortable about [the bootstrap idea] with the elementary calculation of Zachariasen.'

Gell-Mann held a similar view of the importance of the Zachariasen–
Zemach calculation.[24] Figure 6.11 suggests a fairly evident (if crude)
way to begin such a bootstrap calculation. One computes (using
Feyman diagram rules) the amplitude for the exchange of a $J = 1$ ρ
meson in the t channel. This will depend upon the variables s and t. We
then cross to the s channel, project out the $J = 1$, $T = 1$ amplitude, take
its discontinuity on the left cut ($v < -1$), and, from (6.16a), set

$$\text{Im } N(v) \approx \text{Im } A(v). \tag{6.32}$$

Equation (6.17b) yields $D(v)$. Since we had put in a pole in the t-channel
at $t = m_\rho^2$, we require one to come out in the s-channel as

$$\text{Re } D(s = m_\rho^2) = 0. \tag{6.33}$$

We require that Re $D = 0$ here, but not $D = 0$, because a pole *on* the real
(s) axis *above threshold* would make $A \to \infty$ there and that would violate
the unitarity condition, Eq. (6.15). This prohibition against poles *on* the
real axis (above threshold) follows from the *nonlinear* nature of the
unitarity condition. From Eqs. (6.13) and (6.33) we see that near $s = m_\rho^2$

$$A(s) \approx \frac{N(m_\rho^2)}{\left. \dfrac{\text{d Re } D(s)}{\text{d}s} \right|_{m_\rho^2} (s - m_\rho^2) + \text{i Im } D(m_\rho^2)}. \tag{6.34}$$

That is, we have made a Taylor expansion of $D(s)$ about the point
$s = m_\rho^2$.

Therefore, the width of the ρ resonance can be calculated. These
self-consistency conditions determine the width Γ_ρ of the resonance
which is related to Im $D/\text{d}[\text{Re } D]/\text{d}s$ at $s = m_\rho^2$. These same conditions
determine the value of m_ρ^2. Of course, the actual calculation requires a

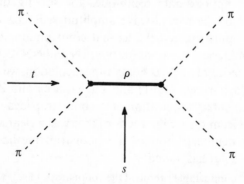

Figure 6.11   The rho bootstrap diagram.

cutoff (i.e., a truncation of the integral) to obtain convergence in the integrals and some judicious choice of parameters, but a fairly good value of m_ρ emerges. Zachariasen and Zemach (1962) then did a coupled-channel calculation in which π, ρ and ω mesons were all treated at once. The result for the mass and width of the ρ got closer to the experimental values. As with all of these numerical bootstrap calculations, it is always difficult to tell how much of the result is produced by the approximations themselves and how much by the surviving dynamics. The approximations are not controlled in the sense of our being able to estimate the error made. They are basically one-shot affairs and you get whatever comes out. We have presented these mathematical details here because it is only through such calculations that one can appreciate just how the bootstrap conjecture was implemented to produce results that commanded assent from the physics community. Chew (1962b) performed a reciprocal bootstrap calculation in which the (3,3) resonance is a composite of the N and π while the N is a composite of the N and π by exchange of the (3,3) resonance. They bootstrapped each other.

Capps (1963) considered the possibility of bootstrapping internal symmetries. In his model an octet (i.e., a set of eight) of vector mesons (ρ, ω, K^* $\bar{K}^*$) is assumed to be bootstrapped from an octet of pseudoscalar mesons (π, η, K, $\bar{K}$). (There are *eight* in each case because some of the particles have more than one charge state: three for the ρ, two each for the K^* and $\bar{K}^*$, for example.) Equality of masses within an octet, conservation of isospin and strangeness (another strong-interaction quantum number) are assumed. The input consists of a set of Born amplitude graphs (as first-order approximations) like that of Figure 6.11. The approximate (matrix) N/D method is applied and the ρ, ω and K^* poles are required as output. The coupling constants, which are the residues of these poles, then turn out to satisfy (as output from the calculation)

$$\gamma^2_{\rho\pi\pi} : \gamma^2_{\rho KK} : \gamma^2_{\omega KK} : \gamma^2_{K^*\pi K} : \gamma^2_{K^*\eta K} = \frac{4}{3} : \frac{2}{3} : 2 : 1 : 1. \qquad (6.35)$$

These are exactly the ratios that would have followed from SU(3) symmetry. Cutkosky (1963a, 1963b), Cutkosky and Tarjanne (1963), Abers, Zachariasen and Zemach (1963), Belinfante and Cutkosky (1965), and Capps (1964, 1965, 1966, 1967) were among the many contributors to this industry of (partially and approximately) bootstrapping internal symmetries. These results were very encouraging, although they did not prove that any of these symmetries are *required* by the S-matrix principles. There were other, more general analytical attempts to bootstrap such internal symmetries (Martin and McGlinn, 1964; Huang and Low, 1965; Blankenbecler, Coon and Roy, 1967; Cushing, 1966, 1969, 1971), but none were successful. In his characteristically careful and critical fashion, Zachariasen (1966) evaluated the status of the bootstrap program as he began a series of summer school lectures.

> These lectures will be devoted to a discussion of the bootstrap hypothesis, its domain of applicability, its limitations, its consequences, and its difficulties as they are known as present ... Stated briefly and qualitatively, the bootstrap hypothesis is that all particles ... are dynamical entities composed of each other and bound by forces produced by the exchange of the particles themselves. Such a hypotheses is intuitively very appealing. (p. 86)

> [O]ur next natural step should ideally be to write down something like this intuitive picture in a precise mathematical form. Unfortunately, at this stage we run into difficulties. ... Let us ignore the fact that we don't really know what we are talking about, and assume that we have made precise what the BS hypothesis is. (We shall frequently use the abbreviation BS for the bootstrap in what follows. The reader is free to draw any conclusions he wishes from this notation.) (p. 87)

> [W]e know nothing firm about either the existence or uniqueness questions in the bootstrap theory. ... If all this discussion of the fundamental aspects, so to speak, of bootstrapism, is too depressing, it is perhaps useful to cheer oneself up a bit by thinking of the colossal rewards to be reaped if it ever becomes possible to make the bootstrap realize its full potential. (p. 88)

Also during the period 1962–65, Weinberg, while at Berkeley, began work on a method for making practical calculations in elementary particle theory (Weinberg, 1963a, 1963b, 1964a, 1964b, 1964c, 1964d, 1965a, 1965b; Scadron and Weinberg, 1964; Scadron, Weinberg and Wright, 1964) and this led to a series of papers on the S-matrix program (Weinberg, 1964e, 1965a, 1965b). Although at one time he advocated 'a point of view midway between that of the classic Langrangian field theories and the more recent approach' (1964b, p. B1318), Weinberg (1964e, p. B1049) later stated:

> It is not yet clear whether field theory will continue to play a role in particle physics, or whether it will ultimately be supplanted by a pure S-matrix theory.

He discussed gauge invariance, the equivalence principle, photons, gravitons and the derivation of Maxwell's and Einstein's equations within the framework of the general S-matrix program. This work appears to have ended abruptly in 1965 when he returned to more standard field-theory problems (Weinberg, 1965c, 1965d). We mention this work of Weinberg's here mainly to indicate how seriously the S-matrix program (in its autonomous form) was taken at the time, even by one who today would be associated in most people's minds with the rival (and apparently more successful) gauge field theory program. Recently, however, Weinberg (1985, 1986a, 1986b) has emphasized that the S-matrix and quantum-field-theory paradigms need not be antithetical to each other and that they may even be limiting case of some larger theory. We return to this in Section 10.5.

6.8 Too much complexity

In his talk at the 1967 Solvay Conference, Chew (1968a) had a very complex situation to review. He began on a general and somewhat philosophical level and then descended into the nitty-gritty of the complexities of the calculational program. He observed (pp. 84–5):

> Properties already identified ... are more than theorists can presently handle. ... [T]here is present in analytic S-matrix theory a superabundance of still unexploited physical content that cannot be ignored.

The problem was that fewer physicists how seemed interested in attempting to exploit it. Fox (1969) reviewed the problems for Regge theory and was not particularly positive. The Pomeron continued to present difficulties (Chew, 1967d) which would only yield to some resolution within the framework of the enormously complex multi-peripheral, multi-Regge exchange formalism of Goldberger (1969b), Chew, Goldberger and Low (1969) and Fubini (1970). The multi-peripheral model we discuss dates back to Amati, Stanghellini and Fubini's (ASF) (1962b) paper. A high-energy collision between two incident particles to produce many outgoing particles is pictured as taking place through a succession of one-particle, low momentum-transfer exchanges between vertices as illustrated in Figure 6.12. The justification they (ASF) offered for taking such a diagram as a reasonable representation of the actual production process was an appeal to the overall systematics of the high-energy experimental data (1962b, p. 897). The idea in the model is that if one knows the matrix element $\langle n|T|i\rangle$ for the process of Figure 6.12, then one can, through unitarity, find the absorptive (or imaginary) part of the two-body amplitude $\langle f|T|i\rangle$ to compute σ_{tot} and $d\sigma_{el}/dt$ as

$$\sigma_{tot} \propto \sum_n |\langle n|T|i\rangle|^2 \propto \text{Im}\langle i|T|i\rangle, \tag{6.36a}$$

$$\frac{d\sigma_{el}}{dt} \propto \left|\sum_n \langle f|T^*|n\rangle\langle n|T|i\rangle\right|^2 \propto |\text{Im}\langle f|T|i\rangle|^2. \tag{6.36b}$$

The behavior of Eq. (6.36b) is consistent with the fact that scattering amplitudes become predominantly imaginary at high energy (near the forward direction). Figure 6.13 represents the multiperipheral approximation for the absorptive part. In this model $\langle n|T|i\rangle$ is written as (an integral over) repeated factors of pion resonance exchanges. In Reggeized versions of this model (Goldberger, 1969b; Chew, Goldberger and Low, 1969), Regge poles (rather than simple pions) are exchanged as indicated in Figure 6.14. This gives the possibility of the Regge bootstrap of Figure 6.15 (Chew, 1968d). Regge cuts as well as Regge poles can, in some approximation, be self-consistently generated in this model. But the approximations *are* really nasty and one can wonder what the results prove. In his Coral Cablel Conference talk on multiperipheral dynamics, Goldberger (1969b, p. 143) observed:

> I'm sure that many of you will notice in the course of this talk many points which although they cannot be defended vigorously must in fact be so defended.

Believable calculations from first principles seemed impossible. Freund (1969) attempted to generate the Pomeron from two-body exchange. Chew and Snider (1970, 1971) argued that these multiperipheral

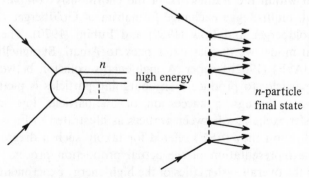

Figure 6.12   A multiperipheral diagram.

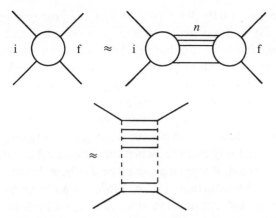

Figure 6.13 Unitarity in a multiperipheral model.

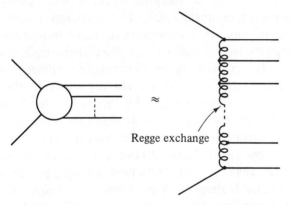

Figure 6.14 A Reggeized multiperipheral diagram.

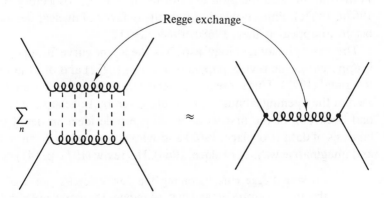

Figure 6.15 A multi-Regge bootstrap.

exchanges could generate the Pomeron. (In fact, the Pomeron problem remains to this day and has never been given a complete and widely accepted solution (Capella, Sukhatme, Tan and Tran Thanh Van, 1985).) Low's (1970a, 1970b) summary of the status of high-energy physics in this framework did not have a simplifying theme to unify it.

At both the practical and the axiomatic levels, the *S*-matrix program had run into the same trouble by the late 1960s. It had been a theory with a lot of promise, had produced some really tantalizing results, but quickly became hopelessly complex when one attempted to refine the theory. As we have seen, Regge cuts were found (Chew, Frautschi and Mandelstam 1962; Mandelstam, 1963b, 1963c) in addition to Regge poles. There were non-Regge terms in other cases as well (Mandelstam, 1965). Gribov, Pomeranchuk and Ter-Martorosyan (1965) generalized Mandelstam's 1963 work on Regge cuts and a Reggeon diagram technique (Gribov, 1968), or Reggeon calculus, was developed to handle multiparticle intermediate states and to attempt to construct a coherent theory of complex processes. A dramatic indication of the contrast between the later complexity and the earlier simplicity can be seen, for example, by looking at Goldberger's (1969b) talk on multiperipheral dynamics or at Low's (1970c) review of the Regge picture and then at the earlier paper by Frautschi, Gell-Mann and Zachariasen (1962) on the Regge-pole formulas for cross sections. It was not a refuted theory in the sense that it made incorrect predictions (Weinberg, 1985, pp. 122–123), but it never got beyond the two-body channels (Capra, 1985, p. 261), although multiperipheral models attempted to do this. It simply collapsed under the weight of its own complexity (Polkinghorne, 1985; p. 25, Chew, 1967a, 1967b, 1968d). Also, some of the pieces of the puzzle never quite fit. Not just the Pomeron, but also the pion did not lie on a Regge trajectory (Chew, 1967a, 1967c). Philosophical arguments in favor of nuclear democracy began to appear (Chew, 1968c, 1968e, 1971).

The 'mere phenomenology' aspect of the Regge curve fitting program is apparent in the review papers of Phillips (1972) and of Phillips and Ringland (1972). This sense of a degenerating research program can be seen in the opening comments of a talk given by Lovelace (1972), who had been one of the first to apply Regge pole ideas in detail to the analysis of data (Lovelace, 1962) and who continued to do so in novel and imaginative ways (Lovelace, 1968). His reviw (1972, p. 141) begins:

> To stop Regge cuts becoming the new epicycles, we need more theoretical constraints and understanding. The most naive view is that they simply correspond to exchange of two reggeons.

One spinoff of the bootstrap program is the sizable and active industry of the 'Paris potential', a semiphenomenological attempt to generate the fundamental nucleon–nucleon potential by applying S-matrix principles. This research is centered in France, mainly in the work of R. Vinh Mau and his co-workers (Vinh Mau, 1979, 1985). We shall later refer to other research programs that had their origins in the S-matrix program. However, as with several of them, we do not pursue the Paris potential here in any detail simply because it is not central to the interest of this case study.

A rival of S-matrix theory, namely, quantum field theory in its resurrected form of gauge field theories, became empirically adequate in the area of a *different* set of problems (Frazer, 1985, p. 5; Weinberg, 1985, p. 123; Pickering, 1984). However, as we shall see in Chapter 8, the S-matrix program did provide the background for duality, which became a field of intense theoretical activity. In the next chapter, though, we first discuss a somewhat radical or revisionary outgrowth of the S-matrix program, one associated mainly with Geoffrey Chew and some of his collaborators. It is in spirit a throwback to the original S-matrix program of Heisenberg.

6.9 Summary

General S-matrix principles, when coupled with the assumption of Regge asymptotic behavior, offered some hope of a workable program independent of QFT and allowed specific testable experimental predictions to be made. At the 1961 LaJolla conference Chew emphatically rejected QFT and advocated an independent SMT program. By no means did all of Chew's collaborators in the SMT project share his unqualified rejection of QFT.

Straightforward use of linearly rising Regge trajectories and crossing symmetry led to the expectation that for the (kinematically) elastic differential scattering cross section, there should be a forward peak that should fall off exponentially in momentum transfer (at fixed energy) and that should have a width that should decrease logarithmically in energy (at fixed momentum transfer). While early data on proton–proton scattering fulfilled this expectation, subsequent results for pion–proton scattering did not. In response to this and other challenges from new experimental data, complicating moves were made to save Regge phenomenology, but at the cost of making that program unlikely to be a fundamental theory.[25]

The core of the *S*-matrix program was the bootstrap (or self-consistency) conjecture. This mechanism was successfully employed to predict (approximately) the mass and width of the ρ meson. Some strong-interaction internal symmetries were also partially bootstrapped. But, tantalizing as these results were, the complicated mathematical structure of the exact bootstrap equations made it impossible to give compelling arguments for the existence of true boostrap solutions or to make convincing higher-order corrections to these 'one-shot' approximations. The situation became so complex that, at both the foundational and calculational levels, the *S*-matrix program ceased to make impressive advances.

7

An autonomous S-matrix program

The inability, in the 1950s and 1960s, of traditional quantum field theory to provide a basis for reliable *calculations* in strong-interaction physics led to quite diverse theoretical responses, just as had the previous divergence problems for QFT in the 1930s. We have already discussed dispersion relations and the mass-shell field theory programs. Another attempt has been the axiomatic field theory approach begun by Arthur Wightman at Princeton. Freeman Dyson has recently even made the claim[1]:

> [A]xiomatic field theory and local observable theory, beginning with Wightman and continuing with Haag and Araki, ... was much closer to Heisenberg's program than the contemporary Goldberger–Gell-Mann–Chew formalism ... .

Wightman (1989) has written an historical outline of this program and a thorough discussion of it would require an entire, highly technical case study of its own. That particular axiomatic approach has not had a strong impact on high-energy theory and we shall not discuss it further. We mention it just as an example of one reaction to the impasse encountered by standard Lagrangian quantum field theory. Our story is of another reaction, the S-matrix program. In this chapter we examine the autonomous S-matrix program some envisaged once Chew had made a radical break with quantum field theory.

Landau had argued that the only directly observable variables were those associated with essentially asymptotically free particles (*before* or *after* a scattering process), such as their initial or final momenta. He felt that the quantized fields which interpolated between these asymptotic states were quite meaningless physically. For him, all but empty or trivial quantum field theories (i.e., those for *noninteracting* particles) were inconsistent and not at all well defined mathematically. By the late 1950s Landau had already advocated a complete break with quantum field theory and supported a program very much in the spirit of Heisenberg's original S-matrix one. He even gave some hint of nuclear democracy.

The subsequent revisionary S-matrix program that Chew favored was formulated for the strong interactions only and had to put aside completely (at least for the time being) the electromagnetic and weak interactions. The basic reason for the exclusion of the latter lay in the severe mathematical difficulties presented for the analytic structure of scattering amplitudes for *massless* particles (such as the photon or the neutrino). The supporters of this autonomous S-matrix program saw in it great scope and potential, both for resolving foundational problems and for calculational effectiveness. As we have already seen, Leibniz's principle of lack of sufficient reason was appealed to as a justification for the analyticity, or maximum smoothness, of the S-matrix as a function of its arguments. (Of course, one might wonder, if anything ought to exist that is not forbidden, why should it be analyticity, rather than a wide class of singularities, that exists.) The photon and electromagnetic interactions were seen as possibly connected with the measurement problem and with our classical concept of a space-time continuum, the latter being expected to go the way of Maxwell's aether. This revolutionary program had a dual appeal to many young workers in the field. It promised a new worldview (to which they would be charter-member subscribers, thus not having to learn the formalism of 'dated' QFT) and it allowed them quickly to become contributing members to the new enterprise by exploiting the considerable calculational power of the program's general principles.

In this bootstrapped world, unitarity, rather than being a constraint or necessary condition any theory must satisfy, became a *sufficient* condition for generating *the* unique list of allowed strongly interacting particles and their properties. In this nuclear deomcracy or partial boostrap (i.e., of the strong, but not of the weak and electromagnetic interactions) self-consistency was the key: only *one* self-consistent 'world' could exist. A *total* bootstrap was recognized as unachievable since everything, even the basic principles themselves, would have to be generated. Rather, a few hopefully uncontroversial principles, such as unitarity, had to be accepted and these were to be constraint enough to provide a unique solution. In fact, as it became calculationally more difficult to extract firm analytical and numerical predictions from the S-matrix principles, the arguments in support of the general program began to shift from empirical adequacy (*vis à vis* QFT) to its great potential. A victory no longer appeared quite so imminent.

An offshoot of the autonomous *S*-matrix program was an axiomatic one whose goal was to provide an independent and mathematically rigorous foundation for the theory. While there was little calculational or empirical advance here, some of the previous proofs and arguments were considerably improved and new areas of (pure) mathematics were even opened up. The highly constrained nature of this SMT program was at once the source of its great appeal and of its downfall (or, at least, fall from general interest). The great complexity of the program gave rise to a vicious circularity of the problems it faced, as SMT left little room for maneuver or free adjustments to *fit* observations (after the fact), whereas QFT provided a general framework within which the systematics of specific observations could be more readily fit and tested for further consequences. Thus, SMT promised *everything* (and uniquely, at that), but solutions to restricted problems became difficult if one could not solve the entire (infinitely) coupled system.

7.1   QFT versus SMT

In his talk at the Twelfth Solvay Conference on Physics, Goldberger (1961, pp. 179–80) gave an overview of the responses to the problems faced by conventional QFT.

> Today's speakers have been invited to throw down the gauntlet and state where the dispersion theorists or if you prefer *S*-matrix theorists are, what they have done and perhaps what they can do. Yesterday Chew expressed a certain amazement at the fact that people failed to recognize until recently the possibility of making a true *S*-matrix theory of the strong interactions and insist on playing with field theory. My own feeling is that we have learned a great deal from field theory as we shall see, even dispersion theory came from it; that I am quite happy to discard it as an old, but rather friendly, mistress who I would even be willing to recognize on the street if I should encounter her again. From a philosophical point of view and certainly from a practical one the *S*-matrix approach at the moment seems to me by far the most attractive.
>
> I have tried to understand some of the reactionary responses I have met in the past about dispersion theory which have been also expressed here or which I have read into the comments of others at the conference. One point which springs to mind is that within the framework of Lagrangian (or old fashioned) field theory, one writes

down a small number of equations such as the field equations, and the Schrödinger equation and it takes only a few lines to do so. It looks then as though one had the whole theory there and one need only start to work. Similarly, in terms of what is called axiomatic field theory, one can produce an infinite set of equations based on only the finest postulates, which with a suitable notation may be written very compactly. This simplicity, which is only superficial, is contrasted with what appear to be rather arbitrary and intuitive procedures on the part of the dispersion theorists who do not seem to express things very neatly.

On the other hand, the axiomatic field theoreticians are very hard pressed to compute the Klein–Nishina formula [for the Compton scattering of photons by free electrons]. The Lagrangian people can do this, but if we allow the discussion at the conference to degenerate to the physics of the strong interactions, they, to say nothing of the axiomaticians, are absolutely helpless. At this level, the challenge of field theory presented to dispersion theory is non existent. Although the axioms if you like, of the dispersion approach have not been stated in what we might imagine to be their final form, no one is really ever in much doubt as to how to proceed. The reason is that the unashamed dispersion theorist has even been willing to resort to experiment to get ideas and to push a little further his understanding of the strong interactions. Thus as Gell-Mann has put it, he is understanding some physics while developing the theory.

It is perhaps correct to say that much of the deeper philosophy of the *S*-matrix approach held by some of us, in particular Chew, who believe that there are no elementary particles, and that there are no undetermined dimensionless constants in the theory has not yet been put to a successful test. I should also hasten to point out that if there are some elementary particles, they may be easily incorporated into the scheme.

Several years later, Goldberger (1969a, p. 86) reiterated his support for the dispersion-theory program as a way to make the general quantum field theory formalism usable.

> I am myself strongly partial to the pure *S*-matrix approach but abhor the extremism with which some proponents and opponents discuss the relative merits of *S*-matrix theory as compared to local quantum field theory.... I know of no conflict between the two theories ... and I feel that they are in fact the same.

Francis Low is even more critical of the possibility that the *S*-matrix program can provide an independent alternative to quantum field theory.

> I did not perceive [in the bootstrap or in the *S*-matrix] either a program or a theory. .... I thought Geoff's conjectures on the absence of underlying degrees of freedom were imaginative and interesting, but did not think the probability of success was very high.[2]

> The *S*-matrix approach replaced explicit field theory calculations largely because nobody knew how to do calculations in a strongly coupled theory. ... I believe that very few people outside the Chew orbit considered *S*-matrix theory to be a substitute for field theory. ...[3]

At a conference in Berkeley in 1966, Low (1967, p. 241) evaluated the options available:

> The historical basis for all our present-day methods of theorizing about particles is the Lagrangian quantum theory of fields. ... During the last 10 years or so, two deviations have grown up among the practitioners of field theory. ... Each wing has abstracted and kept certain aspects of Lagrangian field theory and rejected others. The right wing retains the concept of the local field, but rejects all detailed dynamical assumptions such as equations of motion and commutation relations. The left wing rejects the concept of the local field and insists that the unitary properties of the *S* matrix are sufficient for all calculations. The central position is that there might well be no contradiction between the two sets of axioms and that they might both be right (or, of course, both wrong).

Simiarly, Mandelstam continues to regard these various approaches to fundamental particle theory as belonging to one common fabric.[4]

> I do not regard *S*-matrix theory and quantum field theory as two different theories; I think that they are basically the same theory. Which one one uses should thus depend on which is more likely to give the best results rather than on any question of principle. I was originally attracted to dispersion relations, first because they provided equations which did not contain infinite unrenormalized quantities, and second because, at the time, they appeared to give some hope of obtaining a reasonable approximation scheme. With the success of QCD, my interests turned back to direct field-theoretic methods.

Gell-Mann, in somewhat more colorful terms, has said that he and Goldberger, for instance, did not feel that quantum field theory was wrong but that only some 'nuts' (like Landau) made such a claim.[5] Illustrative of pronouncements which gave rise to such strong reactions are Landau's (1960) remarks below that he made at the 1959 conference in Kiev. As radical as these comments were, they did prove to be a perceptive outline of the basis for an autonomous *S*-matrix program that would follow. Some found Landau's talks an extraordinarily exciting and powerful occasion.

> Accurate measurement of the coordinates of a particle would imply the production of a great multitude of pairs, and an accurate measurement of the momentum in a short time contradicts Bohr's uncertainty relation between energy and time. The only observables that could be measured within the framework of the relativistic quantum theory are momenta and polarizations of freely moving particles, since we have an unlimited amount of time for measuring them thanks to momentum conservation. Therefore the only relations which have a physical meaning in the relativistic quantum theory are, in fact, the relations between free particles, i.e., different scattering amplitudes of such particles. (p. 97)

> [P]si-operators [i.e., the quantized field operators of QFT] contain fundamentally unobservable quantities and should vanish from the theory. As the Hamiltonian can be constructed only with the help of psi-operators, Hamilton's method in the strong interaction theory is dead and should be buried, though pompously, considering its historical merits.

> The new theory should be based on a new diagram technique dealing only with relations between 'free end' diagrams, i.e., such diagrams which represent scattering amplitudes of free particles and their analytical continuations. (p. 98)

> Speaking of the new diagram technique we deal with a new, not yet completely constructed theory. If we speak of the derivation of this technique from the conventional theory, this can be understood only in the sense in which the quantum theory was derived from the correspondence principle some time ago, and this derivation has no bearing whatsoever on the long deceased perturbation theory. In deriving the analytical properties of diagrams and relations between them from Hamiltonian formalism, we actually only find the locality and unitarity conditions in a theory which we have not learned to develop as such. Therefore, I find it very naive to attempt to pose the problem of a rigorous derivation of dispersion relations and other relations that determine the analytical properties of scattering amplitudes. How can one speak of the rigorousness of the derivation of existing relations from non-existent Hamiltonian formalism?[6]

> Because the concept of compound particles is associated with consideration of interacting particles, it may be inferred that this concept cannot have any meaning in the relativistic quantum theory.

Here all the particles should be equal in a sense, being equally elementary and compound. (p. 99)

John Polkinghorne (1979, p. 87) has captured nicely the atmosphere of strong feelings which surrounded the *S*-matrix program in the 1960s.

The development of *S*-matrix theory was characterized by a certain degree of sectarian strife. Some of us who worked in the subject were easy-going eclectics, content to try our hands at any approach to relativistic quantum mechanics which looked promising. Not so some of our senior and more inspiring colleagues. It was not so much a question of its being expedient to be on the mass-shell as of its being sinful to be anywhere else. In particular, they proclaimed the demise of quantum field theory.... [T]he *S*-matrix endeavour looks a good deal less beguiling than it did in those brave early days. This is due both to problems of principle and also to difficulties of practice.

7.2   A conceptual framework for SMT

Let us see what the central tenets of this autonomous *S*-matrix theory program were that raised so many hackles among the more conservative members of the theoretical physics community but that, at the same time, produced a heady sense of a new beginning among other, often younger, colleagues. That a definite break with the field-theory tradition had taken place is evidenced by the last paragraph of Chew and Frautschi's 1962 paper (p. 43):

In conclusion we wish to state our belief that *all* of strong-interaction physics will flow from: (*a*) the principle of maximal analyticity of the *S* matrix, in angular as well as linear momenta, (*b*) the principle of maximum strength, and (*c*) the conservation of *B* [baryon number], *S*[strangeness], and *I*[isospin].

Chew (1962a), in an invited talk before the American Physical Society in New York, gave an extremely optimistic review of the autonomous *S*-matrix program. His paper begins (pp. 394–5.):

In this paper I present an indecently optimistic view of strong interaction theory. My belief is that a major breakthrough has occurred and that within a relatively short period we are going to achieve a depth of understanding of strong interactions that a few years ago I, at least, did not expect to see within my lifetime. I know that few of you will be convinced by the arguments given here, but I would be masking my feelings if I were to employ a conventionally cautious attitude in this talk. I am bursting with excitement, as are a number of other theorists in this game.

I present my view of the current situation entirely in terms of the analytically continued S matrix, because there is no other framework that I understand for strong interactions. My oldest and dearest friends tell me that this is a fetish, that field theory is an equally suitable language, but to me the basic strong interaction concepts, simple and beautiful in a pure S-matrix approach, are weird, if not impossible, for field theory. It must be said, nevertheless, that my own awareness of these concepts was largely achieved through close collaboration with three great experts in field theory, M. L. Goldberger, Francis Low, and Stanley Mandelstam. Each of them has played a major role in the development of the strong interaction theory that I describe, even though the language of my description may be repugnant to them. Murray Gell-Mann, also, although he has not actually published a great deal on the analyticity aspects of strong interactions, has for many years exerted a major positive influence both on the subject and on me; his enthusiasm and sharp observations of the past few months have markedly accelerated the course of events as well as my personal sense of excitement.

Chew (1962a, pp. 400–1) concluded this same talk with a plea for putting aside, at present, some obvious difficulties of principle faced by the new program.

An inevitable question at this point is, 'What about electromagnetism and weak interactions?' I personally have not developed strong convictions on this question, but I do not see how leptons [light fermions; i.e., muons, electrons and neutrinos] and the photon can emerge from the principles enunciated here. One may imagine, in fact, that leptons and weak interactions represent a deficiency in one or more of these principles that will become a major effect in some experimental domain of the future. Had we known of hyperfine structure in the early days of atomic physics, however, it would have been a mistake to insist that any theory should explain the effect. Historically, *all* dynamical theories in physics have had limitations on their domain of validity, no matter how general they seemed when they were proposed. We must not be too greedy.

He motivated the maximal analyticity postulate by an appeal to the 'lack of sufficient reason' (why should it be otherwise?). His 1962 CERN Conference Report (1962c) was in the same vein and he now even expected internal quantum numbers and symmetries to be boostrapped. That the S-matrix principles appeared inapplicable to electromagnetism (because it involves *massless* photons) was to be put aside for the present.[7]

In his 1963 Rouse Ball Lecture delivered at CambridgeUniversity, Chew (1963c) gave a somewhat philosophical analysis of the fundamental implications of the autonomous *S*-matrix program. Several of his points were reminiscent of Landau's remarks (quoted above) at the 1959 Kiev Conference.

> My thesis in this lecture is to suggest that the concept of space and time is playing in current microscopic physics a role analogous to that of the ether for macroscopic physics in the late nineteenth century. (p. 529)

> The capacity for experimental predictions is the only reliable measure of a physical theory. (p. 533)

> The attempt to construct such a theory [based directly upon particle momenta, ignoring the space-time continuum] is gathering momentum at present, the less experienced physicists having an advantage in working with a new framework. (The inverse correlation of productivity with experience in a situation like this is remarkable.) (p. 534)

> If we are frustrated here [in predicting high-energy reactions] it will not be because of lack of content in the analytic *S*-matrix but because the reactions are complicated and human strength is finite. (p. 536)

> *S*-matrix theory is in the unprecedented but agreeable situation at present, therefore, of having more power in a simple principle than theorists know what to do with, without a hint anywhere of a flaw in the principle. (p. 537)

> No suggestion is being made that space and time do not continue to be the basis of *macroscopic* physics; ... Does this mean that there can be no continuous connection between the microscopic and macroscopic worlds? The situation is no more uncomfortable than it has always been for quantum theory, where the conventional explanation of the relation between the classical observer and quantum laws leaves most people feeling queasy. (p. 538)

> As you can see, the new mistress [*S*-matrix theory] is full of mystery but correspondingly full of promise. The old mistress [quantum field theory] is clawing and scratching to maintain her status, but her day is past. (pp. 538–9)

There were several other papers and exchanges in this vein. In comment upon a popular exposition of this new program by Chew (1964a) in *Physics Today*, Stern (1964) criticized it as being simply a (psychologically motivated?) response to frustrations with quantum field theory, rather than a positive and coherent undertaking in its own right.

> Indeed, seldom has there been so much emotion attendant upon the formulation of a new and useful, but abstract, mathematical technique as is embodied in the S-matrix approach to the strong interactions. (p. 42)

> This fundamental inconsistency [of the S-matrix theory] arises from the fact that the formalism does not contain, nor can one build into it, the electromagnetic field. It is inconsistent with the existence of the electromagnetic field. Because the electromagnetic field is involved in our very means of measurement and observation, this is a serious shortcoming. (p. 43)

7.3   A bootstrapped world

In an attempt to give some definite form to these general ideas, let us outline here how the boostrap conjecture might be implemented to form the basis for strong-interaction particle physics (Cushing, 1985). This sketch may help the general reader to understand in somewhat concrete terms the discussion of the following section. A few comments toward the end of this section will depart from a strict historical sequence.

For S-matrix theory the essential ingredient is the unitarity relation, which is a statement about the conservation of probability. Very much as in ordinary nonrelativistic quantum mechanics, the basic probability amplitudes allow predictions of the outcomes of experiments or of observations. The output of this theory consists of these probability amplitudes which correspond directly to measurable quantities. There is no state vector defined at each space-time point (as in quantum mechanics), but these probability amplitudes do satisfy a superposition principle. Macroscopic causality imposes certain analyticity constraints on those probability (or scattering) amplitudes as functions of their continuous variables (such as momentum and energy). Since an analytic function is uniquely determined globally once it has been precisely specified in the neighborhood of any point, there is a (Leibnizian) connection between an individual part and the global whole. However, this similarity is probably merely incidental and we point it out as a curiosity.

The most interesting philosophical aspect of S-matrix theory centers on the unitarity relation which in essence replaces the equations of motion of more traditional theories (Chew, 1974). The reaction amplitude or S-matrix connects the initial, asymptotically free state ψ_i with the final, asymptotically-free state ψ_f as (recall Eqs. (2.31)–(2.32))

$$\psi_f = S_{fi}\psi_i, \tag{7.1}$$

That is, S_{fi} represents the overlap of ψ_f with a given ψ_i as

$$\langle\psi_i|\psi_f\rangle = S_{fi} \tag{7.2}$$

so that the chance, or probability, of beginning in an asymptotic state ψ_i and ending in the asymptotic state ψ_f is

$$|\langle\psi_i|\psi_f\rangle|^2 = |S_{fi}|^2. \tag{7.3}$$

Since the probability of starting in ψ_i and ending up in *some* allowed state ψ, (*any* one at all) must be unity, this conservation of probability requires that

$$1 = \sum_f |S_{fi}|^2 = \sum_f S_{fi}^* S_{fi} = \sum_f (S^\dagger)_{if} S_{fi} \tag{7.4}$$

or, as a matrix equation (cf. Eq. (2.16)),

$$S^\dagger S = I. \tag{7.5}$$

This, as we have seen previously, is referred to as the *unitarity relation*.

Such a restriction upon any theory appears reasonable and innocuous enough. Certainly both quantum mechanics and relativistic quantum field theory should be required to meet this condition. In nonrelativistic quantum mechanics, solutions to the Schrödinger equation satisfy unitarity. However, for a relativistic quantized field theory it has *never* been demonstrated that any realistic interacting system of fields has a unitary solution. While formal power series solutions have been generated that satisfy unitarity to each order of the expansion, the series itself cannot be shown to converge. In fact, there are infinite terms that must be removed from the series in order to obtain finite results for comparison with experiment.

In S-matrix theory the unitarity condition, rather than being taken as a necessary condition which any theory must meet, is taken as providing the central element of a sufficient condition for a *unique* solution (Freundlich, 1980). Because of the possiblity of particle creation as energy increases, the unitarity condition becomes an *infinite* set of coupled nonlinear equations. Such a highly constrained system of equations allows the possibility that it may have just *one* solution. It is through the unitarity equations and crossing (a concept not available to

Heisenberg in the 1940s) that the bootstrap conjecture is implemented in *S*-matrix theory. According to this conjecture, *no* aspect of nature should be arbitrarily assignable, unlike the situation in quantum field theory. This does not mean that the actually existing state of the universe would be uniquely determined. (After all, the theory predicts *probability* amplitudes.) Rather, the masses, charges and other characteristics of all the particles found in nature should be uniquely fixed by self-consistency through unitarity. Given the existence of certain variables, a unique spectrum of particles should emerge.

A complete bootstrap theory would probably be unworkable since everything, including a space-time structure, would have to be generated from the theory (Chew, 1963c, 1971). Chew (1983a, 1983b; Chew and Stapp, 1988) has pointed out that a continuous space-time structure may be only an approximation arising from an ordered, causal event structure in the limit of high complexity. But then, such a continuous structure is ultimately unobservable, much as was the aether of classical physics. *S*-matrix theory requires accurate specification of the momentum $\mathbf{p}$ of a (free) particle, but not of its position $\mathbf{x}$. Operationally, one can determine $\mathbf{p}$ as accurately as necessary but not $\mathbf{x}$ (due to the Heisenberg uncertainty principle *and* the mass–energy equivalence of relativity). Even though the complete bootstrap may not be implementable, the bootstrap conjecture *is*, nevertheless, refutable. For example, if a conventional Lagrangian quantum field theory (having, as usual, arbitrary parameters in it) could be constructed and shown to be unitary, the bootstrap conjecture would be refuted (Chew, 1971).

However, since we must begin someplace in an actual calculation (even if it is only with a language or a set of concepts and quantities that represent certain aspects of reality), only a partial boostrap appears possible. In practice one takes some input from reality and then self-consistently generates other data in addition to those starting facts themselves. The successful generation of the ρ meson from the π meson was an application of this general procedure (Zachariasen and Zemach, 1962). Put very simply, the essence of such a 'bootstrap' calculation is that there are particles – call them A – which can interact to form another type particle, say B, as

$$A + A \rightarrow B. \tag{7.6}$$

For convenience here let us assume that the forces between A and B are such that they do not form a new particle. Suppose though, that two B's can form another particle C as

$$B + B \rightarrow C. \tag{7.7}$$

In general, we would expect particle C to be a type distinct from both A and B. However, since the mass of a 'bound' state is typically less than the sum of the masses of the particles producing it, the mass of C could be the same as that of A. In such a case, we would have, schematically,

$$A + A \rightarrow B, \tag{7.8a}$$

$$B + B \rightarrow A. \tag{7.8b}$$

These processes are described by scattering amplitudes (the S_{fi} of Eq. (7.1)). The unitarity relation (Eq. (7.4)) and the analytic (crossing) properties of the S_{fi} allow one to calculate the parameters of B (say, the mass m_B) in terms of those of A (say, m_A). Furthermore, these masses appear as both the input and the output of the calculation. The requirement of self-consistency (i.e., the numerical values of the input parameters are to equal the numerical values of the output parameters) plus the nonlinearity of the unitarity equation produce (so it is hoped) unique values for m_A and m_B. For the ρ–π case, these numbers agree (approximately) with values obtained from experiment. The actual calculation is not as unproblematic as this sketch would indicate since it is necessary to make mathematical approximations whose reliability cannot be checked. As a practical matter, the π parameters are fed in as fixed values (taken from experiment) and only the ρ parameters are in fact determined self-consistently. So, this is again actually a partial bootstrap.

Now, just as one of the inputs of our partial bootstrap, namely the space-time continuum, may be only a convenient approximation, so may be a description in terms of the reality or existence of identifiable structures that persist in time. Our observation (or stable description) of the reality covering a quantum substructure may depend upon the existence of electromagnetism, the weakly interacting probe that allows us to observe systems without greatly disturbing them. Thus, for the time being, a reasonable place to begin the partial bootstrap is with a space-time continuum and with the electromagnetic interaction taken for granted (rather than as emerging from the partial bootstrap).

7.4 A shift to 'higher' philosophical ground

As new, specific, experimentally observable predictions of the S-matrix program became more difficult to generate, the basis of Chew's argument began to shift to the 'higher' ground of philosophical principles. It is this aspect of the autonomous S-matrix program that has had appeal to some philosophers of science. This has been discussed at length elsewhere (Cushing, 1985)[8]. We shall return to these philosophical issues in Chapter 10. However, some sense of Chew's line of defense can be gotten from his papers over the next several years (Chew, 1964b, 1964c, 1965). He (Chew, 1966a) tried to make the analyticity assumption appear almost self evident.

> It is a commonplace observation that in one form or another almost all physical theories have involved analytic functions. This circumstance is so widespread that physicists tend to forget the exceptional status of analytic functions in mathematics. Fermi used to say: 'When in doubt expand in a power series.' This statement reflects the belief, shared by most of us, I am sure, that natural laws are likely to depend analytically on any physical parameter which is continuously varied. (p. 369)

> [A]ll strongly interacting particles seem to be composite structures. Using Gell-Mann's phrase, we are dealing with what appears to be nuclear democracy. (p. 374)

He continued to stress the broad aspects of his program (Chew, 1966b, 1966c, 1966d). Also, Chew (1966e, p. 10) attempted to undermine the consistency of the entire concept of an elementary particle, which had been the basis of paradigms in high-energy theoretical physics.

> Ultimately, however, science must answer questions of 'Why', as well as 'How?', and it is difficult to imagine answers that do not involve self-consistency. Why is space three-dimensional? Why are physical laws relativistically invariant? Why do physical constants have values that make it possible to understand sub-classes of natural phenomena without understanding all phenomena at once? In other words, why is science possible in the first place?

> The bootstrap idea for hadrons does not directly touch on these profound questions but there is a similarity of spirit in concern for the notion that the laws of nature may be the only possible laws. ... Perhaps the hadron dilemma is the precursor of a new science, so radically different in spirit from what we have known as to be indescribable with existing language.

At a conference in Brussels, Chew (1968a, p. 85) placed the S-matrix program in the context of a long-range enterprise.

> Although one must anticipate periods of theoretical frustration over these enormously difficult questions, we may be confident that the theory will not stagnate so long as experiments are continued.

In 1971, after the autonomous *S*-matrix program had somewhat run out of steam while gauge field theories had gained much momentum, Chew (1971) published a rather philosophically oriented article on the scientific status of *S*-matrix theory. He now advocated a partial versus a complete bootstrap.

> [T]he concept of measurement, on which hard science is based, is admissable only because of certain special attributes of nature, attributes that a complete bootstrap theory would have to explain as necessary components of self-consistency. It is in this sense that the idea of a complete bootstrap, while not obviously foolish, is intrinsically unscientific. (p. 2330)

That is, at the practical or calculational level one must be satisfied with a partial or hadron bootstrap with certain facts about the real world (e.g., space-time structure, 'separability', the existence of electromagnetism) accepted as *givens*, while the hadron (or strong-interaction) properties *alone* would be bootstrapped. *The* basic conjecture now became phrased as:

> Quantum superposition, when expressed through a nontrivial *S* matrix, can achieve compatibility with the real (classical) world in only one possible way – close to the way exhibited by nature for hadrons. (p. 2330)

Furthermore, the origin of electromagnetism is circumscribed as being beyond the bootstrap.

> [There is a] basis to believe that it would be futile to seek a scientific bootstrap theory of electromagnetism. (p. 2332)

Finally, in that same paper (Chew, 1971) a criterion is given for judging the success of this partial bootstrap.

> [C]an one even imagine what might constitute 'verification' of the hadron bootstrap hypothesis? ... [I]ncreasingly remarkable theoretical correlations of experimental facts about hadrons may come to be accomplished purely through general properties of the analytic *S* matrix.... [I]t may gradually become plausible that the only uniquely necessary input is the requirement of self-consistency. (p. 2335)

There had been a considerable retreat from the sense of impending and total victory prevalent a decade earlier. In a curious paper, Allen (1973) even attempted to argue that the structure of formal logic favored the bootstrap over the more conventional program involving fundamental fields. Of course, the 'demonstration' was inconclusive.

7.5   Axiomatic SMT and its offshoots

A program of axiomatic *S*-matrix theory emerged in the early to mid-1960s as well. Although not everyone who contributed to this was necessarily a supporter of the autonomous *S*-matrix program, the thrust of much of the work was toward constructing an *S*-matrix theory independent of (even if not actively opposed to) quantum field theory. It is for that reason that we mention it in this section. The axiomatic approach was associated mainly with Stapp, Olive, and later, with Iagolnitzer. Before he began working on the *S*-matrix program, Stapp had been interested in the analytic properties of scattering amplitudes within the framework of axiomatic field theory and in the theorem on the connection between spin and statistics and CPT invariance in that theory. He had also made ultraviolet divergence corrections using dispersion relations. His calculations involved only (analytic extensions of) measurable quantities, simultaneously avoiding both divergences and unphysical quantities. In this on-mass-shell framework he was able to prove the CPT-spin-statistics theorem with assumptions about the analytic and boundary-value properties of these amplitudes. He then felt that these assumptions plus unitarity and Lorentz invariance (with suitable asymptotic bounds on the amplitudes) could in principle allow one to reconstruct the Feynman perturbation solution.[9] Results of the CPT theorem work appeared in Stapp (1962a, 1968) and in general reviews of the axiomatic program (Stapp, 1962b, 1965b). An important motivation for this axiomatic *S*-matrix approach appears to have been the inability of the formal field-theory program to make contact with experiment. For example, we read (Stapp, 1962b, p. 390):

> It has been the common practice to consider field theoretic axioms as the proper basis for rigor in physics This is evidently due more to the lack of any satisfactory alternative rather than to their obvious merit. For although the axioms of field theory provide a basis for rigorous mathematics, there is considerable doubt that they are of relevance to physics. ... [A]xiomatic field theory seems in fact very distant, if not totally disconnected, from most practical calculations.

On the other hand, some in the axiomatic field-theory school felt that Stapp had assumed more than he had proved, especially in the analyticity properties he used. In a strongly critical comment on Stapp's original (1962a) proof of the CPT theorem (about a general invariance property that any theory should satisfy), Jost (1963, p. 81) states:

> Now Henry P. Stapp has recently proposed a true axiomatization of

the Chew *S*-matrix theory in which he attempts to derive, as a first application, the CPT theorem and the connection between spin and statistics. It appears to us that Stapp, who spared no pains in his criticism of the other [field-theory] approach, has, on this particular point, also failed to observe that same self-criticism and clarity of formulation which one is accustomed to demand in a physical investigation. ...

The main objection to Stapp [is that] he has not really thought about the problem, but has slavishly copied the well known method of proof, whereby he obviously hopes that the result would also be included.[10]

In his text on quantum field theory a few years later, Jost (1965, pp. xiv–xv) returned to the same theme:

Recently other attempts to axiomatize theories of as yet unknown physical significance have been made. The quotation of one of the proposed postulates will suffice to show that they have little in common with the work presented here.

'*Physical connection.* Physically interpretable functions obtained by analytic continuation from functions describing physical phenomena also describe physical phenomena; they are not mere mathematical chimeras. Specifically, the *M* functions [a form of the *S*-matrix elements particularly useful for analysis] at all physical-type points of a physical sheet correspond to processes actually occurring in nature. Regarding interpretation, if a simple connection can be set up permitting a consistent interpretation of the quantities appearing in the theory, and also those that could be obtained by analytic continuation, then this interpretation accords uniformly to reality if it accords at all.' [Stapp, 1962a, p. 2141]

In all fairness to the author one has to point out that 'the mathematical forms' (of the postulates) 'will be introduced as they are needed in the proofs'. We feel that this is, depending on the standpoint, a rather dangerous or an extremely economical procedure: economical in the sense that you quote things exactly where you need them, and there they mean what they need to mean, neither more nor less.

In his first proof of the connection between spin and statistics, Stapp (1962a) assumed that the magnitudes of linear combinations of particle and antiparticle amplitudes could be observed. This assumption has no physical foundation. It is contradicted, as Stapp later pointed out (1962c), by superselection rules (Wick, Wightman, and Wigner, 1952) which decompose a Hilbert space into orthogonal subspaces such that the relative phase of the components of a state vector in these subspaces is not measurable (or, equivalently, such that no observable can connect states in these different subspaces). For example, charge conservation is a superselection rule. Stapp (1962c) was able to give another proof of the spin-statistics theorem based, essentially, on the cluster decomposition property (a type of factorization of multiparticle scattering amplitudes) and on the Cutkosky rules for the S-matrix. Lu and Olive (1966) gave an independent proof of that theorem. Froissart and Taylor (1967) published a complete version of Stapp's (1962c) proof, using the assumptions of cluster decomposition, crossing symmetry and hermitian analyticity (a particular type of analytic continuation). Stapp (1968) showed that macroscopic causality implied the scattering amplitudes are analytic except possibly on the Landau surfaces (i.e., those surfaces containing the Landau singularities).

Eden, Landshoff, Olive and Polkinghorne (1966) presented a detailed, largely heuristic, perturbation-theory based approach to S-matrix theory. Many of the subsequent axiomatic papers, for example Iagolnitzer and Stapp (1977) on pole factorization (the factorization of the residue at a pole) and Iagolnitzer (1980) on macrocausality and physical-region analyticity, tended to give more mathematically elegant proofs of previous results without bringing the enterprise much closer to calculational reality. Iagolnitzer's (1978) book presents an abstract, more mathematically rigorous approach which, like its counterpart in axiomatic field theory, did not lead to any empirically testable advances. However, for Iagolnitzer and some other practiners, the focus of interest became not calculational adequacy but, by conscious choice, the appreciable new developments of a conceptual and mathematical nature.

> [M]ost of the works on ASMT [axiomatic S-matrix theory] in the 2nd part of the sixties and in the seventies (and in particular all my own works) did *not* lead to progress in [actual calculations and 'empiric' physics], but turned out to be concerned with more conceptual and more mathematical aspects. As far as I (and some other people) were involved, this was *deliberate*.[11]

While the problems posed by the S-matrix theory of the 1960s were not solved by this axiomatic approach, a new field of *mathematics* was opened up. This has become an area of considerable interest and activity for mathematicians. Several important problems were treated, including the status of space-time in S-matrix theory (Stapp, 1965a; Iagolnitzer, 1969), a discussion of the implications of macroscopic causality on the S matrix in Chandler (1968), in Chandler and Stapp (1969) and in Iagolnitzer and Stapp (1969), an S-matrix derivation of the Cutkosky rules in Coster and Stapp (1969, 1970a, 1970b), and pole factorization in Iagolnitzer and Stapp (1977). Stapp also considered, within the S-matrix framework, the general interpretation of quantum theory (1971, 1972b) and the measurement process (1972a). Once again, we shall not pursue this topic in the present case study, because that would divert us from our discussion of the development of the physical theory under consideration.

7.6   Enormous complexity again

In 1962, Gunson, in a widely referenced University of Birmingham preprint, presented an S-matrix formulation purely in terms of on-mass-shell ($p_j^2 = m_j^2$) quantities, using the assumption of maximal analyticity (i.e., only those singularities *required* by unitarity). (This work was finally published in Gunson (1965a, 1965b).) From unitarity he proved the existence of poles and was able to establish factorization of the S-matrix at these poles. While attempting to use the Yang–Mills theory (a QFT based on local gauge invariance) to construct a renormalizable theory of weak interactions, Olive read something on dispersion theory and S-matrix theory, became intrigued by some circular arguments of Goldberger and caught some of Chew's enthusiasm. Chew visited Cambridge in 1962–63, where Olive was then studying. Gunson, Heisenberg and Cutkosky also lectured there. Landshoff and Polkinghorne, then at Cambridge, influenced Olive with their work on the S-matrix program. Olive prepared a translation of Heisenberg's early S-matrix papers. The highly constrained nature of the S-matrix theory attracted him. Olive (1964) made a first attempt at a systematic S-matrix theory. He stressed the highly involuted nature of the singularity structure of the S-matrix, one that might prove impossible to untangle in a purely axiomatic approach (Olive, 1964).

> [I]t is logically impossible to deduce the complete singularity structure without the results we are trying to prove. A suggested resolution of this difficulty is to set up a scheme of successive iterations in singularity structure to be justified by self-consistency. Then our work is the first step in such a scheme. (p. B745)

> All these approaches make assumptions involving either the unphysical unitarity relations or the antiparticle theorems or both and are therefore not immediately applicable in view of the fact that it is these relations which we wish to deduce. ... It is these very complications [the multiparticle aspects of the theory] that enable us to do anything,

whose sheer complexity at present prevents us from writing down our arguments in any but the simplest cases.

We find two main conclusions:

(A) The validity of the unphysical unitarity relations [i.e., unitarity constraints applied outside of or beyond the physical region] and the antiparticle theorems depends upon certain multiparticle features of the physical unitarity equations and upon certain plausible topological properties of the singularity structure which are satisfied in the case of a crude model of the S-matrix singularity structure consisting only of normal thresholds.

(B) Since one cannot deduce the complete singularity structure of the S matrix from the physical unitarity equations without using the theorems mentioned above, a vicious circle develops in the attempt to simultaneously deduce both the singularity structure and the fundamental theorems. We suggest that it may be possible to overcome this critical difficulty by proving that the singularities must possess some sort of hierarchical structure which will enable one to unravel the powerful consistency requirements of the theory. In this case we outline a program for generating the remaining singularities and completing our proof. (p. B746)

It does not appear possible to deduce much of the S-matrix singularity structure from the physical unitarity equations without introducing the unphysical unitarity relations and antiparticle theorems. Since the specification of these results must involve paths of continuation passing through regions where there may lie singularities we cannot yet deduce, a severe difficulty arises which is characteristic of S-matrix theory rather than our particular method. (p. B759)[12]

A unified summary of unitarity-based analyticity arguments was later given by Bloxham, Olive and Polkinghorne (1969a, 1969b, 1969c; also, Pham, 1967). Intractability, at both the calculational and axiomatic levels, remains a characteristic of the S-matrix program.[13]

Recently, Stapp gave this summary of the present status of the (autonomous, axiomatic) S-matrix program:[14]

The necessary extension to multiparticle processes brought in complexities. It was difficult to formulate simple-looking dispersion formulas even for 5- and 6-particle processes, and no one had the strength needed to set up concrete recipes for specifying exactly what the dispersion relations would be in the general $n \to m$ case. Without such a recipe the S matrix program remained a program rather than a complete theory.

The 'most telling weakness' of S-matrix theory, apart from the fact that the needed dispersion relations were never specified, is the fact that it seemed impossible in practice to derive the incredibly complex particle structure now being revealed in high energy experiments directly from merely the requirements of unitarity, Lorentz invariance etc. In field theory one can put in structure pretty much by hand to reproduce observed data. And symmetry principles can be postulated, rather than having to be derived. Thus field theory provides a tool much more facile for coping with the apparently very complex particle structure that is gradually being revealed.

In order to have any hope of *deriving* this structure, rather than merely just feeding it in, in response to the data, one needs, in addition to the basic S-matrix principles, at least some interim principles such as symmetries or topological requirements. Whether these interim principles will eventually be found to follow from the basic S-matrix principles remains to be seen.

Notice that Stapp here points out one of the great practical advantages that QFT has shown to help insure its own survival: *flexibility* or *adaptability*.

Chew wrote a foundational paper on S-matrix theory (1970) and later a general article (1974) on the impasse for the elementary-particle concept. These stressed the negative aspects of traditional field theory and the hope of eventual rewards from the autonomous S-matrix program. During most of the 1970s this S-matrix program, as applied to physics, languished. But as we shall see in the next chapter, some work continued, largely through the duality program, and by the end of that decade a topological version of the S-matrix program emerged.

7.7 Summary

Many of the central contributors to the dispersion-theory and S-matrix theory program saw no necessary conflict between QFT and SMT, but rather looked upon dispersion theory as *a* way of making the QFT formalism useful for applications. However, Landau and Chew adopted a radically *anti*-field theory stance. Chew hoped that S-matrix priciples – mainly unitarity and analyticity (especially crossing) – would make workable a bootstrap mechanism to generate (uniquely) all the observable consequences of strong-interaction physics. Electromagnetism remained an anomaly for the analytic S-matrix program. As this axiomatic S-matrix theory encountered serious conceptual and calculational difficulties because of its great mathematical complexity, its supporters shifted the line of argument to the great *potential* of the project and to the level of philosophical principle. The axiomatic *S*-matrix program never attained a precise formulation in terms of a well-defined set of principles from which one could convincingly derive equations having wide and unambiguous empirical support. While the axiomatic program was never refuted, it did not succeed either.

8

The duality program

In this chapter we examine the origins of a program that emerged from the S-matrix formalism but that soon took on a quite independent existence. Its greatest interest for present high-energy physics is that it led to superstring theories. This evolution is also of interest methodologically, as an example of an abandoned research program giving rise to a possibly 'correct' theory that might otherwise never have been formulated.

The concept of duality was an outgrowth of the Regge program. The simple Regge-pole form for the scattering amplitude arises from the exchange (in the crossed channel) of a Regge pole and is valid at *high* energy. When that form is extended (or extrapolated) to low energies, it gives the same result as the average value of an amplitude generated by the exchange of resonances in the direct channel (see the 'pictoral' representation of Eq. (8.1) below). More specifically, when sum rules (an example of which we have seen in Eq. (6.31)) that are integrals over differences of total cross sections are evaluated once numerically using experimental data and once 'theoretically' using the simple Regge form (at *all* energies), the results agree. This type of equivalence between direct-channel resonance and crossed-channel Regge exchange became known as *duality* and was a self-consistency form of the bootstrap. Veneziano offered a specific and simple analytical model that provided a concrete instantiation of the duality conjecture. This example was important both for subsequent theoretical progress and for immediate phenomenological applications. His remarkable model showed how several of the S-matrix principles, along with Regge asymptotic behavior and duality, could be incorporated analytically and consistently to form the basis of a workable bootstrap program. Subsequently, Veneziano suggested a perturbation type of expansion of dual scattering amplitudes, not in terms of a coupling constant (or smallness parameter) as had been familiar from QFT, but rather in terms of the topological complexity of various dual amplitudes (or diagrams). This gave some understanding of the origin of the Pomeron, which had remained a mystery in previous S-matrix theory. Duality supported a successful program of phenomenology that was able to correlate a large body of data.

Topology was incorporated into the S-matrix program and gave it new life and promise. A specific topological property of these dual amplitudes – 'order' – was conjectured to hold in nature. This order is a formal mathematical property of the topology of certain dual amplitudes and is not present (in exact, unbroken form) in nature. Nevertheless, the assumption of order did allow explanation of several important and known, but previously mysterious, regularities in the systematics of hadron phenomenology. This is not only an obvious example of hypothetical-deductive reasoning to justify an assumption (here, order). Even better, it is a case of formalism leading physical intuition, since it is unclear what physical property of an elementary particle could correspond to this notion of order. It seems to be a purely formal (even if quite useful) device. Furthermore, in terms of the hierarchy of this topological complexity, some mathematical objects are more fundamental (or 'elementary') than others (much as were Gell-Mann's 'mathematical' quarks in his own early formulation of the quark model; cf. Section 6.7 above). That is, compatibility was now seen as a possibility between the S-matrix principles and a quark picture (of elementary entities). Much of the new topological SMT was even formulated in terms of the language of its competitor (quark-gauge-field theory). Again, though, as with earlier versions of the S-matrix program, the theory was so complex mathematically that, after some initial and promising success, further progress became difficult.

Nevertheless, the duality program itself did lead to string theory when Nambu recognized that the formal structure of dual amplitudes could also be obtained from a field theory for strings. Again, we have an important example of mathematical formalism leading physical intuition. Not only did SMT and its descendant, duality, provide the origin for string theories, but these string models (in quantized form) require a certain number of space-time dimensions for mathematical consistency (including, possibly, a removal of the old renormalization infinities of QFT). This would be the ultimate bootstrap. As a bonus, which was neither sought nor even contemplated in the early days of duality and string theory, current superstring theory may include gravity (as well as strong and electroweak forces) and be truly the theory of everything (TOE). Such a unique, universal theory was the bootstrapper's dream. There are indications that SMT and present quantum field theories are either equivalent or limits of a larger theory. It is also possible that SMT may bootstrap superstrings. Whether or not any of these conjectured scenarios is realized, it does remain that superstring theories owed their formulation to duality and to the *S*-matrix program.

8.1 *S*-matrix origin of duality

In a previous chapter (6) we saw that Igi (1962) used Regge asymptotic behavior for the πN reaction amplitudes to derive the sum rule of Eq. (6.31). In 1966 de Alfaro, Fubini, Furlan and Rossetti discovered a similar idea of superconvergence relations. They first derived these relations from current commutators in quantum field theory using a technique employed by Gell-Mann for weak interactions. De Alfaro *et al.* (1966), as well as Dolen, Horn and Schmid (1967), also related these superconvergence relations to a limit of ordinary dispersion relations. Schwarz (1967) obtained superconvergence relations from helicity (or spin) amplitudes and an assumed Regge asymptotic behavior.[1] Igi and Matsuda (1967) applied these to the πN charge exchange reaction. Dolen, Horn and Schmid (1967, 1968) developed the use of finite-energy sum rules (FESR) to determine Regge parameters by connecting high-energy Regge behavior with integrals over low-energy data. The dual representation or duality of an amplitude in terms of resonances at low-energy and Regge behavior at high energy provided a bootstrap or consistency condition imposed by analyticity. Here Regge and resonance behaviors are both contained as different approximations to the same exact amplitude and they cannot both be added together independently (Schmid, 1970). Symbolically we can express this as (Mandelstam, 1974b)

$$\sum_{\substack{1 \\ 1}}^{3 \ 4} = \sum_{\substack{1 \ 2}}^{3 \ 4} = \sum_{1}^{3} \!\!\!\!\! \xrightarrow{\hspace{1cm}} \!\!\!\!\! <_{2}^{4} . \tag{8.1}$$

A simple example of how one obtains a FESR is provided by considering (Dolen, Horn and Schmid, 1967, 1968) an amplitude F that is odd in the variable[2]

$$v = \frac{(s-u)}{2M} \tag{8.2}$$

and that satisfies an unsubtracted dispersion relation

$$F(v) = \frac{2v}{\pi} \int_0^\infty \frac{\text{Im} F(v') \, dv'}{v'^2 - v^2}. \tag{8.3}$$

FESR'S are based on the assumptions that $F(v)$ is analytic and that the asymptotic behavior is given by the Regge form

$$F(v) \underset{v \to \infty}{\approx} \frac{\beta(t)[1 - e^{-i\pi\alpha}]}{\sin(\pi\alpha)\Gamma(\alpha + 1)} v^\alpha \equiv R(v). \tag{8.4}$$

If the leading Regge term has $\alpha < -1$, then the $v \to \infty$ limit of Eq. (8.3) requires the superconvergence relation

$$\int_0^\infty \text{Im} F(v) \, dv = 0. \tag{8.5}$$

If *only* the leading Regge trajectory has $\alpha > -1$, then we can subtract R from F and obtain the superconvergence relation

$$\int_0^\infty \text{Im}(F - R) \, dv = 0. \tag{8.6}$$

Assuming that the Regge approximation of Eq. (8.4) is valid for $v > N$, one obtains from Eq. (8.5) the FESR

$$\int_0^N \text{Im}[F(v)] \, dv = \sum_\alpha \frac{\beta N^{\alpha+1}}{\Gamma(\alpha + 2)}. \tag{8.7}$$

This relation should be valid at large, but *finite*, energy.

An early and important use of FESR'S (Dolen, Horn and Schmid, 1967, 1968; Freund, 1968) was a new type of bootstrap (Chew, 1986b), or self-consistency, condition to determine the parameters of the ρ trajectory from pion–nucleon charge-exchange data (recall Figure 6.9). However, for our purposes here, this πN case led to an important observation (Schmid, 1968a, 1968b, 1970). Only the ρ trajectory is above $\alpha = -1$ (at $t = 0$) so that Eq. (8.6) becomes

$$\int_0^N \text{Im}[F(v) - F_\rho(v)] \, dv \approx 0 \tag{8.8}$$

where $F_\rho(v)$ is the one-pole ρ Regge amplitude (Eq. (8.4)). Here Im $F(v)$ can be related (through the optical theorem) to the experimentally measurable $[\sigma_{\text{tot}}(\pi^- p) - \sigma_{\text{tot}}(\pi^+ p)] \equiv \Delta\sigma(v)$. Numerical evaluation of the integrals verified Eq. (8.8). This meant that, *on the average*, the high-energy Regge amplitude, extrapolated to low energies, gave the same average behavior as the resonances which dominate the low-energy region. This is illustrated in Figure 8.1. This equivalence of the average of the resonances and of the Regge term is known as FESR duality. This duality is the meaning of the symbolic Eq. (8.1) above. If no approximations were made, this equality or equivalence should be exact. As applications of the FESR bootstrap, other resonances (denoted as the ρ, f, and g ones) in the direct $\pi\pi$ channel were used to generate the ρ Regge pole in the t channel and to calculate $\alpha_\rho(t)$ (Schmid, 1968a) and the Regge πN charge-exchange amplitude was projected into partial waves to reproduce the low-energy πN resonances observed experimentally. Chew and Pignotti (1968) used duality to justify a Regge multiperipheral model at all energies, as in the production reaction $\pi N \rightarrow \pi N\rho$.

8.2 The Veneziano model

Veneziano's (1968) model was of central importance for the subsequent development of the duality program (and finally, indirectly, for string theory). Since the mathematics involved is fairly simple, we outline the general features of his model below.

Veneziano (1968) constructed a particularly simple model that exhibited this duality property explicitly. Let us take as an example the (somewhat fictitious) case of the elastic

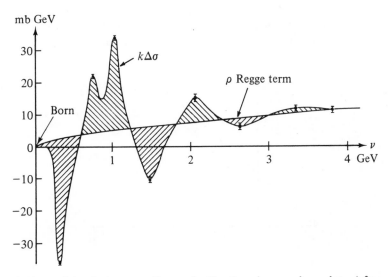

Figure 8.1   Resonance–Regge duality for pion–nucleon data (after Schmid (1970), p. 114).

scattering amplitude for identical scalar (spin zero) particles, $s + s \rightarrow s + s$. The complete amplitude $F(s, t, u)$

$$F(s, t, u) = A(s, t) + A(s, u) + A(t, u) \qquad (8.9)$$

is given in the Veneziano model as the 'narrow' (actually, zero-width) resonance approximation as

$$A(s, t) = \frac{\Gamma[-\alpha(s)]\Gamma[-\alpha(t)]}{\Gamma[-\alpha(s) - \alpha(t)]} = \int_0^1 dx \, x^{-\alpha(s) - 1}$$
$$\times (1 - x)^{-\alpha(t) - 1} \qquad (8.10)$$

where $\Gamma(z)$ is the usual gamma (or factorial) function and the Regge trajectory $\alpha(s)$ is assumed to be linear

$$\alpha(s) = \alpha_0 + \alpha' s. \qquad (8.11)$$

(Recall that $s + t + u = 4m^2$.) Here α' is positive to represent a rising Regge trajectory (corresponding to physical particles of positive spin) as shown in Figure 8.2. Incidentally, that these *exactly* linear trajectories correspond to the *zero*-width resonance approximation can be seen from the dispersion relation for $\alpha(s)$ (Mandelstam, 1974b)

$$\alpha(t) = a + bt + \frac{(t - t_0)^2}{\pi} \int \frac{dt' \, \mathrm{Im} \, \alpha(t')}{(t' - t)(t' - t_0)^2}, \qquad (8.12)$$

once we recall that $\mathrm{Im} \, \alpha$ is proportional to the width of a resonance (cf. Section 5.4 and Eq. (6.34)).

Since $\Gamma(z)$ has simple poles when $z = -n$ (i.e., at the negative integers), the amplitude $A(s, t)$ of Eq. (8.10) has poles at those values of s for which $\alpha(s) = n$, $n = 0, 1, 2, \ldots$.

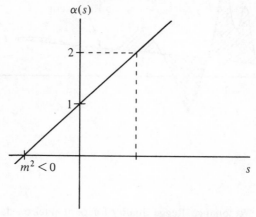

Figure 8.2   A linear Regge trajectory in a Veneziano model.

This is the expected resonance behavior in the direct (here, s) channel. Furthermore, the Stirling approximation for $\Gamma(z)$ implies

$$\frac{\Gamma(z+a)\Gamma(z+b)}{\Gamma(z+c)\Gamma(z+d)} \xrightarrow[|z|\to\infty]{} z^{a+b-c-d} \tag{8.13}$$

so that[3]

$$A(s,t)\xrightarrow[s\to\infty]{} \Gamma[-\alpha(t)][-\alpha(s)]^{\alpha(t)} \approx s^{\alpha(t)}. \tag{8.14}$$

which is the desired asymptotic behavior in the s channel. The full amplitude of Eq. (8.9) has the asymptotic form (Schwarz, 1973)

$$F(s,t,u)\xrightarrow[s\to\infty]{} \Gamma[-\alpha(t)][1+e^{-i\pi\alpha(t)}][\alpha(s)]\alpha^{(t)}. \tag{8.15}$$

Notice that, when $\alpha(t)=$ odd integer, the second term in square brackets vanishes. This is the signature factor and it implies that a given trajectory contributes a pole to the physical amplitude only at alternate integers.[4] In the Veneziano model we can avoid a recurrence at $\alpha=1$ as follows. From Eqs. (8.9) and (8.10), as well as the standard recursion relation

$$\Gamma(z+1)=\Gamma(z), \tag{8.16}$$

we find that the residue of $F(s,t,u)$ at $\alpha(s)\equiv 1$ is proportional to $[-2-\alpha(t)-\alpha(u)]$. We can arrange to have this residue vanish if

$$\alpha(t)+\alpha(u)=-2. \tag{8.17}$$

Combined with $\alpha(s)=1$, this constraint implies (for the linear trajectories of Eq. (8.11))

$$\alpha(s)+\alpha(t)+\alpha(u)=-1. \tag{8.18}$$

The 'bootstrap' condition that the Regge trajectory pass through zero (for a spin *zero* particle) at $s=m^2$

$$\alpha(m^2)\equiv\alpha_0+\alpha'm^2=0 \tag{8.19}$$

then requires that

$$\alpha'm^2=-1 \tag{8.20}$$

or that

$$\alpha_0=1. \tag{8.21}$$

This corresponds to $m^2<0$ which is a tachyon (i.e., a particle that travels faster than the speed of light).[5] This unphysical property proved to be one of the persistent problems for dual models and we shall return to it later. In fact, Veneziano (1968) in his original paper considered the case $\pi\pi\to\pi\rho$ and specifically avoided the case $s+s\to s+s$ because of the unphysical result of Eq. (8.21).

Lovelace (1968) made a very nice application of Veneziano's model to the data on decays of the type $p\bar{n}\equiv X\to 3\pi$. In order to fit experiment, Lovelace introduced, *ad hoc*, an imaginary part, Im $\alpha=\text{const}\sqrt{s-4m^2}$, to the linear Regge trajectory $\alpha(s)$ of the Veneziano model. The phenomenological fit to the data was quite good, although the model was no longer wholly consistent since Veneziano's results obtained only for *linear* trajectories. Veneziano's model had been for a 'four point' scattering amplitude (i.e., two particles, or 'lines' in and two out, for a total of four lines). Bardakci and Ruegg (1968) generalized Veneziano's integral representation of Eq. (8.10) to the five-point amplitude (e.g., two lines in and three out) and Virasoro (1969b) also gave a similar result. Chan Hong-Mo (1969), Goebel and Sakita (1969), and Koba and Nielsen (1969) found a generalization of Veneziano's model for the N-point function. Of course, since such a model violates unitarity, it can at best be considered only a zeroth-order approximation to the true scattering amplitude. That is, from the list of desirable properties for scattering amplitudes (e.g., Lorentz invariance, unitarity, analyticity, crossing, Regge asymptotic behavior), the duality program began by giving up unitarity and then attempting to implement that requirement 'perturbatively'. More complicated generalizations (e.g., Virasoro, 1969a), including integral representations of Veneziano's model, were proposed and extended to include arbitrary values of the isotopic spin for the particles (Paton and Chan, 1969). An entire operator formalism (Mandelstam, 1974a) was generated to guarantee the imposition of subsidiary conditions, such as factorization of residues and the elimination of ghost states (resonances with negative decay probabilities). We shall return later to this and its successor – the quantized string theories – after we have followed a different offshoot of the duality program.

8.3   A topological expansion

Harari (1969) and Rosner (1969) independently suggested a graphical representation to exhibit the duality property of scattering amplitudes. They assumed, in terms of (the group theoretical) SU(3) quark terminology, that *all* 'legal' particles (external to a diagram) and poles (internal to a diagram) are either baryons (which are fermions) consisting of three quarks (B→qqq) or mesons (which are bosons) consisting of a quark–antiquark pair (M→qq̄).[6] However these duality diagrams are 'cut' in a given channel, only 'legal' states should appear. Thus, Figure 8.3(*a*) is an allowed duality diagram, whereas Figure 8.3(*b*) is not. The contribution from the Pomeron is not given by these diagrams, but the rest of the scattering amplitude in any channel is given by the sum of single-particle states. These duality diagrams can equally well be viewed as a sum of resonances in either channel, as indicated in Figure 8.4. All of the allowed duality diagrams are connected and planar, in the sense that the 'quark' lines do not cross one another (i.e., they all lie uncrossed in a plane). When these planar dual diagrams (e.g.,

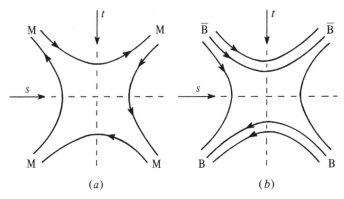

Figure 8.3   Harari–Rosner diagrams.

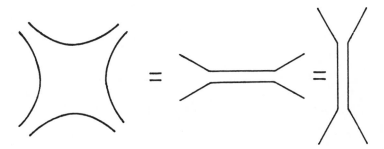

Figure 8.4   Duality diagrams.

Figure 8.3(*a*)) are combined *via* unitarity, one obtains higher-order diagrams, both planar (Figure 8.5(*a*)) and nonplanar (Figure 8.5(*b*)). Duality diagrams could be seen as a first (Born) approximation.

These duality diagrams, which began largely as a phenomenological set of rules, bear some resemblance to the Okubo (1963)–Zweig (1964)–Iizuka (1966) (or OZI) rule, also termed (Okubo, 1977) the quark line-rule (or QLR). Although Okubo (1963) originally derived a phenomenologically successful rule from a 'nonet hypothesis' for SU(3), Zweig (1964) and Iizuka (1966) stated the rule in terms of connected *planar* quark diagrams. Only processes that can be represented by these connected planar diagrams, such as figure 8.6(*a*), are allowed. A process like Figure 8.6(*b*) is not allowed, according to the OZI rule. This rule, which had no independent theoretical basis, proved quite successful in that reactions violating it were strongly suppressed. Its origin remained somewhat of a mystery, although its interpretation was straightforward enough in terms of the quark model (Freund, Waltz and Rosner, 1969).

The concept of planar diagrams as the lowest-order terms in an

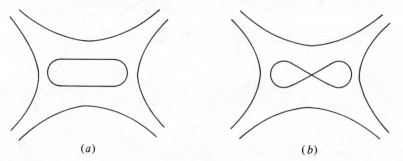

(*a*)   (*b*)

Figure 8.5   Planar and nonplanar diagrams.

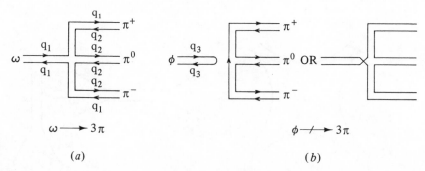

(*a*)   (*b*)

Figure 8.6   The Okubo–Zweig–Iizuka (OZI) rule.

expansion of the physical scattering amplitude became important. Lee (1973) gave an indication of how dual resonance amplitudes in a multiperipheral model could bootstrap the Pomeron. Veneziano (1974a) suggested the expansion and unitarization of dual amplitudes in a hierarchy of topological complexity, beginning with the planar at the lowest level.

> If one sums ... 'per topology' ... , a new perturbation expansion emerges (Veneziano, 1974a, p. 365)

> which we name the 'topological expansion'. (Veneziano, 1974b, p. 220)

That is, rather than expanding in terms of a coupling constant (or smallness parameter), Veneziano now suggested expanding in terms of topological complexity. The hope was that the topologically simpler diagrams would dominate the expansion.

> [W]e have proposed a new 'order of summation' for implementing unitarity in planar dual models. (Veneziano, 1974b, p. 220)

> [We] hope to associate the ... equation

> $$\text{Planar} \times \text{Planar} = \text{Planar} + \text{Non-Planar}$$

> with another equation

> $$\text{Reggeon} \times \text{Reggeon} = \text{Reggeon} + \text{Pomeron}.$$

(Veneziano, 1974a, p. 367)

The unitarity sum generates nonplanar output terms from planar input, just as the Pomeron may be generated through unitarity by Reggeon exchange. This led to calculations, by Rosenzweig and Veneziano (1974) and by Schaap and Veneziano (1975), of Regge couplings (i.e., a vertex joining two particles to a Regge pole as in Figure 6.7) and Regge trajectories with this planar duality and unitarization scheme. An early effort to calculate diffractive scattering (associated with the Pomeron) was made within the framework of a dual Regge picture (Chan, Paton and Tsou, 1975) and the results produced reasonable fits to a wide range of experimental data. A Reggeon bootstrap was attempted (Chan, Paton, Tsou and Ng, 1975) with a dual model.[7] The results were encouraging but not conclusive. In 1974 Chew, while visiting CERN where Veneziano was, became interested in Veneziano's topological expansion. Rosenzweig, who had been working with Veneziano, began a collaboration with Chew and then went to Berkeley. The origin of the Pomeron was clarified as a correction to the planar level in the topological expansion (Rosenzweig and Chew, 1975; Chew and Rosenzweig, 1975), a bootstrap origin of the Okubo–Zweig–Iizuka (OZI) (Okubo, 1977) rule was found (Chew and Rosenzweig, 1976) and the relation between G parity (another conserved quantum number) and Regge-trajectory exchange degeneracy (for certain pairs of trajectories as, for example, the ρ–A_2 and the f–ω) was studied (Chew and Rosenzweig, 1977). Balázs (1976, 1977a, 1977b, 1977c, 1977d, 1977e) used a dual multiperipheral model to make approximate numerical calculations of the Pomeron parameters, triple Regge couplings (a vertex at which three Regge trajectories meet) and a Regge trajectory giving a linear meson spectrum in agreement with experiment. Capra (1977) and Stapp (1977) made some tentative suggestions for incorporating baryons into this topological expansion scheme.

8.4   The topological S-matrix program

In the process of writing a review paper, Chew and Rosenzweig (1978) put forward as an hypothesis the concept of *order* in the topological S matrix. This gave some understanding of several previously observed regularities of high-energy scattering amplitudes. For example, we have already seen that it had long been known that as a matter of principle there must exist Regge cuts (which are moving singularities other than poles), in addition to Regle poles, in a relativistic scattering amplitude. Nevertheless, the effect of these cuts is remarkably absent from much experimental data. That is, good empirical fits can often be obtained using only Regge poles without any Regge cuts. This fact remained a curiosity for which there was no compelling explanation. Furthermore, the empirical OZI rule, which is simple to state and quite accurate, had no theoretical explanation either in QFT or in SMT. So, both the absence of Regge cuts and the OZI rule appeared to be accidents.

It was found that by adding one more postulate to SMT, these facts, plus many predictions that had been characteristic of the quark model in QFT as applied to meson–meson scattering, can be explained (Chew and Rosenzweig, 1978). This additional property is *order* and can best be explained here by means of a specific example. Consider the scattering process (or reaction) in which particles a and b scatter into (or become) particles c and d as

$$a + b \rightarrow c + d \tag{8.22}$$

The analytic scattering amplitude (or S-matrix element) describing the process (8.22) equally well describes the reaction

$$a + b \rightarrow d + c. \tag{8.23}$$

The order of the particles in the final state (or in the initial state or in both) is immaterial. This is found to be the case in nature. However, we could begin by postulating an ordered S-matrix for which the amplitudes corresponding to the ordered diagrams of the theory have the property

$$A_0(a + b \rightarrow c + d) \neq A_0(a + b \rightarrow d + c). \tag{8.24}$$

Such an S-matrix can be consistently relativistically invariant, unitary, crossing symmetric and analytic. In other words, order is a perfectly acceptable S-matrix principle. By demanding self-consistency (a key to any bootstrap program) when unitarity is imposed, one can show that only a certain (unique) subset of graphs (from among all conceivable ordered connected graphs) is allowed (Capra, 1979). The only difficulty is that the actual physical S-matrix is not ordered. One can attempt to remedy this by defining a physical, unordered scattering amplitude as[8]

$$A_p(a + b \rightarrow c + d) = A_0(a + b \rightarrow c + d) + A_0(a + b \rightarrow d + c). \tag{8.25}$$

Unfortunately, this so-called planar S-matrix now lacks just one of the properties required of an S-matrix – unitarity. If we leave this drawback aside for a moment, we find that the planar S-matrix has several nice properties, among which are the OZI selection rule, exchange degeneracy, and lack of any need for Regge cuts. (That is, there is no inconsistency if Regge cuts are absent).

One can then set up an iterative scheme in terms of the topological structure of these amplitudes and of the unitarity relation to generate corrected amplitudes that do satisfy unitarity. It appears plausible that there may be a dynamical mechanism which suppresses the higher-topology terms to produce convergence of the topological expansion (Chew and Rosenzweig, 1978, p. 293), but such a convergence has not been proved. The hope is that the planar diagrams provide the dominant contribution to the physical amplitudes so that many remnants of order do survive in the physical world. Incidentally, one can in retrospect 'see' in Veneziano's paper on planar dual models (1974a, Fig. 1, Eqs. (3.1) and (3.2)) what is essentially the ring diagram of Chew's later program, even though order does not appear to have been in the air at that earlier date. This dual topological unitarization (DTU) program required the existence of baryonium states (or exotic mesons) which, in terms of the quark model language, would represent qqq̄q̄ states (Chew and Rosenzweig, 1978, p. 320). Lack of firm experimental evidence in the available data (Nicolescu, Richard and Vinh Mau, 1979; Pickering, 1984, pp. 312–315; Nicolescu and Poénaru, 1985) for these high-mass narrow resonances (with zero baryon number) was *the* central problem for DTU and order. After the 'November Revolution' of the discovery of the J/ψ particle (Aubert *et al.*, 1974; Augustin *et al.*, 1974), there was a loss of interest among the theoretical physics community in the problem of the topological program in favor of the gauge-field-theory program. If the idea of order had come a few years earlier, perhaps there would have been a wider interest in the concept.[9]

It was in his (1979a) paper on the bootstrap theory of quarks that Chew began to give this entire program a highly sophisticated formulation in terms of a topology that could handle baryons as well and he has claimed (1979b) that the weak and electromagnetic interactions may also be able to be accommodated. In fact, electromagnetism now plays a central role in the basic measurement process for Chew (1985). However, as ever in the S-matrix approach, calculations remain difficult. (Some sense of these complexities can be gotten from the papers of Balázs, Gauron and Nicolescu (1984); Balázs and Nicolescu (1980, 1983); Jones and Uschershon (1980a, 1980b); Nicolescu (1981); Gauron, Nicolescu and Ouvry (1981).) In a paper setting forth the axioms of the new topological S-matrix theory (TSMT), Chew and Poénaru (1981) acknowledged the effectiveness of the rival QCD approach and reiterated their belief that the two approaches might indeed be compatible.

> Bootstrap theory developed slowly because of its essential nonlinearity and lost favor when the capabilities of the seemingly opposite quark approach, eventually formalized within quantum chromodynamics (QCD), became recognized. It was, however, never established that conflict exists between quark and bootstrap ideas. This paper describes a bootstrap theory which explains quarks and their properties on the basis of S-matrix consistency. (p. 59)

Some workers in TSMT continue to see signs of this equivalence:[10]

> [W]hen we repeatedly said [in TSMT] that 2-dimensional surfaces are the essence of hadron dynamics, people were more or less laughing. Now they take very seriously into account this idea (for all particles) simply because it comes from superstrings.

An example of what Nicolescu may have in mind here is the importance that Schwarz (1982a, p. 12) assigns to topology in the current superstring theories.

> The two-dimensional world sheets may be classified by their topological complexity – the number of holes and the handles – which is a convenient way of organizing the perturbation expansion.

Nicolescu also appeals to an overarching principle as a motivation for pursuing the TSMT program[11] (see also, Nicolescu, 1985, p. 63):

> For me it is very important that the bootstrap principle succeeds to unite 'world-view' and 'scientific theory' in a single approach.

An exposition of this topological S-matrix theory involves complex concepts quite different from any we have discussed thus far and would add relatively little of new conceptual interest to our case study, aside from the possibility that a bootstrap may be possible even with (topological) hierarchies. (A review can be found in Capra (1984) and in Nicolescu and Poénaru (1985).) This model, like the duality and string models (to which we turn next), began as a mathematical 'toy' model for the mesons but, once it had been extended to include spin and baryons (Chew and Poénaru, 1980, 1981), became for its proponents a candidate for a theory of all particle interactions. Even though first formulated for the hadrons, the TSMT now includes also the electroweak interactions. This theory predicts exactly three 'quarks' in a baryon, four as the maximum number of quarks and antiquarks in an elementary hadron, and just four hadron, generations (for a total of eight flavors) (Finkelstein, 1985). The electroweak interactions may be required in order to produce a consistent topological theory for the hadrons and the fine structure constant ($\alpha \simeq 1/137$) may be fixed in the process (Chew, 1983a). The standard (Salam–Weinberg) electroweak model has been seen as a natural outgrowth of TSMT (Chew and Finkelstein, 1983). Also, *separate* T (time reversal), C (charge conjugation) and P (parity) invariances for strong interactions in TSMT have been deduced (Jones and Finkler, 1985; Finkler and Jones, 1985; Jones, 1985). This brings us up to some of the latest claims of TSMT. Capra's review article (1984) can be consulted for recent further developments in that program. Let us leave this topic for now and turn to the operator formalism and quantized string program that grew out of the duality models.

8.5 The emergence of quantized string theories

Shortly after generalizations of Veneziano's (1968) original model had been proposed, Fubini and Veneziano (1969) considered the factorization problem for the N-point duality amplitude. If an N-point amplitude A, such as that of Figure 8.7(a), is to exhibit duality in every channel, then it must be representable as a sum of resonance terms in *any* channel. For example, in the s channel it must reduce to a sum of terms (cf. Figure 8.7(b))

$$A \xrightarrow[s \to s_m]{} \frac{R(p,q)}{s - s_m} \tag{8.26}$$

and the residue $R(p,q)$ must factorize as

$$R(p,q) = F(p)F(q) \tag{8.27}$$

for a single level, or at least as

$$R(p,q) = \sum_{i=1}^{d_n} F_i(p)F_i(q) \tag{8.28}$$

for d_n levels. This simultaneous factorization in *all* channels is an extremely stringent requirement. It must be demonstrated explicitly that such is the case, since an arbitrary function with the proper pole (or resonance) structure will not usually possess this property. Furthermore, if ghosts (states of negative norm) are to be avoided, the sum in Eq. (8.28) must be positive-definite.[12] Fubini and Veneziano (1969) were able to show that the duality amplitudes factored, but the residues were not necessarily positive definite. However, they demonstrated that because of a set of invariance transformations of these amplitudes, many (but not necessarily all) of these ghost states were effectively decoupled from the physical states. Using a representation of the N-point function due to Koba and Nielsen (1969), Fubini, Gordon and Veneziano (1969) were able to express these amplitudes in terms of

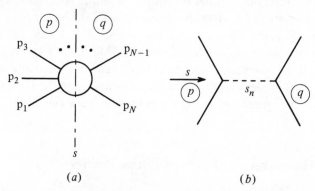

(a)   (b)

Figure 8.7   Residue factorization at a pole.

expectation values of functions of creation $(a^\dagger)$ and annihilation (a) operators in such a way that this factorization property was manifestly guaranteed. This operator formalism was further explored by Fubini and Veneziano (1970) (Alessandrini, Amati, Le Bellac and Olive, 1971). Nambu (1970) also handled factorization by means of an operator formalism in which the basic operators a and $a^\dagger$ satisfy the 'harmonic oscillator' commutation relations[13]

$$[a, a^\dagger] = 1. \tag{8.29}$$

On the basis of a formal analogy for the energy of an intermediate state, Nambu (1970) observed that the internal energy of a meson can be interpreted in terms of the normal modes of vibration of a string of finite length. It is interesting to note that, although string theories *could* (logically) have been formulated decades ago, they might never in fact have been considered had it not been for SMT *via* the duality models (Mandelstam, 1985). John Schwarz (1975, p. 64), one of the central figures in the construction of superstring theories, has made a similar observation.

> The theory of light strings could have been investigated 40 years ago, but there was little motivation for doing so until the connection with dual models was suspected.

Goddard, Goldstone, Rebbi and Thorn (1973) showed that the Lagrangian in the string theories had a large invariance group which generated the Virasoro (1970) gauges to eliminate the ghosts. Scattering of strings can be pictured as strings breaking and joining at the ends (Mandelstam, 1973a, 1973b).[14] Thus, dual resonance models can be treated either as S-matrix theories (Veneziano, 1974c) or as a system of interacting strings (Rebbi, 1974), although the string formulation can be made manifestly ghost-free (Mandelstam 1974b; Frampton, 1974; Jacob, 1974). We see that the string theory can be looked upon as a natural outgrowth of S-matrix ideas. The existence of ghosts (which involve negative probabilities, the sum of which need not equal unity – a violation of unitarity) can be seen as a conflict between the demands that the theory satisfy the principles of both (special) relativity and quantum mechanics (Schwarz, 1975). In the simplest string model, the space-time dimension must be 26 for these simultaneous demands to be met! (Lovelace, 1971; Cremmer and Scherk, 1972; Olive and Scherk, 1973a). A model allowing both fermions and mesons (Neveu and Schwarz, 1971a, 1971b; Thorn, 1971) can be proven consistent for a space-time of 10 dimensions (Olive and Scherk, 1973b). The tachyon problem (violation of causality) remained.

All of these string models contained a massless spin-2 particle. This was considered undesirable for a theory of *strong* interactions (Schwarz, 1982a), since there is no such strongly-interacting particle known. However, Scherk and Schwarz (1974a, 1974b, 1975) proved that this particle interacted as a graviton would. They suggested using string models as a basis for *all* particles, not just for the hadrons. Gliozzi. Scherk and Olive (1976, 1977) interpreted the string model as a supersymmetry[15] in which fermions and bosons are represented in the same state. Such a supersymmetry must be spontaneously broken since bosons and fermions do not occur in nature with the same mass. Superstring theories have no tachyons (Green and Schwarz, 1981; Schwarz 1982b). At low energies, some of these superstring theories have the possibility of reducing to the conventional gauge field theories.

These 10-dimensional superstring models are candidates for a quantum theory of all interactions in which all calculated quantities are finite (Schwarz, 1982a; Green and Schwarz, 1984a, 1984b, 1984c, 1985). The internal degrees of freedom of the string (as opposed to a structureless *point* particle) allows a *single* string to have an infinite number of states with masses and spins that increase without limit.[16] It appears that conventional (point) field theories cannot remain finite, or even renormalizable,[17] when the gravitational interactions are included. Extended (e.g., string) field theories, in a severely restricted number of cases, appear to have the possibility (likelihood?) of being renormalizable (or even finite) when gravity is included (Green, 1985). These latter theories may, in fact, be (nearly) *unique* (Schwarz, 1984, 1985; Gross, 1985).

> This was a unique state of affairs: the gauge symmetry of a theory being imposed by the requirement of quantum consistency. (Schwarzchild, 1985, p. 19).

A similar observation has been made by Richard Capps, a theoretical physicist who was prominently and actively involved in the dispersion theory and *S*-matrix bootstrap program.[18]

> [R]ecently I have been struck by the similarity of the motivation of the bootstrap idea and the motivation of modern string theory. ... The modern consistency condition is based on anomaly cancellation [a type of inconsistency that must be removed or prevented from occurring in a theory], rather than *S*-matrix analyticity, but the two approaches are more similar than most people realize.

Spontaneous compactification (a type of 'curling up' of certain dimensions), *via* an old idea due to Kaluza (1921) and Klein (1926), has been suggested to reduce the (observable) number of space-time dimensions from 10 down to the 4 of the real world. (Of course, this is an *ad hoc* theoretical move.) These theories even suggest the existence of a 'shadow world' of matter that can interact with ordinary matter *only* through the gravitational field (but not through the strong or electroweak ones so that it is 'invisible' to them) (Kolb, Seckel and Turner, 1985). This could offer a possible resolution of the 'missing mass' problem of cosmology (i.e., the discrepancy between the observed mass of the universe and that required by the observed motion of stars near the galaxies in the universe). Here, too, we have come to some of the latest developments in superstring theory, an area of much activity at the current time (Schwarzschild, 1985; Neveu, 1987).

8.6 Superstrings – the ultimate bootstrap

Let us make two observations about the relationship of string field theories (and gauge field theories like QCD) on one hand and S-matrix theory (including TSMT) on the other. Frazer (1985) has reiterated the possibility that the current rage – QCD – may be, like a previous theory of great interest – SMT – not really a *fundamental* theory. Both SMT and QCD have had successes in *different* domains of high energy physics (Pickering, 1984) and SMT is certainly not a theory that has been proven incorrect. Frazer (1985) summarizes Weinberg's (1985) two 'folk theorems': (1) that a quantum field theory with the most general Lagrangian is without content, being simply a way of implementing S-matrix theory; and (2) that, at low enough energies, a simple effective Lagrangian field theory can always be found. Frazer (1985) has termed this latter a 'harmonic oscillator theorem' (presumably in analogy with the result from classical mechanics that 'any' small oscillation about a stable equilibrium point is essentially a harmonic oscillation, no matter how complex the mechanical system).

The possibility certainly exists that SMT and QCD could be equivalent and/or 'low energy' limits of some larger theory (such as superstrings or TSMT). When one looks at the topological diagrams and expansions of superstrings and TSMT, one cannot help but be impressed by their (perhaps only formal and superficial) resemblance. In a recent paper, Balázs (1986, p. 1759) attempts to bootstrap superstrings from a self-consistent S-matrix formation.

> It is proposed that the superstring and its spectrum may not be fundamental but may themselves correspond to an approximate

solution of a self-consistent dynamical theory built from the basic principles of S-matrix unitarity, analyticity and crossing. In particular, it is shown that a simple approximate dual unitary scheme based on these principles does generate linear Regge trajectories and selects closed (rather than open) strings.

This would imply that superstrings would be *demanded* by the S-matrix principles. On this ambiguously optimistic note we end our technical discussion of the S-matrix program.

8.7   Summary

Within the framework of the S-matrix program of the late 1960s, finite energy sum rules were discovered and a duality was recognized which allowed one to consider a scattering amplitude either in terms of resonances at low energy or in terms of Regge behavior at high energy. Resonances in one channel were used to bootstrap a Regge pole in the crossed channel. Veneziano in 1968 constructed a simple mathematical model which explicitly exhibited this duality behavior and satisfied several S-matrix postulates *except* unitarity. This model was generalized to the N-point function and Veneziano in 1974 suggested a topological expansion to unitarize these amplitudes. This led to an independent duality program for calculations in strong-interaction physics.

In the late 1970s and early 1980s the S-matrix program was resuscitated in the form of a topological S-matrix theory. But, as regards novel empirical content, it was largely a camp follower to the highly successful gauge field theories. Matters of principle continue to motivate a relatively small band of present-day S-matrix theorists.

Nambu in 1969 discovered a connection between duality models, which had emerged from the S-matrix program, and the field theory of a string. Interest in quantized string theories arose because of the duality program, which had been rooted in S-matrix concepts. Although quantized strings were plagued for many years with consistency problems and with undesirable features, Scherk and Schwarz in 1974 recognized that they could be interpreted as representing gravitational as well as strong interactions. Green and Schwarz have recently found indications that all quantities calculated in these theories may be finite and that this requirement may yield a *unique* theory of all fundamental interactions.

9

'Data' for a methodological study

In the previous eight chapters we have presented a fairly detailed case study of the S-matrix program. Along the way, we often pointed out the relevance of certain developments to methodological issues in theory construction and theory choice. In the next chapter the 'data' from this and several other episodes in modern physics will be used as a basis for drawing some methodological lessons about theory selection in current science. This chapter is devoted to summary sketches of methodologically relevant aspects of several programs in modern physics.

We begin (Sections 9.1 and 9.2) with an overview of the 'factual' content of the S-matrix case study itself and some relevant methodological issues because that may help focus the reader's attention on points important for the argument of the next chapter. Also, those readers most interested in the discussion of methodological issues in Chapter 10 can use the present chapter as an easy and fairly self-contained review of the pertinent historical materials. (Recall, too, that this book has a glossary of technical terms for quick reference.) To demonstrate that those features are not peculiar to S-matrix theory alone in modern physics and, thereby, to broaden the foundation upon which subsequent arguments are made, Section 9.3 provides a summary of some other major episodes in modern physics. This includes a review of certain aspects of the old quantum theory, of early quantum field theory, of quantum electrodynamics after World War II and of gauge field theories.

Section 9.4 illustrates similar general features of theory creation, selection and justification in an episode of smaller scope and of a less revisionary character than the SMT and QFT programs. This case of the compound nucleus model of nuclear reactions shows how formal analogies were used to suggest a physical mechanism that could account for the observed systematics in data for the scattering of a particle by a nucleus. This led to a general framework and a language with which to describe these phenomena. Finally, an argument was fashioned that made the model appear almost self-evidently *required* by general principles. In Section 9.5 we consider a causal interpretation of quantum mechanics, one that is a rival to the standard, acausal 'Copenhagen' version accepted by most scientists. The interest, for us, in this story is that it provides an example of an empirically adequate, fertile and viable theory that has been rejected (in fact, scarcely even considered) by the theoretical physics community. The reasons for this appear to have more to do with historical accident or contingency than with rational criteria wholly internal to and peculiar to science itself.

9.1   An overview of this case study

The story of the S-matrix program must be seen against the backdrop of quantum field theory if we are to place the events of this case study in any reasonable perspective. Both initially, in the mid 1940s, and again in the late 1950s, S-matrix theory was turned to as a hoped-for means of coping with, or perhaps of just avoiding, what some perceived to be intractable difficulties and inconsistencies in the quantum-field-theory program. If QFT had encountered no serious difficulties, it is unlikely that SMT would have come to exist. However, theorists were able to recast QFT to overcome divergence problems through renormalization and, two decades after that, to formulate a local gauge principle and a mechanism of spontaneous symmetry breaking that allowed QFT to cope with the strong interactions. Each time that QFT made a comeback, the overwhelming majority of theorists returned to the fold. Even grand unified field theories have not proven wholly satisfactory, though, as fundamental theory. A new creation, superstring theory, is seen by many as offering hope of providing *the* unique fundamental theory for all interactions – strong, electroweak and gravitational.

Although the S-matrix program does not now have, among most theorists, the status of an actively pursued and viable research program, it has played an important role in bringing us to superstring theories. Heisenberg's early S-matrix papers of 1943–44 helped to focus attention on possible uses for analyticity and causality in strong-interaction theory. These concepts proved useful in nuclear physics in providing some explanation for the Breit–Wigner resonance formula. Analyticity properties of scattering amplitudes were combined with perturbative quantum field theory by Rohrlich and Gluckstern in the early 1950s in a calculation of photon scattering by the Coulomb field of a nucleus. Goldberger and Gell-Mann saw this work and sought some fundamental underpinning for these dispersion relations. Thirring provided an essential ingredient when he suggested implementing the causality condition, in some sense of a first-signal principle, by requiring that the commutator vanish for two field operators acting at space-like separations. Kramers–Kronig type dispersion relations were derived and successfully applied to correlate experimental data. The style of this dispersion-theory enterprise was a pragmatic one in which needed analytic properties were often simply *assumed* to hold, before they could be justified (if ever they were) on the basis of fundamental principles or of a coherent theory. By the mid 1950s Gell-Mann had suggested how, in principle, certain general properties of the S-matrix *might* be sufficient to recover all of the results of perturbative quantum field theory, but without the need of specifying in detail any particular Lagrangian. He did not see this possibility of a mass-shell quantum-field-theory program as in any way opposed to the conventional approach, but rather as an alternative.

Although Chew and Goldberger had been graduate students at the University of Chicago just after World War II and had heard of Heisenberg's S-matrix paper from Wentzel, they did not directly pursue research in that program at that time. Only much later did they, and Gell-Mann, appreciate the relevance of their own work to Heisenberg's earlier program. Goldberger and Gell-Mann arrived at the S-matrix through dispersion-relation applications of quantum field theory, but Chew's route was quite different. While still a graduate student, he was very much impressed by Fermi's interest in practical applications of theory. In particular, previous work by Fermi on the scattering of neutrons by a bound state (here, the deuteron) was generalized by Chew into the impulse approximation. At this point in his career, Chew was heavily involved with the analysis of high-energy scattering data, in which directly measurable quantities were related to each other. Extrapolation and some effective use of analyticity were important for these applications. This work generated in him a belief that there should be no arbitrariness in the fundamental equations of a theory. He felt uncomfortable about standard quantum field theory in which a largely arbitrary Lagrangian was allowed. However, at that time (say the early 1950s), this was only a sense of uneasiness and Chew had no concrete suggestions for an alternative formulation. It was while at the University of Illinois that he collaborated with Low and became an expert in applications of the QFT static model to pion–nucleon phenomena. The Chew–Low model resulted and impressed upon Chew the importance of analyticity and of crossing (which he had learned about from Gell-Mann and Goldberger). It was the confluence of the Chew–Low model and of the Goldberger–Gell-Mann dispersion-theory approach that gave rise to the S-matrix program of the 1960s.

Several technical developments at this time, such as Mandelstam's conjecture of his double dispersion relation and Regge's use of complex angular momentum in his investigation of the analytic properties of scattering amplitudes, provided plausible means for implementing a calculationally effective S-matrix program. Chew and Mandelstam discovered an unexpected, self-sustaining solution to the S-matrix equations for pion–pion scattering. This was the origin of the bootstrap conjecture for strong-interaction physics. At the 1961 La Jolla Conference, Chew proposed cutting any connection with quantum field theory and pursuing an independent S-matrix program instead. While Chew's attitude was certainly *anti*-QFT, by no means all of those who contributed centrally to the SMT program shared this view. The S-matrix program had considerable initial success in correlating a large body of experimental data and in making specific predictions that were subsequently confirmed. However, as more data came in and the program was pressed for greater detail, the situation became overwhelmingly complex, at both the conceptual and calculational levels. Rapid forward progress was no longer possible. As SMT encountered this impasse, QFT made an impressive comeback in the form of spontaneously broken gauge field theories. Interest in the theoretical physics community shifted to this new area of activity. Proponents of SMT now argued for their program largely on the basis of principle and of its hoped-for scope.

Nevertheless, one offshoot of SMT, namely the duality program, did prove influential for subsequent developments in mainstream particle theory. In 1968 Veneziano constructed a model of a crossing-symmetric, Regge-behaved scattering amplitude. Even though this model was unrealistic, since it violated unitarity, it did suggest how one might implement a calculationally adequate scheme for strong-interaction physics. The duality program did allow analysis of a considerable body of data and also was responsible for the introduction of topological methods into this area of particle physics. The S-matrix made a minor comeback in the late 1970s and early 1980s in the form of topological SMT, but no convincing program emerged. The important development for particle physics was Nambu's observation (by way of a formal and quite indirect mathematical analogy) in 1969 that these duality amplitudes could be associated with a field theory of strings. Work on string theories by Schwarz and others has led to superstring theories, which are the current candidates for a unique, finite theory of all the fundamental interactions, including gravity.

9.2 Hallmarks of this episode

Let us now focus on certain characteristics of these developments in modern high-energy physics in order to highlight factors that are especially relevant for proposed methodologies of science. In this section we are particularly concerned with the origin and pragmatic selection of theories.

The recycling of analogies (Pickering, 1984) and the use of models (Cushing, 1982) have played important roles in the developments we have discussed. The Breit–Wigner nuclear resonance formula and Heitler's theory of radiation damping for strongly-interacting mesons were both based upon an analogy with the familiar example of the scattering of light by a (classical or quantized) atomic system. This dispersion of light by atomic systems had proven central to the development of quantum physics. Then the concept of 'virtual oscillators' proved useful when the result (cf. Eq. (A.80)) of the classical dispersion theory of Drude (1893, 1900) (based on the scattering of light by a classical charged harmonic oscillator) was reinterpreted in terms of virtual quantum oscillators by Ladenberg (1921), by Slater (1924), by Kramers (1924) and finally by Kramers and Heisenberg (1925) (cf. van der Waerden, 1968, pp. 9–16). The result had been better than its derivation. Similarly, Kramers (1927)–Kronig (1926) dispersion relations, based on classical electromagnetic theory and the causality principle, were justified in the context of relativistic quantum field theory when Thirring (Gell-Mann, Goldberger and Thirring, 1954) suggested that 'microcausality' (i.e., a first-signal principle assumed to be valid even over subatomic dimensions) be implemented *via* the vanishing of the commutator of the field operators (Eq. (3.1)). Here one had a result that was believed correct and a means had to be found to justify it within an accepted general framework (in this case, quantum field theory). The concept of causality, which in classical electromagnetic theory had a clear and precise meaning in terms of the first-signal principle, was reinterpreted (or redefined) in terms of microcausality – the vanishing of field-operator commutators (or anticommutators) at spacelike distances. The *exact* physical meaning of this commutator condition is not clear. As we indicated earlier, this is underscored in the 1957 LSZ paper in which an operator $A(x)$ satisfying Eq. (3.1) for spacelike separations is *defined* as a *causal* operator – 'without discussing here [or anywhere in that paper] the physical interpretation' (Lehmann, Symanzik, and Zimmermann, 1957, p. 323). It seems *plausible* that it may have something to do with first-signal-principle causality, but the equivalence of these two definitions of causality is not established. The main justification of this new criterion for a causal theory is that it produces a desired result (dispersion relations), while not obviously conflicting with a more direct notion of causality. We return to the evolution of the term 'causality' in a later section.

The dispersion-theory and S-matrix-theory programs of the late 1950s and early 1960s had great appeal initially because they *worked* (i.e., they successfully related many directly measurable experimental quantities to each other). Of course, some of this success was 'arranged' (or greatly aided) since needed results (such as dispersion relations for massive particles and for nonforward directions, Regge asymptotic behavior, etc.) were assumed long before they could be proved (and many never were). But, then, some of the best people in the business (such as Goldberger) seemed to 'know' the correct answer in advance and would leave the detailed proof to someone else (similar to Heisenberg's formal use of scattering-theory limits, which would be justified years later.) The Chew–Low model had some solid foundation in a limited regime, but since it suggested the possibility of determining the πN coupling constant by extrapolation (which worked), it led to the inspired guess of the general Chew–Low extrapolation scheme for multiparticle amplitudes. Mandelstam's double dispersion relation can also be seen as inspired guessing, guided by field-theory examples. These programs were characterized by a desire to 'get on with things', to 'do something'. Cini (1980) and Pickering (1989a) have stressed the pragmatic aspect of these approaches and Schweber (1989) has suggested that this was a hallmark of much of theoretical physics after the Second World War (as contrasted to the period before the War).

The central ideas for these programs – such as microcausality, the bootstrap, Regge poles, Veneziano's duality model – came quite directly from the mathematics and from the successes of the equations used. These seminal concepts had their origins in detailed and often highly technical applications, rather than in some grand methodological principle that had been subscribed to in advance. These are also examples of physical thinking being led by (at times perhaps replaced by) the mathematics, rather than *vice versa* as in an earlier era (Cushing, 1983a). (Zahar, 1976, has also written about the heuristic value of mathematics in developing physical theories.) After all, what would be the ontological status of a Regge pole? In a sense, the autonomous *S*-matrix program can be seen as a logical extension of this (mathematically-motivated) 'guessing game': just assume what you need and use empirical success as the justification. Indeed, considerable initial success (such as the shrinkage of the diffraction peak and the correlation of nucleon resonances by a common Regge trajectory) stimulated a great interest in the program.

The *S*-matrix program of the 1960s consisted of a loose set of postulates coupled with some fairly specific rules for calculation which lent themselves to introducing young workers into a field of great activity. The project was ideal for generating PhD theses in theoretical physics. However, later, after it had encountered calculational difficulties, the appeal of the *S*-matrix program (in its autonomous and topological versions) was based on its (grand) promises or hopes. This program was never refuted (by logic or by experiment), but it failed to fulfill its promise (that is, to succeed in a strongly positive and overwhelmingly convincing sense). In describing the criterion he used for selecting theories in his own construction of the quark (or aces) model, Zweig (1980, p. 35) expressed very nicely the value scientists place on 'usability'.

> Theories which lacked predictive power, like Heisenberg's nonlinear spinor theory of matter, were discarded, not because they were necessarily incorrect, but because they were operationally useless.

At this same time that the predictive fertility of the *S*-matrix program waned, it continued to have considerable philosophical appeal (Cushing, 1985). In fact, to the extent that it has a solid empirical base, *S*-matrix theory can be largely equated to its dispersion-theory content. Without great (apparent) forward progress, the number of active workers in the field reduced sharply. The duality and superstring models also became theory-driven, having little direct contact with experiment (Schwarz, 1975, p. 67; 1982a, p. 7). Consistency, potential scope, and hoped-for contact (in a limiting regime) with an empirically adequate theory (such as QCD) remained the major motivations for pursuit.

Chew's role in incorporating Regge theory into the *S*-matrix program is a nice illustration of what Pickering (1984) has termed 'opportunism in context'. (There is no negative connotation implied in this phrase.) A good theorist grabs what is needed and puts it to use, exploiting those skills and techniques he has already developed and for which he has a particular knack. We have referred several times in earlier sections to Goldberger's attitude that one should use any tool to solve a problem and that there need not be any high-minded, 'theological' guiding principle behind this.[1] Chew showed an extraordinary ability to achieve a goal by orchestrating his own ideas and those of others (e.g., Fermi, Low, Goldberger, Gell-Mann, Mandelstam, Regge, Frautschi, Veneziano, Poénaru). That goal, over a long period of time, we have characterized as *no arbitrariness* in the fundamental laws of nature. He also had exceptional influence upon a very large number of extremely productive students (Frazer, 1985) (including F. Salzman, W. Frazer, J. Ball, J. Fulco, L. Balázs, C. Jones, J. Stack, J. Schwarz, J. Finkelstein, J. Sursock) who propagated the dispersion- and S-matrix-theory pro- grams. Goldberger recalls that Chew always had a program, even when he was doing fairly routine things like fixed-source calculations, that Chew had a 'school' (e.g., dozens of students who walked and talked like him) and that Geoff always knew what he was going to do when he went into the office.[2] Geoff Chew was a very appealing, perhaps almost romantic, figure – 'charismatic' is not an inappropriate term. Polkin- ghorne (1985) has characterized Chew as a man you *would* buy a used car from. Sustained growth in a field depends upon a large, active work force. This certainly does show an influence of sociological factors at work in the scientific enterprise. Chew's personality and the relative ease of entry into the forefront of this field were important for the wide acceptance of the program. That is, the early S-matrix program depended for its actual calculations upon relatively simple mathematics (by the standards of modern theoretical physics). This made it possible to recruit and train fairly quickly a sizable cadre of young contributors. Cini (1980) has written on the role of such factors in the history of modern dispersion theory and has argued that they were the *dominant* forces in that history. Such a position does not appear defensible when one considers in detail the rise and fall of the S-matrix project. It ceased to be the major area of activity for theorists because it did not continue to produce new fields of applications at a time when a rival (gauge field theory) was able to do so. A more interesting question is the relevance of a 'Forman-type thesis' (Forman, 1971). The *Forman thesis* refers strictly only to a claim (Forman, 1971) that the intellectual milieu in post World War I Germany conditioned scientists to accept and even, possibly, to fashion the concept of acausality (and, eventually, quantum mechan- ics). We use it in the more general sense of social factors being a determinant of the acceptance of a scientific theory. One can ask why the (radical) *S*-matrix program of the early 1960s had its most solid core of originators located in Berkeley, a center of (radical) social and political activity at the time. Perhaps one *could* make a case for such a connection, but it does seem far-fetched to see that social atmosphere determining the content of high-energy theory. However, the *appeal* of such a program in the Berkeley atmosphere is a more plausible thesis.

Pickering (1984) has presented the process of choice (or judgment) as a largely social exercise. In the tradition of the radical relativist-constructivist program in the sociology of knowledge, he has attempted to show that not only the *form*, but even the very *content*, of scientific knowledge is sociologically determined. His vehicle for this has been a case study of the transition from what he terms the old (pre-1974) physics, which was dominated both in theory and in experiment by soft-scattering (that is, low transverse momentum transfer) hadron (meson–baryon) phenomena, to the new physics, which is concerned mainly with hard-scattering lepton–quark phenomena. It is important to distinguish between the possibility that there could exist an equally successful theory that would be very different from whatever one is currently in vogue, and the thesis that *any* theory can be made to work (Lakatos, 1970, pp. 187–8). It is not really clear just where Pickering places himself between these two points. However, he (1989b) has recently argued for a pragmatic realism in which not just anything goes. He does make a good case that experiment did not *force* a particular theory choice on physicists and that other avenues might have been viably pursued (cf. the comments on this in Cushing, 1986b). We have seen instances of this in our *S*-matrix case study and shall return to this theme in later sections. Galison (1987), in his study of the change in experimental practice in high-energy physics during the twentieth century, argues convincingly that it is not by deductive reasoning alone that scientists pass from the raw data of an experiment to a conviction that an effect has been seen. Nevertheless, he does stress how tight all the constraints become so that there is not a large degree of arbitrariness in the final, sustainable position (once, of course, one has accepted a given general theoretical framework).

9.3 A review of several developments in modern physics

In order to enlarge somewhat the sample upon which we base our conclusions about methodology in science in the next chapter, we now turn to a set of examples of how theories have been selected by scientists. First, there will be several brief and summary illustrations from modern physics, using material from the present case study as well as from others' case studies. A fair number of cases, presented here without great detail, may be of the most use in attempting to see common characteristics of how theories are appraised. We do, however, give in the next section a more detailed specific example from theoretical nuclear physics.

Let us begin by summarizing an earlier attempt (Cushing, 1982) to compare and contrast Kuhn's and Lakatos' models of science by studying developments in the modern theoretical physics of this century. (We have given a brief statement of Kuhn's and of Lakatos' positions in the Preface of the present book.) This was an historical outline not subscribing to either methodology. (Detailed references can be found in Cushing, 1982, and we do not repeat them in this summary.) A brief sketch (see Sections 4–7 of Chapter 1) of this history, from 1930 on in terms of quantum field theory (QFT) and S-matrix theory, indicates a reasonable *prima facie* case for competing research programs in the sense of Lakatos' methodology of scientific research programs (Cushing, 1982, p. 7). After we have highlighted several points from this study, we return to the question of Kuhn versus Lakatos.

We start with the development of the old quantum theory from classical mechanics. In the formulation of these interim theories, empirical success alone was not enough, as illustrated by Haas' 1910 derivation of an expression relating the electron's charge (e), the electron's mass (m), Planck's constant (h) and Rydberg constant (R). His argument applied only to the ground state of hydrogen (Hermann, 1971, p. 94), while Bohr's famous 1913 semiclassical model for the atom had immediate extension to any allowed orbit for the electron. Here we see the importance of scope in a theory and of its ability to lead to further work. This characteristic of *fertility* in a successful theory has been stressed by McMullin (1976b), who has himself used Bohr's model to illustrate such fertility. McMullin distinguishes two types of fertility: P-fertility and U-fertility. From him it is the *proven* fertility (or P-fertility) over the long run that is important for the acceptance of a theory. That is, the theory by then has established a good track record of generating new and successful avenues of research. On the other hand, *unproven* fertility (or U-fertility) is the potential for avenues of fruitful research that the theory promises or suggests. Nickles (1987) has referred to a similar concept as generative potential, the ability to generate new problems and areas of further research. It is U-fertility, or generative potential, that is important during the developmental phases of a theory. If researchers do not perceive this promise in a theory, they may never pursue it. The relevance of P-fertility will not arise. This Haas–Bohr example illustrates two theories with greatly different U-fertilities. Such fertility confers practical advantage on a theory for its own survival.

Bohr's quantization, in his 1913 paper, of the angular momentum of the electron was not the radical break with accepted physics of the time that it is often portrayed to be. In fact, his first argument in that paper has nothing to do with quantizing the angular momentum, but rather quantizes the energy of the emitted light, in a somewhat *ad hoc* manner, yet still very much in the spirit of Planck's by-then-accepted program. Only later in this same paper did Bohr show that this first procedure is equivalent to the now-familiar quantization of the angular momentum. This episode illustrates the importance of argument by analogy, the need to cement even new departures into a presently accepted background and the high degree of continuity in what may appear retrospectively as a discontinuous development. The rationality of such developments, or of moves by creators of new theories, lies largely in the continuity with accepted practice and ideas. Similarly, the Sommerfeld–Wilson quantization integrals for canonically conjugate variables were accepted not just because they were successful but also because they appeared reasonable in terms of Ehrenfest's work on adiabatic invariants (a set of quantities that remained constant under a specified set of modifications of mechanical systems), a concept previously developed by Rayleigh within the framework of classical mechanics.

In Section 4 of Chapter 1 we outlined the important role played by analogy and model building in the Heisenberg and Schrödinger developments of quantum mechanics (see Cushing, 1982, for more details). We now highlight some of the developments of quantum field theory. A prominent feature of the QFT story is the degree to which new ideas were suggested by the mathematical formalism itself. That is, in large measure, the mathemtics (or mathematical formalism) now led the physics (or physical intuition), just the reverse of what often occurred in previous eras. The central importance of one person, Dirac, is evident throughout the early development of quantum electrodynamics (QED), which Dirac himself formulated in 1927.[3] Soon the theory was driven not by any direct demands of empirical adequacy but by the need to make the theory internally consistent; namely, to rid the theory of the infinite quantities to which its calculations led. It was the presence of these infinities or divergences that motivated Heisenberg in the early

1940s to propose an alternative theory, the S-matrix theory (SMT), which he hoped could be based only on observable quantities.

Heisenberg's original motivations were largely 'theoretical' as opposed to being experiment driven. Properties (such as analyticity) found in specific models formed the basis of principles that were taken to hold generally. After the War, Heisenberg seems to have lost touch with major developments and could no longer command the respect (even fear) among colleagues that he had previously. Heisenberg's interest in the program waned, both because he could not find a way to overcome a difficulty SMT had with a general principle (causality) in specific models and because of the great calculational success of the rival QFT program once renormalization had been made a working tool of QED. But, even though the S-matrix program was soon dropped as a subject of active research, one of its offshoots was important for providing a background for the highly successful dispersion-theory program of the late 1950s and the 1960s. This in turn was instrumental in leading to a revival of S-matrix ideas in the 1960s. This SMT program had largely died out by the 1970s, not because it had been proven wrong, but because it no longer led to new avenues of experimental research, whereas rival gauge field theories did. SMT had become a camp follower. Nevertheless, it was instrumental in leading to duality models and to today's superstring theories.

The entire S-matrix and dispersion theory program of the late 1950s and 1960s is characterized by an extremely pragmatic attitude in which motivations for key assumptions were often based on specific models and can be seen in terms of an ability to 'get on with things' in the sense of allowing the correlation of more data or the suggestion of a new experiment. There was a strong interplay between theory and experiment, providing good examples of what Pickering (1984) has termed the symbiosis between theory and experiment (see also, Hones, 1987, who has examined in detail the methodological relations between theory and experiment in the case of the neutral weak currents). Furthermore, from the point of view of the philosophy of science, the radical bootstrap program of the 1960s S-matrix theory, associated largely with Geoffrey Chew, provides a fascinating example, in terms of the type of explanation science could offer, for what could have been a significant change in the *methodology* (or even possibly the goal) of science (Cushing, 1985), a point we return to in Section 10.2. The bootstrap conjecture was an expression of the hope that the S-matrix equations, based largely on the principle of unitarity (or conservation of probability), might be so restrictive that, once partial input had been given from experiment, only *one* solution (for *all* physical properties) would be consistent. The universe (in its *constituents*, not in its *history*) would be unique. It would be the way it is because it *could not* be otherwise. On this same point of a new methodology, let us just mention the anthropic cosmological principle (Gale, 1986; Barrow and Tipler, 1986; Wilson, 1989). Some forms of this principle attempt to explain the past evolution of the universe *in terms of* the eventual appearance of intelligent life. These are relevant for the claim that methodology and goals, as well as practice, are in principle open to change in science (Laudan, 1984a; Cushing, 1984).

Let us return now to the triumphant QED program just after World War II. Again, there was stimulus for the renormalization program provided by new experimental measurements. Renormalization, which is a program for eliminating or ignoring the infinities in a quantum field theory (Teller, 1988, 1989a), was discovered in QED (a specific model) and then *required* as a general guide to discovering possibly admissible theories. It is quite clear that the renormalization program was accepted mainly because it allowed one to make calculations which led to incredibly accurate agreement with experiment. The history of renormalization 'proofs' is interesting, since later proofs cover 'holes' in earlier ones. Nevertheless, the methods of renormalization were used immediately, even before the 'good' proofs existed. The techniques *worked* and the proofs allowed one to believe that the whole procedure might make sense. The details of the proofs are so complicated that it is unclear how many people who use the techniques could claim to understand these proofs completely (not just in outline or in principle). Furthermore, these renormalization proofs can be carried out only in perturbation theory and there were known since the 1950s a class of nonperturbative (Landau) singularities that should invalidate the perturbation expansion at high enough energy. In practice, this difficulty was put aside and the (highly successful) perturbation calculations were carried on. It is only lately, with the advent of superstring theories which may provide a way to eliminate these Landau singularities, that such nonperturbative singularities are now much in evidence and heralded as a victory for superstrings (Cushing, 1988). That is, a problem was forgotten until it became a success. Furthermore, the real values of superstring theory may lie in its resolution of fundamental difficulties such as these rather than in an ability to make direct, experimentally verifiable predictions.

By the late 1950s it had become clear that standard QFT could not be used to make believable calculations for the strong interactions. It was then that the dispersion and S-matrix theory programs enjoyed much success. However, by the 1970s, QFT made a strong comeback in the form of gauge field theories, both as the unified electroweak model and as a quantum chromodynamics (QCD) for the strong interactions. In 1954 the gauge principle had been put forth by Yang and Mills for theoretical reasons (not driven by experiment). This theory predicted strongly interacting *massless* particles, of which none are known to exist experimentally. For years this theory remained an elegant, but apparently useless, curiosity. Once a clever mathematical trick (the Higgs mechanism) had been found to get rid of the massless particles and these gauge theories had been proved renormalizable, gauge field theories became the center of activity. This construction of usable gauge field theories is more readily represented in terms of creation by scientists rather than of discovery. Interestingly enough, QCD, which by general consensus is the theory of choice (over its vanquished rival SMT), accounts for a set of experimental results *different* from those that SMT was able to handle. That is, the successor theory has not captured all the successes of the previously favored theory. Pickering (1984) has argued that this entire process of constructing gauge field theories illustrates scientists' convincing themselves of the reality of entities (e.g., quarks) as a result of the (experimental) practice they chose to engage in (i.e., the types of events they studied), and not *vice versa*. Experiment did not serve as the wholly impartial arbiter of theory. The symbiosis between the two was important. We might add that the tremendous momentum of practice (i.e., 'getting results') in 'big' science today may be exacerbating certain undesirable aspects of this dependence, thus producing an essential change in the scientific enterprise (Cini, 1980; Schweber, 1989a).

Certainly, the history of the development of and the competition between QFT and SMT can be *fitted* with Kuhn's or with Lakatos' descriptions of change in science (Cushing, 1982). However, if one sees as essentially important the contributions of the few creators of scientific theories, rather than those of the many followers, then the spirit of Lakatos' model may seem a bit more faithful. Still, though, neither model is *refuted* by such a case study. In fact, each seems to stress too little the complexity of actual science. There does not appear to be a fixed framework within which to fit scientific practice. The means used to establish a given theory vary from one episode to another and it is unclear that a (finite and small) list of means can be given which fairly characterizes *all* of science. Whatever is necessary and acceptable (to the community) is done. In this section we have offered only a sketchy recapitulation of a competition between two major modern theories. Perhaps a little more detail about a theory of more modest proportions would be helpful.

9.4 An illustration: the compound nucleus model[4]

As a specific illustration of some of these mechanisms of theory construction, selection and justification, let us consider the development of the compound nucleus model in a period which spanned roughly the mid 1930s to the mid 1950s. This is not a theory of 'revolutionary' proportions, but neither is it (during the developmental period under discussion) a fully articulated theory characteristic of 'normal' (or routine) science. It is an intermediate case, an example of a fairly specific theory being fashioned and tested within the framework or tradition of (essentially nonrelativistic) quantum mechanics, but with much input from experimental results and with appeal to general principles, some of which (e.g., the causality condition as yielding an upper bound on the speed of propagation of influences) are not strictly required by the theoretical basis provided by the nonrelativistic Schrödinger equation. A very readable review of the main ideas involved is given by Wigner (1955b) and more details, both historical and technical, can be found in another paper (Cushing, 1986c). We use a modern, uniform notation, corresponding fairly closely to that found in Blatt and Weisskopf's (1952) textbook, *Theoretical Nuclear Physics*.

We restrict our discussion to a class of two-body reactions of the form

$$a + X \quad \rightarrow \quad Y + b \quad . \tag{9.1}$$

α channel $\quad \beta$ channel

Here the reaction is initiated in the entrance channel (α) when a beam of projectile particles (a) is incident upon a target consisting of nuclei (X). The reaction products, a recoil nucleus (Y) and an emitted (or scattered) particle (or photon) (b), appear in the exit channel (β). This is illustrated in Figure 9.1. A specific example would be a neutron (n) projectile incident upon an aluminum (Al) target. An experimentally measurable quantity of great importance is the cross section ($\sigma_{\alpha\beta}$) for the reaction of Eq. (9.1)

$$\sigma_{\alpha\beta} = \frac{\text{number of events in channel } \beta \text{ per unit time per nucleus X}}{\text{number of incident particles a per unit area per unit time}} .$$

$$\tag{9.2}$$

The total cross section σ_t is the sum of the elastic scattering $(a+X\rightarrow a+X)$ cross section σ_{sc} and the reaction (or inelastic) cross section σ_r as

$$\sigma_t = \sigma_{sc} + \sigma_r. \tag{9.3}$$

Bohr (1936; Bohr and Kalckar, 1937) suggested that the reaction (9.1) be pictured as taking place in two distinct and *independent* stages: (i) the formation of a compound nucleus C by a and X of the entrance channel α, (ii) the subsequent decay of this compound nucleus C to produce the products b and Y of the exit channel β. The nuclear force is assumed to have strictly finite range (r_0) and the nucleus itself is treated as a 'black box' of size r_0. The 'Bohr assumption' (i) and (ii) is formalized by the *Ansatz*

$$\sigma(\alpha, \beta) = \sigma_c(\alpha)G_c(\beta) \tag{9.4}$$

where $\sigma_c(\alpha)$ is the cross section of the formation of the compound system C and $G_c(\beta)$ is the probability for C to decay into the products of channel β. For a large class of nuclear reactions (but not *all* nuclear reactions), the cross sections exhibit resonances, or peaks, at various bombarding energies E of the projectile as shown in Figure 9.2. That is, the compound nucleus C is formed only in certain discrete energy states, corresponding to its own energy levels.

This resonance, or 'bump', structure of the cross sections was the occasion for an important analogy in the history of this model. It had long been known classically (Drude, 1900) and had more recently been shown (in the now classic papers) by Weisskopf and Wigner (1930) and by Weisskopf (1931) that a charged atomic system interacting with an electromagnetic field (e.g., a light wave) emits radiation (or photons) whose intensity $I(\omega)$ is distributed according to the resonance formula

$$I(\omega) = \frac{\text{const.}}{(\omega - \omega_0)^2 + \gamma^2/4} \tag{9.5}$$

where ω_0 is the resonant frequency of the atomic system and γ is the

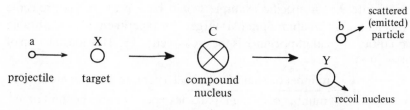

Figure 9.1   A scattering event *via* the compound nucleus.

width of the line or peak in the spectrum. This is illustrated in Figure 9.3. When the cross sections for the scattering of neutrons by nuclei were found to have a resonance structure, it was natural to associate these peaks with nuclear resonances. The basic analogy made was that the incident neutrons are like the photons and that there are quasi-

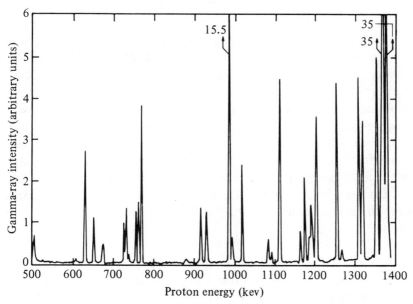

Figure 9.2   Nuclear cross section resonances (after Blatt and Weisskopf (1952), p. 486).

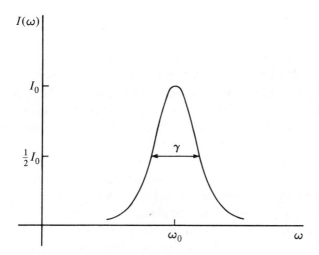

Figure 9.3   A typical resonance curve.

stationary energy levels of the system (nucleus plus neutron). On the basis of this analogy Breit and Wigner (1936) wrote down a formula for the cross section for the transfer reaction from channel α to channel β. It had the structure of Eq. (9.5) and proved very successful for fitting experimental data.

However, the basis of the Breit–Wigner formula was certainly open to question, since the original system (photon–electron) is a weakly interacting one (*via* the electromagnetic force) and this assumption had been essential for obtaining the Weisskopf–Wigner (1930) result, whereas the neutron–nucleus system is a strongly interacting one (*via* the nuclear force) where a similar approximation in the calculation could not be made. Kapur and Peierls (1938) and Siegert (1939) used the one-particle Schrödinger equation to provide the derivation of a resonance formula without the assumption that the interaction is weak. This derivation did not *require* pronounced peaks in the cross sections, but it did *allow* them in a natural and general way. In terms of the nuclear wave function $\psi(r)$, all of the cross sections could be obtained from the derivative function $R(E)$ defined as

$$R(E) \equiv [\psi(r)/\mathrm{d}\psi(r)/\mathrm{d}r]_{r=r_0} \qquad (9.6)$$

evaluated at the nuclear surface $r = r_0$. This $R(E)$ characterizes *all* of the effects of the nuclear force upon the measured cross sections. This derivative function was used by Wigner and Eisenbud (1947) and was shown, on the basis of the Schrödinger equation, to have the *exact* expansion

$$R(E) = \sum_\lambda \frac{\gamma_\lambda^2}{E_\lambda - E} \qquad (9.7)$$

where the γ_λ are real constants independent of the energy E. This structure for $R(E)$ implies a Breit–Wigner resonance structure for the cross sections. Schützer and Tiomno (1951) indicated that the resonance form of Eq. (9.7) might follow from a causality condition (i.e., a first-signal principle) and Wigner (1955a) made this connection apparent (see Section 2.6). Here we have an example of justifying a theory by making it seem *required* by what we already accept. Nickles (1987) has termed this 'discoverability'. The strong appeal of such (after the fact) arguments would appear more psychological than purely rational since such 'proofs' have a large heuristic element in them.

What are some of the relevant lessons we can draw from this brief summary of one episode in modern physics? Well, there is certainly little reason to speak of the compound nucleus model as having been *justified* by the empirical data. After all, the resonance structure of the data *suggested* the analogy with resonance phenomena in light scattering and this led to the (*ad hoc*) phenomenological Breit–Wigner fit to the data. This blurring of the distinction between discovery and justification has been emphasized by Galison (1983b) in his study of modern high-energy experimental physics. This (physical) analogy and formal manipulation of the Schrödinger equation (an already accepted theoretical framework) provided a plausible, but not compelling, basis for this resonance structure. At the same time, the compound nucleus model provided a *language* with which to discuss, organize and interpret more data, while the resonance model was extended to the situation of overlapping resonances to allow a fit to (or an 'explanation' of) data which exhibited a smooth (monotonic) energy variation rather than any pronounced resonance structure (cf. Wigner, 1955b, for a discussion of this point). Here we see how an initially successful model provides a *vocabulary* for any general discussion of the relevant phenomena. It becomes difficult to separate the 'raw' (uninterpreted) experimental data from a 'picture' (language) that these data seem to support. Another example of a successful theory *defining* the terms of discourse, even for its competitors, was the topological *S*-matrix theory (Section 8.4) being formulated in terms of 'quarks' which underlay a hierarchy. In the present situation, appeal to a generally accepted principle (causality) was then used to justify the expectation of a resonance behavior (in principle, even if not always in a detectable way). A tightly knit interplay among experiment, theory and general beliefs had been partly arranged, partly 'discovered', to cement this model into a stable, accepted configuration. This is an example of bootstrapping knowledge (Shapere, 1984; Nickles, 1987). The bootstrapping process outlined here is much less formal than that suggested by Glymour (1980) is made to appear in his work and our goal is far less than his '... ambition to locate general, content-free principles determining the relation of evidence to theory' (p. 8).

The status of causality in this episode, and in modern theoretical physics generally, provides a nice example of the evolution of the meaning of a scientific term. Although causality in the simple sense of determinism ceased to be a requirement of fundamental physical theories after the acceptance of quantum mechanics, a modified concept of causality continues to be an important constraint on acceptable physical theories. We do not have in mind here Planck's (1949, p. 136) observation in his essay on 'The concept of causality in physics' that '... the wave function ... is fully determined for all places and times by the initial and boundary conditions ... .' In this view it is simply the connection between this determinate entity (the wave function) and the observables (say, position and momentum) that no longer yields the old causal picture for the time evolution of these observables. Rather, the causality that has survived is that of special relativity, which we may simply identify with a first-signal principle. That is, if two observers are separated by a distance l and if the speed of light is c, then no event at one of these positions can produce an effect at the other position before a time l/c has elapsed. This is some type of necessary condition for a causal theory. It is not *a priori* obvious that such a requirement would *have* to be consistent with *nonrelativistic* quantum mechanics. In fact, there is current debate among philosophers of science (Shimony, 1978; Redhead, 1983; Cushing and McMullin, 1989; Teller, 1989b) whether such a 'peaceful coexistence' of relativity theory and quantum theory is possible. Nevertheless, usage of 'causality' by the scientific community has assumed that no such conflict exists.

As we have just seen, the early stages of the development of the compound nucleus model were strongly guided by the experimental data and by the analogy these data suggested to previous models of scattering phenomena for very different physical systems. This initial epistemic justification of the compound nucleus model was provided by the empirical adequacy of that model. But that still left unexplained *why* the model was so successful – that is, the *foundation* of the model itself. Wigner (1955b, pp. 376–7) described this development as follows:

> First, the ephemeral and the general elements of the theory are separated. Second, the general elements are formulated in such a general way that they can serve as a framework for the description of a wide variety of phenomena. This, unfortunately, also means that the physical content of the framework becomes so small that almost any experimental result can be fitted into it. The interest, therefore, shifts away from the framework to the body of information which can be described in the language given by the framework. Third, an attempt is made to deduce from very general principles the physical content which remains in the framework. The story of almost every physical theory goes along this way; some run through it fast, the more important ones less rapidly.

This last, but important, phase of the justification of the compound nucleus model had nothing to do with empirical adequacy. This move is an attempt to make the theory (or model) appear 'obvious' and, in fact, *required* by plausible or nearly self-evident principles. The more convincingly this can be done, the better we *feel* about the security of the theory. This element of theory justification is not peculiar to the present model but is characteristic of theories of broad generality. Chew attempted to do this for the S-matrix program by appealing to a broad philosophical principle ('lack of sufficient reason') to underpin the key assumption of analyticity. In the context of quantum field theory, Steven Weinberg (1977, p. 34) described the goal of physical theories in terms of seeking *necessary* explanations:

> After all, we do not want merely to describe the world as we find it, but to explain to the greatest possible extent why it has to be the way it is.

Similarly, a great appeal of the S-matrix program in particle physics was its promise of providing a unique and necessary description of fundamental processes. An important element in this program, as we have seen, was analyticity. The connection between causality (in the sense of a first-signal principle) and analyticity is an old idea dating from the work of Kramers (1927, 1929) and of Kronig (1926) for classical electromagnetic theory. The important extension to relativistic dispersion relations was made when Gell-Mann, Goldberger and Thirring (1954) showed how the desired analytic properties could be obtained from the so-called microcausality condition which requires that the commutator of the quantum field operators $\phi(x)$, $\phi(y)$ vanishes when the space-time points x and y are spacelike separated (i.e., $|\mathbf{r}_1 - \mathbf{r}_2| > |t_1 - t_2|/c$),

$$[\phi(x), \phi(y)] = 0, \quad (x - y)^2 < 0. \tag{9.8}$$

Although we might *expect* that the causality condition (in the sense of a first-signal principle) could be implemented through Eq. (9.8) as a reflection of the independence of observables at spacelike distances, a logically compelling argument for the connection between causality and the commutator condition does not seem to exist. (In fact, much of the recent discussion about nonseparability and instantaneous influences (but not controllable effects) in quantum mechanics would seem to raise doubts about the usual reasons given for Eq. (9.8); cf. Cushing and McMullin, 1989, for various views on the separability question.) The main justification of this new criterion for a causal theory is that it produces a desired result (dispersion relations), while not obviously conflicting with a more direct notion of causality.

The story of these episodes in modern physics illustrates that the pragmatic meaning of the term 'causality' has changed significantly in the context of the applications made of the principle of causality. Rather than meaning that every effect has an identifiable cause whose speed of influence is bounded by the speed of light, it has actually become a statement of non-acausality. Appeal to this principle has been important in providing epistemic support for theories (or programs), even though it is not really clear what the logical status of causality should be in a quantum theory. Indeed, there has at times been the suggestion that the source of the divergence problems in quantum field theory may be the requirement of microcausality. Still, recourse to some, even vauge, notion of causality continues to have great appeal and effectiveness in providing support for a theory. It is not just the logical structure of a final theory that is the basis for acceptance or rejection (assuming, of course, its empirical adequacy has already been established). One obvious function of appeal to such an overarching principle is to provide support for the necessity or uniqueness of the law (a *result* that would be very conducive to a realist interpretation of the law). Grounding a theory on such a principle helps to cement the result into an accepted, stable framework. Such a (relatively) stable framework is the knowledge that the scientific enterprise gives us. Even if in the process we do learn something about the real (micro) world, how significant this is for supporting a (scientific) realist position with regard to these laws remains quite an open question. That is, the success of a theory and its conforming to our general, overarching expectations (such as causality) do not drive us into the realist camp.

Nersessian (1984) has also illustrated in a very different context how a scientific theory develops its own concepts and the meanings of the terms it employs. Her concern is the way in which scientific theories and the meanings of their central concepts are *constructed* (not discovered) in actual scientific practice. Her vehicle for addressing this issue is a detailed study of the development of one key construct in the physics of the nineteenth and early 20th centuries: the concept of the electromagnetic (and, to a lesser extent, the gravitational) field. The modern field concept began with Faraday's intuitive and quite graphic notion of lines of electric and magnetic force. Maxwell gave a mathematical formulation of lines of force as states in a mechanical aether. Although Lorentz did not accept Maxwell's literal picture of the aether, he did take the concept of an aether quite seriously as necessary for understanding fields. By the time Einstein had done away with the aether, physical fields had an existence of their own. The concept of 'field' had evolved into something that would probably have been incomprehensible to Faraday. It has undergone equally dramatic changes to encompass the quantized fields of today's theories. As Nersessian herself puts it (1984, p. 143) '... scientific concepts do not arise in some "creative leap";

rather, they are constructed from a changing network of beliefs and problems'. This outlook is consistent with the view of science that emerges from our present examination of modern physics. Scientific theories are created and constructed (as opposed to being discovered). They are developed, selected and justified within an evolving framework that is not uniquely demanded by nature.

9.5 Causal quantum theory

As a plausible case for the existence of empirically equivalent but not equally well accepted scientific theories, we outline here an interpretation of quantum mechanics that differs radically from the standard, 'Copenhagen' one. As we shall see, it can arguably be taken as a candidate for a viable, fruitful theory. One must ask why it has not been seriously studied by the physics community since there exists such a causal interpretation of quantum mechanics, one version of which is empirically indistinguishable from the (nonrelativistic) Schrödinger equation. That theory requires considerably less drastic departures from or revisions of the epistemology and ontology associated with classical physics. This more nearly traditional theory can be traced to its origins in the mid 1920s.

In 1927 Louis de Broglie proposed a 'principle of the double solution', according to which he suggested a synthesis of the wave and particle nature of matter. At the Fifth Solvay Congress in 1927, he presented (1928) some of these ideas in a form he termed the pilot-wave theory. Here a physical particle is pictured as being guided by the pilot wave. In discussion at that Congress, Wolfgang Pauli (1928) criticized de Broglie's theory on the basis of the example of the inelastic scattering of a plane wave by a rigid rotor. Although de Broglie felt he understood the general outlines of a suitable response to Pauli's objection (de Broglie, 1960, p. 183; 1973; Bohm and Hiley, 1982, p. 1003), he in fact did a poor job in attempting to rebut Pauli at the 1927 Congress so that his response appeared *ad hoc* and was not convincing then (Jammer, 1974, pp. 113–4). In addition to Pauli's negative reaction to de Broglie's paper, neither Einstein nor Schrödinger gave positive support to de Broglie's ideas: Einstein because he did not like the nonlocal (or nonseparable) nature of the theory and Shrödinger because he wanted a theory based only on waves (not on waves *and* particles). Furthermore, de Broglie had the reputation of being an unorthodox theoretician (MacKinnon, 1982, p. 216) and that did not condition the scientific community to consider his ideas seriously. Also, while the German physics community may have been prepared 'socially' to abandon causality (Forman, 1971), the French scientific community had isolated the Germans for reasons of nationalism after World War I (Mehra and Rechenberg, 1982, pp. 578–604). In addition to these factors, the people who were producing results (e.g., Heisenberg and Born who spoke at the 1927 Solvay Congress) strongly favored the indeterministic or noncausal picture (de Broglie, 1973, p. 16). Bohr was for a long time against the concept of the photon (Slater, 1973, p. 23), so that de Broglie's ideas had never spread rapidly in the Copenhagen school. The Institute at Copenhagen was a very closed community and those invited there were identified as the 'respectable' theorists. De Broglie was never a member of this group. By 1930 when he wrote a *very* standard quantum-mechanics book, de Broglie had himself changed his mind about the pilot-wave theory: 'It is not possible to regard the theory of the pilot-wave as satisfactory' (1930, p. 7). In the same book he rehearsed other arguments against the pilot-wave theory, both general conceptual ones (pp. 119–21) and a specific thought experiment involving the reflection of light from an imperfect mirror (pp. 132–3; cf. de Broglie, 1960, pp. 183–4 as well). Von Neumann's (1955) 1932 impossibility 'proof' for hidden variables theories further confirmed de Broglie's position against his own previous theory (de Broglie, 1962, p. 99).

There matters essentially stood until 1952 when David Bohm published two papers on a causal interpretation of quantum mechanics. Initially, de Broglie was against Bohm's ideas (which were similar to his own pilot-wave theory of 1927) and he raised the same objections against Bohm's theory that had been raised against his own (Bohm and Hiley, 1982, p. 1003 and pp. 1014–15). In an appendix to his (1952) papers, Bohm explicitly replied to Pauli's and to de Broglie's objections and showed them to be specious (cf. Bohm and Hiley, 1982, p. 1014–15). Not only did Bohm (1952, 1953) produce a causal version of quantum mechanics – one capable of a *realistic* interpretation with a largely classical (micro) ontology – but he and Jean-Pierre Vigier (1954) produced an explicit hydrodynamic model to underpin the Bohm formalism and interpretation. Their model has a certain resemblance to much older work by Madelung (1926). They (Bohm and Vigier, 1958) subsequently extended this model to a relativistic fluid to encompass the (relativistic) Dirac and Kemmer equations for particles with 'spin' (in the standard formulation of quantum mechanics). Bohm's work of the early 1950s reconverted de Broglie to his former ideas (de Broglie, 1962,

p. vi). For de Broglie the issue at stake was not (classical) determinism, but rather the possibility of a precise space-time representation for a clear picture of microprocesses (pp. vii–viii). In this, his expectations were similar to Einstein's. He felt (1973, p. 12) that classical Hamilton–Jacobi theory provided an embryonic theory of the union of waves and particles, all in a manner consistent with a realist conception of matter. With the concept of the quantum potential (de Broglie, 1970, p. 8), one could, as Bohm and Vigier had actually done, provide a *model* for 'elementary' particles.

Bohm's (1952) basic idea is the following. Beginning with the Schrödinger equation (which is *accepted, not derived*, there)

$$ih\frac{\partial\psi}{\partial t} = \frac{-\hbar^2}{2m}\nabla^2\psi + V\psi \tag{9.9}$$

one defines two *real* functions R and S as

$$\psi = R\exp(iS/\hbar). \tag{9.10}$$

Substitution of Eq. (9.10) into Eq. (9.9) and separation of the real and imaginary parts yields

$$\frac{\partial R}{\partial t} = -\frac{1}{2m}[R\nabla^2 S + 2\nabla R\cdot\nabla S], \tag{9.11}$$

$$\frac{\partial S}{\partial t} = -\left[\frac{(\nabla S)^2}{2m} + V - \frac{\hbar^2}{2m}\frac{\nabla^2 R}{R}\right]. \tag{9.12}$$

The *quantum potential* U is defined as

$$U = \frac{-\hbar^2}{2m}\frac{\nabla^2 R}{R}. \tag{9.13}$$

With the definition $P \equiv R^2 = |\psi|^2$, Eq. (9.11) can be rewritten as

$$\frac{\partial P}{\partial t} + \nabla\cdot\left(P\frac{\nabla S}{m}\right) = 0. \tag{9.14}$$

If U were identically zero, then Eqs. (9.12) and (9.14) together would represent a continuous 'fluid' of particles of momentum

$$\mathbf{p} = \nabla S \tag{9.15}$$

following well-defined classical trajectories. The quantum potential U of Eq. (9.13) introduces highly non-classical, nonlocal effects. However, as is evident from the construction of these equations from the nonrelativistic Schrödinger equation (Eq. (9.9)), the predictions of Bohm's theory must be empirically indistinguishable from those of the standard ('Copenhagen') formalism of quantum mechanics. But, Bohm's theory is based on the existence of an objective (observer-independent) physical (micro) reality. There is no in-principle 'cut' between the micro and the macro (or between the system and the observer) and no reduction (or 'collapse') of the wave packet. Still, the physics community has not shown interest in or sympathy toward the causal interpretation of quantum mechanics. 'Copenhagen' got to the top of the hill first and, to most practicing scientists, there seems to be no point in dislodging it. Furthermore, senior members of the majority school even have apocryphal versions of Bohm's own convictions about quantum mechanics (Ne'eman, 1980, p. 267; Bohm and Hiley, 1982, pp. 1014–15).

Edward Nelson (1966, 1967) argued that particles subject to Brownian motion, with a diffusion coefficient $(\hbar/2m)$ and no friction, and responding to imposed forces in accord with Newton's second law, $F = ma$, obey (*exactly*) the Schrödinger equation. Although there is randomness, a radical departure from classical physics is unnecessary so that the resulting theory is probabalistic in a *classical* way (Nelson, 1966, p. 1084). William Lehr and James Park (1977) generalized Nelson's work to the relativistic case. With this, with Bohm–Vigier and with Vigier's program to provide an understandable (picturable) model for quantum phenomena, the aether had returned! (There are interesting parallels between Lorentz versus Einstein with special relativity and Bohm–Vigier versus Heisenberg–Schrödinger with quantum mechanics.) This alternative program has thus shown a great deal of fertility for generalization within its own resources, not just as *ad hoc* moves. That is, *if*, say in 1927, the fate of the causal interpretation had taken a very different turn and been accepted (over the 'Copenhagen' one), it would have had the resources to cope with spin and other generalizations essential for a broad-based empirical adequacy. We could today have arrived at a *very different* world view of micro-phenomena. If someone were then to present the (merely) empirically equally as adequate Copenhagen version, with all of its own counterintuitive and mind-boggling aspects, who would listen!

That is, a highly 'reconstructed' but entirely plausible bit of history could run as follows (all around 1925–27). Study of a classical particle subject to Brownian motion leads to a 'classical' understanding of the already discovered 'Schrödinger' equation, which is then given a 'deBroglie–Bohm' realistic interpretation. A 'Bohm–Vigier' model underpins this interpretation with a visualizable physical model of micro-phenomena. These models and theories are generalized to include relativity and spin. The program is off and running! It is essential to appreciate that this 'story' is neither *ad hoc* (in the sense of these causal models having as their sole justification an origin in successful results of a rival program) nor mere fancy, since all of these developments exist in the physics literature.[5]

David Bohm himself (1987, p. 39) recently speculated on why a causal theory of quantum phenomena, in spite of being just as empirically adequate as the standard Copenhagen one, has not been actively pursued by a larger part of the scientific community.

> These proposals did not actually 'catch on' among physicists. The reasons are quite complex and difficult to assess. Perhaps the main objection was that the theory gave exactly the same predictions for all experimental results as does the usual theory. I myself did not give much weight to these objections. Indeed, it occurred to me that if de Broglie's ideas had won the day at the Solvay Conress of 1927, they might have become the accepted interpretation; then, if someone had come along to propose the current interpretation, one could equally well have said that since, after all, it gave no new experimental results, there would be no point in considering it seriously. In other words, I felt that the adoption of the current interpretation was a somewhat fortuitous affair, since it was affected not only by the outcome of the Solvay Conference but also by the generally positivist empiricist attitude that pervaded physics at the time.
>
> In addition, it was important that the whole idea did not appeal to Einstein, probably mainly because it involved the new feature of nonlocality, which went against his strongly-held conviction that all connections had to be local. I felt this response of Einstein was particularly unfortunate, both during the Solvay Congress afterwards, as it almost certainly 'put off' some of those who might otherwise have been interested in this approach.'

Finally, there is a paradoxical (even slightly inconsistent) aspect to the attitude of many physicists in the following regard. Heisenberg–Schrödinger is preferred over Bohm–Vigier because the former got there first, so that the latter is *ad hoc*, being empirically indistinguishable from the first and 'merely' offering the bonus of an understandable (or, really, less puzzling) picture of the world. Yet, the formalism of quantum field theory (QFT) is claimed to be more picturable (especially with the artifice of Feynman diagrams) and to have a richer ontology than a former rival, the *S*-matrix program

(SMT), which is a more nearly instrumentalist program. But, in the opinion of most physicists, QFT works better (i.e., has had a larger domain of successful empirical applications) than SMT (which, however, is *not* a refuted program). So, in fact, being empirically adequate is most important to scientists and being first seems only somewhat less crucial for the entrenchment of a theory.

9.6   Summary

In this chapter we have focused on methodically relevant issues in several developments in modern theoretical physics. There has been a back and forth between the paradigm of quantum field theory (in which the field concept is primary) and the S-matrix one (in which particles are the natural basic entities). It was the failure of QFT (in the late 1930s and early 1940s) that opened the door for SMT as a possible alternative. Each subsequent shift from one of these paradigms to the other was occasioned by serious difficulties encountered by its competitor. The physics community, by and large, stayed with the successful program and shifted only when forward progress became difficult. Empirical success and fertility for future work were both necessary requirements to be met. In spite of its ups and downs, the S-matrix program has continued to play an important role in the development of theoretical high-energy physics leading, through the S-matrix duality program, to today's superstring theories.

A characteristic common to all of the episodes we have discussed is the origin of new (physical) ideas in highly technical, mathematical developments. Examples of these are Chew's belief in no arbitrariness (in the analytic scattering amplitudes) as an outgrowth of his being impressed early on by Fermi's old model of particle scattering from a compound system; the Mandelstam representation; Chew and Mandelstam's bootstrap mechanism; Regge poles; Veneziano's model that led to the duality program; Nambu's recognition of the connection between dual amplitudes and string field theory; and Green and Schwarz' discovery that superstring models could include gravity. A recycling of formal analogies was often central to such exercises. General principles were then sought to underpin these successes. We have illustrated how causality was used to underwrite dispersion relations and the Breit–Wigner resonance formula, just as Chew appealed to a general philosophical principle ('lack of sufficient reason') to justify the analyticity assumption of an S-matrix program that would be independent of quantum field theory.

The dispersion-theory and *S*-matrix theory programs were character-ized by a very pragmatic attitude in which a premium was placed on immediately useful techniques and models of the analysis of experi-mental data. The relative ease (compared to much of QFT) of many of the early calculations required for application provided rapid entry into an active area of research for many young workers. This was an important sociological factor in the explosive development of the program. Chew's personal influence on the field is another instance of the crucial role of a few *individuals* (as opposed to the collective effort of large groups) for the initiation and development of key ideas, just as had been the case with Dirac for the early days of QFT. There was an ebb and flow in which theory sometimes led the way and at other times experiment.

Dominant models (e.g., the compound nucleus model, QFT versus SMT today, the Copenhagen interpretation of quantum mechanics versus the Bohm–Vigier causal one) establish the language employed in discourse. That language itself further entrenches the favored theory, colors the interpretation of data for its own support and must often be used even by competitor theories that have radically different concep-tual frameworks. It is through the *use* of a successful theory that a language and the meanings of terms used (such as causality) are constructed. In this complex process, the construction and justification of a theory become inextricably linked, both temporally and logically. A study of empirically successful, fertile and viable theories that are rejected points to a role for factors external to science in theory selection. This is not meant to imply that such theory choice is irrational, capricious or wholly arbitrary. We discuss this issue in the next chapter.

10
Methodological lessons

Here we present some philosophical lessons drawn from, and discussion of, questions raised by this case study. The previous chapters ought to be able to be taken or left independently of this chapter, but not *vice versa*. That is, some familiarity with the facts of the case study is necessary for an appreciation of the argument of this chapter. We point out specific features of our story that are relevant to constructing and evaluating a methodology of scientific practice. The discussion is not based only upon or restricted exclusively to the *S*-matrix case study. Several methodologies currently on offer are reviewed in the light of that case study and of the other episodes in modern physics reviewed in the previous chapter. We argue for a multifaceted and evolving description or model of scientific practice.

Since there is a good deal of over and back between the specific case studies and general methodological issues in this chapter, it may be helpful to outline here the structure of the argument to be followed. Some general characteristics of science, as identified by modern philosophers of science, are first outlined (Section 10.1) as a preliminary to comparing them with those features of scientific practice found in the SMT study and in our summary of other major episodes in modern physics in Chapter 9. In Section 10.2 the recent works by Pickering (1984) and by Galison (1987) are used to illustrate the role sociological factors have played in scientific practice in high-energy physics (HEP), both theoretical and experimental. Then (Section 10.3) characteristics common to such episodes (as, for example, SMT, QFT and the compound nucleus model) are identified that mitigate against any straightforward belief in the discovery of fixed, universal criteria unique to science and that speak in favor of an ever-expanding and changing set of criteria. Specific candidates for changes in the methods and goals of science are identified in the S-matrix program, in the cosmological anthropic principle and in the type of explanation that quantum mechanics allows us (Section 10.4). In Section 10.5 we question the plausibility of the reliability of pictures of the world given by our 'successful' theories, once we have seen their basic underdetermination at a foundational level. But, in spite of this fundamental underdetermination, scientific practice and opinion *do* converge and an explanation for this must be offered (Section 10.6). Finally, we arrive (in Section 10.7) at a rather skeptical view of the degree of trust we should put in the pictures (epistemological and ontological) of the world accompanying our 'best' physical theories.

10.1 Some general characteristics of science

In order to bring the observations of the last chapter to bear on the question of a model of scientific practice, let us begin by discussing in summary some previous and more current pictures of science. An initially appealing view of science can be caricatured as presenting it as basically an unproblematic rational enterprise in which objective truth is sought and discovered. Of course, there are social factors present, but these are not relevant to the essence of science. Such factors may be unavoidable since humans are involved in science, but surely they are undesirable for and extraneous to the scientific enterprise. These influences are to be eliminated from the final, objective product of scientific knowledge. The interaction of the individual with others is an accidental or practical necessity for scientific activity but certainly, *in principle*, science could be done by a lone computer of sufficient intellectual or logical capacity. In spite of this basically hostile view toward social factors prevalent among scientists, there is an uneasy awareness that all might not be well with this folklore version of science. This is reminiscent of the revulsion that was once experienced on knowing that Newton, the embodiment of the age of reason, had shown a genuine and prolonged interest in subjects as irrational and mystical as alchemy and an unorthodox interpretation of the *Book of Revelations*. John Maynard Keynes (1951) in his essay, 'Newton, the man', tells us that Newton had kept these and other unorthodox writings in a large box and that: 'After his death Bishop Horsley was asked to inspect the box with a view to publication. He saw the contents with horror and slammed it shut. A hundred years later Sir David Brester looked into the box. He covered up the traces. ...' (Keynes 1951, p. 317).

Much of the scientific community is genuinely uncomfortable with what it sees as a dark side of science. This is especially true of those engaged in the routine, less creative aspects of science. Today, more reflective scientists do not subscribe to such an uncritical view of science. Still, there is a strong tendency to maintain a separation between the sociological and the intellectual. This is one reason Medawar (1979, pp. 91–4) finds the Popper–Lakatos school an appealing matrix within which to anchor the scientific enterprise. No matter how tangled and messy the path to scientific knowledge may be, the end product is objective and pure (that is, rational).

Nevertheless, attacks on traditional, foundationist approachs to the philosophy of science have come from several quarters. The basic criticism is that purely philosophical programs, such as logical positivism and its descendants, which study rational reconstructions of how science *ought* to function, have simply not been successful in explicating the nature and methods of science as it historically has been practiced and as it continues to be pursued. Scientific models and theories, which result from the interplay of disciplined, creative imagination with the hard boundary conditions provided by natural phenomena, have proven to be so much richer (almost unimaginably so) than unbridled abstract conjecture would ever have led us to believe plausible. *A priori*, apparently self-evident, foundational principles (such as the possibility of causal explanations or a belief in a purely deductive choice among theories) often assumed in the philosophy of science have proven to be essentially incapable of coping with the diversity of the real world. As obvious examples, from the history of science, of conceptual changes, we have non-Euclidian geometries, the possibility that space-time may not be a fundamental structure in nature, but only an approximation, and, most remarkable of all, inherent indeterminacy and the nonseparability of many spatially distant systems, as demanded by quantum mechanics and supported by experiment. We have been forced to learn these hard lessons, the common message of which is that all concepts must be open to revision. There can be no givens that are beyond question. We shall argue that this applies not only to the laws and entities of science, but that it extends also to the rules and methods of evaluating theories. It is this repeated failure – and by so wide of the mark – of foundationist approaches, rather than clever abstract counter arguments, that tells most heavily against them and that recommends strongly that we examine actual scientific practice. Scientific knowledge, as well as knowledge about science, must be bootstrapped, starting from what may seem a plausible position, but always keeping open the possibility (often the likelihood) of fundamental revisions. Part of the difficulty, of course, is that the *method(s)* of science must also be determined in the process. Harvey Siegel (1985), in a recent defence of the traditional grounding of the rationality of science in its method, has taken this method to be a commitment to evidence. That is certainly a major part of the story, but not obviously the entire story. Also, a criterion such as 'commitment to evidence' is both too vague to be of much use and so general as to be of little assistance in distinguishing science from everyday common-sense reasoning. After all, the question is not just what are the characteristics of science, but rather what are the *distinguishing* hallmarks of science that demarcate it from other rational activities.

Not surprisingly, this message – that the methods of the philosophy of science can no longer even in principle be independent of or autonomous from the methods of its subject (science) – has not been welcomed by everyone in the philosophical community. After all, it claims (Gale, 1984; Cushing, 1984) that much of the philosophy of science must look to science for its central concepts, especially as regards epistemology and ontology (as has been the case in quantum theory and as could have happened in a dominant SMT worldview). Charges brought against such a program of naturalized philosophy of science are that it may lead to a total philosophical relativism and that an historical case-study approach may generate only disconnected accounts of various episodes in the history of science which yield no common themes or general insights into the nature of science. While it is *in principle* possible that this *could* turn out to be the case, it does not necessarily follow that it will (a point also made by Shapere, 1984). As with any undertaking in science, one cannot guarantee the outcome prior to an examination of the actual facts. Although it might be more comforting to be able to insure in advance that a certain conceptual matrix must remain invariant, it simply has not proven possible to do this. That observation applies just as well to cherished beliefs about how scientific theories must 'surely' be evaluated as it does to the forced abandonment, say, of a 'sacred' field-theory paradigm in favour of an S-matrix one (and *vice versa*). We do the best we can in regard to maintaining such stability, but our best often turns out to be so much less than we had hoped for. So, we must do the case studies and see what comes of it all. As in science, we must learn how to learn (Shapere, 1984).

Losee (1987) has examined the relation of philosophy of science to the history of science. In particular, he is concerned with assessing how reasonable it remains for 'high' philosophy of science to be, on the basis of evaluative principles, prescriptive of what constitutes 'good' scientific practice and to be a meta-enterprise over and above (mere) descriptive history and/or philosophy of science. While he illustrates that historically philosophers of science have in fact traditionally been prescriptive, he argues that attempts to provide justifications of such standards, either by appeal to necessary principles or to further metalevels, have largely failed. So, in the end, his position appears to be not only that the history of science is necessary for the philosophy of science but even that the philosophy of science is essentially and irreducibly historicist in nature. He is led to a position closer to that of Shapere and of Laudan than to any of the others he criticizes. Such a message is certainly consonant with the position we are advocating.

Nevertheless, it is important to appreciate that the *a priori* or foundationist approach to the philosophy of science has yielded many valuable insights and useful starting points for further investigation. For example, Popper's (1963) emphasis on the method of hypotheses, Lakatos' program of meta-methodologies (research programs and heuristics) (1970, 1976), Salmon's (1984) focus on causal structure in the world and, in a very different vein, Pickering's (1984) thesis of the overriding role of social factors in science have all identified important characteristics of actual scientific practice and argument. The mistake comes when one attempts to use exclusively just one of these characteristics as representing *the* (invariant) hallmark of science (as opposed to nonscience) or of scientific explanation. Scientific practice is multi-faceted and dynamic, so that no fixed, single-component view is sufficient to represent that practice. Several authors have stressed this many-faceted and evolving nature of the scientific enterprise. For reference, later, let us recapitulate a few of these.

Dudley Shapere (1984, pp. xiii–xv) characterizes a central weakness of the logical empiricist program as its focus on the formal logical structure of scientific theories to the exclusion of the process by which these theories were constructed, thus ignoring the possibility of fundamental changes in the nature of science itself. He has stressed the importance of formulating a view of science based on an accurate description of actual scientific practice, which includes attention to how the meaning of a scientific term is rooted in this evolving matrix of practice. Janet Kourany (1982) has attempted '... to lay the foundations for a purely empirical method for establishing a theory of science' (p. 526). In arguing that the *a priori* has no place in such an approach and in responding to the charge that such an empirical method cannot produce an absolutely fixed set of criteria, Kourany (p. 546) reminds us of Popper's observation about another empirical enterprise and suggests that we apply this same admonition to our expectations for an empirical method of constructing a theory of science itself.

> The empirical basis of objective science has nothing 'absolute' about it. Science does not rest upon solid bedrock. The bold structure of its theories rises, as it were, above a swamp. It is like a building erected on piles. The piles are driven down from above into the swamp, but not down to any natural or 'given' base; and if we stop driving the piles deeper, it is not because we have reached firm ground. We simply stop when we are satisfied that the piles are firm enough to carry the structure, at least for the time being. (Popper, 1959, p. 111)

No one method is used exclusively in science to justify the knowledge generated. Thomas Nickles (1987) has long been an advocate of the central epistemic role of the way a scientific theory is discovered and formulated in providing a warrant for the acceptance of that theory. Nickles (1987, p. 48) points out that '... a theory is justified by seeing how to derive it (usually in a highly reconstructed way) from what we already "know" about the world.' That observation is particularly relevant for the principle of causality and its use in justifying dispersion relations. We have examined this at length in our case study and summarized it in Section 9.4. Larry Laudan (1984a) has presented a case for a coupled, triadic model of scientific practice in which the various levels or components of thought and activity are coupled in their evolution so that no level is immune from changes produced at other levels. He also refers to this as a *reticulated* model. None of these components is forever fixed, but each is influenced, and ultimately determined, by practice. In later sections we shall illustrate the multicomponent nature of this practice by discussing the various mechanisms that produce convergence of scientific opinion as that opinion or position evolves over time. Rather than attempting to build a house, we are building a ship, and the structure of that ship is continually changed as the ship sails. Some recent studies of scientific practice in high-energy physics illustrate this.

10.2   The role of sociological factors in high-energy physics

Pickering (1984) has shown the large sociological component present in the symbiotic relation between theory and experiment. The case study he uses is about the transition from what Pickering terms the old (pre-1974) physics, which was dominated both in theory and in experiment by soft-scattering (that is, low transverse momentum transfer) hadron (meson–baryon) phenomena (i.e., those of our *S*-matrix era), to the new physics, which is concerned mainly with hard-scattering lepton–quark phenomena. The old physics is that prior to the November 1974 Revolution, the new that thereafter.

In opposition to what he terms the scientist's account, which presents experiment as being the supreme and compelling arbiter of theory, Pickering stresses the role of the scientist's *judgment* in choosing between (or possibly even among) competing underdetermined theories. He places great emphasis upon the scientist as an active and creative agent in generating scientific knowledge; he opposes this view to a more traditional one of the scientist as a passive observer of nature. This is another attack on the scientific 'textbook' practice of evaluatively rewriting the history of science from the vantage of currently accepted 'correct' scientific knowledge. The circularity he is criticizing is nicely put (p. 18, note 6):

> If and when a consensus is reached, ... particular theoretical constructs become regarded as preexisting attributes of the natural world, independent of the particular arguments and practices implicit in their establishment ... [T]he reality of the theoretical constructs is used to explain the validity of these arguments and practices, and to indicate the invalidity of any arguments or practices which support an alternative construction of reality.

The old physics was dominated by the strong interaction rather than by the weak one and hence hadron beams, typically protons or pions, were much more commonly employed than were lepton (usually electron) beams. The old physics had been faced with a proliferation of strongly interacting ('elementary') particle states and the constituent quark model was an attempt to cope with this proliferation of states by interpreting them in terms of basic constituents, the quarks. Pickering sees one of the great appeals of this model for physics to lie in its recycling of the atomic analogy with which physicists were already familiar from their training. This model allowed theorists to make calculations and hence provide experimentalists with motivation to explore in great detail the low-energy resonance scattering of hadrons.

Pickering focuses on the role of choice when physicists singled out the relatively rare hard-scattering data, which could be obtained only by rejecting the immensely more common background of soft-scattering data. These relatively esoteric hard-scattering events were those for which the gauge field theory could make fairly reliable predictions, whereas the soft-scattering events could not be accounted for in detail by these theories. On the other hand, S-matrix theory had been quite successful with (a different class of) soft-scattering data, but could not make reliable predictions for hard-scattering data. Of course, it is not new to the history of science that experiments and observations concentrate on those phenomena and situations about which current theory can make the clearest statements. In classical gravitational theory, early efforts centered on the one-body (or the equivalent reduced two-body) problem rather than on complex, several-body problems. Even though this is an obvious and rational stratagem for scientists to use, it does still make the point that certain phenomena derive much of their importance (relative to 'uninteresting' phenomena) from the theory under consideration, while the theory in turn is supported by the important or relevant data. This should not be taken as saying that some fraud is being perpetrated by theorists and experimentalists collusively scratching each other's backs. But the choices are there and are essential to the process. This leaves the possibility that things *could* go differently.

The November 1974 Revolution refers to the observation of the J-ψ particle, a narrow, long-lived resonance at 3.1 GeV, interpreted as a $c\bar{c}$ (charm-anticharm) state in the e^+e^- channel which decays into lower-mass hadrons. The exceptional stability of the state was taken to reflect a conserved quantum number c (charm) whose 'naked' charm states were too massive for the J-ψ to decay into directly. Pickering (1984) sees the decision to follow the predictions of the QCD charmonium model of J-ψ as crucial to the eventual dominance of that theory. Thus we read (p. 266):

> If on the contrary, experimenters had oriented their strategy around some other model the pattern of experiments around the world would have been quite different, and the outcome of this whole episode might well have been changed.

And, again (p. 273):

> The key to charm's success lay in the social and conceptual unification of HEP practice which was achieved during the November Revolution. And, as we shall see in the following chapters, progressive social unification in the common context of gauge theory was intrinsic to the establishment of the new physics in its entirety.

Pickering can certainly be read as advocating an 'anything can be made to go' brand of social construction of scientific knowledge, although some of his later writing (e.g., Pickering, 1987, 1989b) considerably tones down this aspect of his picture of scientific practice and attempts to defend him against a charge of total relativism. Nevertheless, our purpose in discussing Pickering's position here is that it is one, based on a study of modern physics, that definitely establishes the important role played by sociological factors in theory construction and selection. It is not necessary that we decide just now exactly how radical his position is. He has at least made a case that things *might* have gone very differently from the course taken after the 'November Revolution'. Earlier in this book, we have presented a similar case for SMT versus gauge QFT and, at least in outline, for Bohm's causal quantum theory versus the Copenhagen interpretation.

While Pickering emphasizes the role of choice in accommodating theory to experiment, Galison (1987) has argued convincingly that modern experimental practice is not parasitic upon theory since experimentalists have their own stratagems for judgement which do not depend upon the details of a theory. (Franklin (1986) has made a similar point.) Galison effectively demolishes both any tight, uniquely compelling, purely logical representation of how choices are made in actual scientific practice and an image of science as a free-floating, anything-can-go activity of largely arbitrary choices. In the process, he again puts lie to the old and artificial dichotomy between the context of discovery and the context of justification of scientific theories. These two contexts are intextricably linked, not just temporally or historically, but more importantly, in the logical or actual process of constructing scientific knowledge. Scientists eventually become persuaded of the reality of an effect, but not by purely deductive argument from their laboratory observations. To lay out how actual experiments are conducted and terminated, and how scientific practice has changed in essential ways throughout the twentieth century, Galison begins with a summary of experimental practice in the nineteenth century, but then concentrates on three episodes in the experimental physics of this century. The first is the series of experiments by Einstein and de Haas, begun in 1914 and subsequently modified by others into the 1920s, to measure the gyromagnetic ratio of the electron. Next is what eventually became the discovery of the muon, work that covered the 1930s and 1940s. Finally, and most extensively, comes the story of how the CERN and Fermilab groups separately established the existence of weak neutral currents and of the W boson in the early 1970s.

By comparing and contrasting experiments on three very different scales (of cost, of complexity, of sheer physical size), Galison illustrates how marked quantitative changes of this kind have produced qualitative changes in the type of argumentation employed to establish the reality of the phenomena the experiments sought to find. In his sociological study of science, cognizance is taken of the complex of factors responsible for producing a convergence of scientific opinion. The presuppositions (both experimental and theoretical) of scientists are often important for deciding when to end an experiment, but they are not uniquely determining nor singularly overriding. Theoretical expectations or guides are crucial, at least in providing order-of-magnitude estimates of effects (either to be looked for as the principal phenomenon or to be ignored as unimportant background). As the size and cost of experimental equipment in modern high-energy physics has increased several fold (even orders of magnitude), varying the equipment and the experiment itself (during execution) has become unfeasible. Hence, much of the argument–counterargument now goes on during the reduction and analysis of the raw data, often long after completion of the actual run of the experiment itself. It is a hunt for an effect that can be buried in or mimicked by unwanted background. The art of data analysis is to peel away the background, to rule out 'faked' events and to produce a tight network of constraints and arguments that result in a stable physical effect. In spite of initial biases and expectations (sometimes contrary to the final, accepted outcome), certain effects can't just be bent to conform to predilections (even of so strong a personality as a Millikan) and can't be made to go away in any consistent fashion. In an assessment of the CERN neutral-current experiment, we read (Galison, 1987, p. 193):

> Thus, using a plethora of approaches, techniques, heuristic arguments, prior data, theories, and models, the members of the collaboration persuaded themselves that they were looking at a real effect.

In this episode of the 'discovery' of the weak neutral current, a longstanding judgment (based on much other, earlier data) against the existence of such neutral currents had to be *reversed*.

There are various levels of constraint (and of theory commitment) that interlock and that must at times be (mutually) adjusted to produce a stability issuing in the verdict that an effect has been 'seen'. This is in some ways reminiscent of the 'consilience of inductions' familiar from an earlier era in the philosophy of science. It is a mistake to have an image of science in which theory is always primary, just as it is equally incorrect to take experiment as primary (Galison, 1988). There is an essential interplay between the two. Some episodes are dominated by theory, others by experiment. Galison (1987) argues that one must hold the floodgates against a tide of relativism that could swamp the objectivity of science. For him, sociological factors are important, but he does come down much closer to the scientific objectivity end of the spectrum than does Pickering. For us, the important point is that there *is* a spectrum of positions each emphasizing in varying degrees the magnitude of the role played by sociological factors in producing accepted scientific knowledge. What does become apparent, though, is that the importance of an inextricable sociological component cannot be denied.

A theme common to these case studies is that any single-component model of scientific practice is inadequate to represent accurately – even in first approximation – the actual functioning of the scientific enterprise. The components are coupled and nothing need remain forever fixed. Schweber (1989) has made a case that can be interpreted as showing that the very nature of scientific practice changed significantly with the advent of 'big science' after the Second World War. This becomes especially apparent in some of the recent large collaborative efforts in high-energy physics in which experiments may come to be performed just *once*. Some indications of the danger inherent in unrepeated experiments is evident in the discovery of the weak neutral currents by the groups in Europe and in America (Galison, 1987). If the American group had been the *only* one doing the experiment and analyzing the data, the final verdict could have been very different from what is accepted today.

To establish the plausibility of, or really the need for, such a model or representation of a scientific enterprise with many interlocking but distinct epistemic factors requires, at a minimum, a detailed study of significant episodes in the history of science. In the present monograph we have examined several features of this evolving scientific practice. In our case study, we have found support for various aspects of scientific practice identified by several philosophers of science and summarized in this section. As one illustration, the program of Regge phenomenology discussed in Chapter 6 is a good example of a degenerating research program in the most pejorative sense of Lakatos' term (Lakatos, 1976; Zahar, 1976). It became so complicated and involved so many (in practice) arbitrary parameters that it lost any defensible claim to be theory and became (mere – though not easy) curve fitting. Some fundamental, unifying insights were provided. True, the program did continue (in part due to the commitment and accumulated expertise of its practitioners who pressed on with this phenomenology), but this was not a 'research program' as an undertaking that might reasonably be expected to lead to a basic theory. So, it is surely an overstatement to claim that just *any* theory can be made to work (in any interesting sense of that term).

A theme of this section has been a retreat from epistemic certainty in our view of science and of the knowledge it gives us. This retreat has been forced upon us by the hard facts of experience. Scientific practice has turned out to be a much more complex and less certain enterprise than it was once perceived to be.

10.3   Structures and dynamics in methodology

Thus far in this chapter we have argued for a representation of science that takes cognizance of and that is faithfully based upon actual scientific practice. This requires an admission that no methodological principles can be prescribed ahead of time. They must be discovered or recognized in science as it is practiced in the real world. Even after one has made this concession, though, one may hope or expect to find an invariant (ahistorical) characterization which represents *the* essence of science and of knowledge construction in science. We have already recommended as a starting point for a model of science Laudan's (1984a) ('reticulated') triadic scheme which organizes scientific activity in terms of practice, methods and goals. Nevertheless, we argue below that any level-metalevel distinction is largely a false dichotomy. Paradoxically, Laudan and his collaborators (Laudan *et al.*, 1986; Donovan, Laudan and Laudan, 1988) have recently undertaken a project of amassing a huge range of case studies (across several centuries and across many fields) in an attempt to abstract from all of these the common characteristics of scientific practice. We say 'paradoxically' because, given the types of change in an evolving methodology which we have found and which Laudan's own (1984a) triadic model suggests, it is by no means evident that one should expect to find *one* description that will fit all eras and all fields. Why need there be *a* method of science or *a* (fixed) set of modes of change and argument in science? In this section we discuss why such a straightforward inductivist approach to constructing a methodology is unlikely to succeed. We also identify several important characteristics in those cases of scientific practice that we have considered.

It is entirely possible, as Fine (1986, p. 174) has recently suggested, that science is an historical entity (his expression) whose practice, methods and goals are *contingent* (by which we mean here not fixed by logic or necessity). There may not be *a* rationality that is the hallmark or the essence of science. Shapere (1984, 1986) has made similar suggestions. If this view of science is correct and if, as Laudan *et al.* (1986, p. 142) claim '... ["post-positivist"] models of scientific change and progress ... were based upon and supported by empirical study of the workings of actual science, as against the logical or philosophical ideals of epistemic warrant emphasized in the positivist tradition', then such methodologies derived from the historical record of science will reflect the characteristics of the episodes from which they have been abstracted. And, all too often, these methodologies are mired in the scientific practice of previous centuries. One can well expect an evolution of these methodologies themselves. In a given era certain factors or criteria will be of major importance, in another episode others. We have seen that Laudan (1984a) has outlined just such an evolving and coupled network or matrix of practice, methods and goals. Shapere (1986) has discussed the distinction between external and internal factors in science and has argued that science has successively internalized certain once-external factors and criteria as they have proved themselves useful for the progress of science. On this historical-entity view one can at best hope to generate a loose enough framework to accommodate a *description* of this change. This would make seeking a general theory of scientific change analogous to asking for a general theory of historical change. Would any historian attempt such? Would we expect a project of that type to succeed? Perhaps Toynbee's (1935–1961) *A Study of History* could serve as such an attempt. But, even if we accept Toynbee's stimulus–response theory of historical change, have we gained anything of predictive power or of understanding, or simply a framework that can be fit, *post hoc*, to various episodes from the historical record? Does it allow us to *understand* why history developed as it did? After all, *we* impose much of the structure, or inform the facts, later, once the events have occurred. Laudan's (1984a) reticulated model gives a coherent framework within which to tell a narrative of scientific activity, in terms of practice, methods and goals. This includes the ability of science to bring under scrutiny (i.e., to internalize) factors once held to be external (or beyond modification) (e.g., the requirement of causal explanation; cf. Shapere, 1986).[1] By this process of internalization, 'transcendent' values become a part of science and, thereby, subject to revision. Just as the belief in a need for causal explanation was once internalized by science, so it has now been significantly modified (arguably even abandoned; Cushing, 1989c) by the requirements of quantum theory. Similarly, we have seen that an *S*-matrix paradigm for science could have modified significantly the nature of scientific explanation (cf. Section 7.3). Another important characteristic of actual science is the pyramid structure of its work force, with the apex occupied by a very few creative people who are able to generate the new theories for testing by the rest of the community (Cushing, 1984). For the *S*-matrix program such a list would contain the names Chew, Goldberger, Gell-Mann and Veneziano. This structure is a key factor in producing stability and convergence of scientific opinion (Cushing, 1984). An overall characteristic of successful scientific practice we have examined might be fertility or the ability to 'get on with things' (progress?), but that is pretty vague.

Earlier in this chapter, we have given reasons supporting this view of science. Let us tentatively accept this picture representing science as a coupled, evolving, three-tiered structure which internalizes by assimilation or modifies or even rejects external factors (e.g., generally-accepted epistemological criteria) when it is productive to do so and which has a relatively small elite who generate the new theories for test and elaboration and see what follows concerning reasonable expectations for a theory of scientific change. Here we refer again to Laudan *et al.* (1986). Its authors observe (p. 146):

> Despite this historical orientation, however, all members of this school [i.e., the 'historical approach' to the philosophy of science] perceive (except Feyerabend) their task as one of enunciating normative principles of scientific inquiry which will show *what the nature of scientific rationality is.*

If science is a contingent historical entity, we need not be able to identify a universal (fixed) type of rationality for all of science (Fine, 1986, p. 174), although it *could* (fortuitously) turn out that such a universal characteristic exists (a hope often hinted at by Shapere, 1984). That is, even if one eschews an *a priori* foundationist belief in necessary and therefore timeless criteria at the metalevel, it remains a logical *possibility* that fixed (or permanent) meta-criteria could emerge from the history of science. While this *could* happen, we are arguing that it *does not*, in fact, happen. Quantum theory and the *S*-matrix program have each provided examples of actual or possible changes produced at the metalevel by success at the level of practice. Empirical adequacy has become a generally acknowledged hallmark of the epistemic validation for scientific knowledge, but this, too, is a learned criterion and its relative importance varies with the episode considered. If there were a universal rationality for science, then it might be revealed by an empirical investigation of the historical record. But we have no *a priori* assurance that such a universal characteristic exists. How reasonable is it even to expect that *one* 'description' of science will prove adequate for *all* times and fields? By way of analogy, would anyone ask for *the* way that scientists make measurements? There is no *one* way. We saw in the previous section that Galison (1987) has provided a detailed discussion of this in modern high-energy experimental physics. As an 'epistemological shortcoming' of some methodologies based on the 'historical approach', Laudan *et al.* (1986) list (p. 148):

> ... restriction of the range of cases considered to a handful of five or six favored revolutions (usually those associated with Copernicus, Galileo, Newton, Lavoisier, Darwin, and Einstein) ...

But, if it is true that only relatively few of the best people can *create* new theories (the 'pyramid' structure), then there will be a quite restricted list of examples to select from for *major* theory choices. Their paper also criticizes (p. 152) professional historians of science for not seriously considering the claims made by (philosophers') models of scientific change and for not using these models to inform their own (the historians') interpretations of science. But, if these philosophers' models are so particular that they lack real generality or are merely general at the price of vagueness, then why should an historian (or anyone else) feel that it is a wise investment of time and effort to use the models as guides? Heilbron (1982, p. ix), in an historical study of electricity in the seventeenth and eighteenth centuries, notes that:

> Our case history shows that the metaphysics of the paradigms and research programs supposed to guide scientists are seldom close enough to experimental work and theory to order them in useful ways.

Why does it follow that (p. 154 of Laudan *et al.*, 1986) '... to grow in a strong and enduring manner it [the history of science] must once again become deeply involved in the development of a general theory of scientific change.'? One may simply not exist. Case studies are criticized because many (p. 159) '... are not "tests" of the theory in question at all; rather, they are applications of the theory to a particular case.' However, a perfectly good means of testing a theory (or description) of change is to see whether it fits naturally various episodes in the history of science. We have illustrated this with specific examples. Further, historians of science are taken to task for not having examined (p. 159)

'... how general patterns of development have characterized all episodes [of the scientific enterprise] ...' Again, though, we do not *know* that such a general pattern does exist for science. (Unless, of course, we choose the existence of such a pattern as a *definitional* criterion for science itself.) There may exist only clusters of episodes that share some common characteristics of development.

The point to be made with these several citations of Laudan *et al.* (1986) is that the authors seem to take as given the existence of universal criteria for change and evaluation in science. After all, philosophers of science *do* have a vested interest in finding such universals. Otherwise, philosophy of science might reduce largely to a 'mere' collection of historical episodes. We have argued, by citing instances of change (e.g., quantum mechanics and causality; SMT and explanation) in the case study of the earlier chapters and in those outlined in the present chapter, that such an assumption (of the existence of universal criteria) is by no means obviously supported by the data. (By 'data' we mean here case studies of episodes in the history of science.) Our general representation of the functioning scientific enterprise is that of activity within a matrix of evolving constraints having both internal and external components. Furthermore, as we have emphasized previously, the structure of the scientific community is that of a pyramid, the apex being occupied by the relatively few creative people who can invent and sustain successful theories. They ultimately make the rules of the game. The method of argument and presentation has been historical and case studies (both the *S*-matrix program as well as others summarized in the previous chapter) have provided the empirical base for the claims made. Even though the preceding comments have been critical of some of the assumptions on which Laudan *et al.* (1986) appears to be based, our basic position is in complete agreement with the *spirit* of that paper – to subject theories of scientific change to test. However, the piecemeal test of separated (and isolated) components of these theories may not be the most productive way to proceed, since one can generate lots of supporting evidence, as well as many counterexamples, for specific theses (an extensive list of which can be found in Cushing, 1989f). One must look for general (often elastic) characteristics (or parameters) that can be fit to various episodes of scientific practice.

There is no lack of suggestions for the criteria used by scientists in making their choices. Among those philosophers of science who attempt to identify a particular characteristic of theory choice, we find Popper (1963), who stresses falsifiability as the hallmark of scientific theory; Kuhn (1970), who sees a successful scientific theory as providing a paradigm for further investigations or puzzle solving within a fairly well articulated set of ground rules; Lakatos (1970, 1976), who emphasizes the research program as a sequence of theories guided by a largely fixed heuristic (or set of principles) that must be respected as theories are modified in light of experimental results; Laudan (1977), who focuses on the problem-solving ability of a successful theory; Giere (1983), who attempts to explicate the rationality of science in terms of decision-theoretic models of choice between competing theories; and van Fraassen (1980), who argues for empirical adequacy as the main justification for our acceptance of a theory. More recently, as we have already pointed out, Laudan (1984a) has suggested viewing scientific debates as conducted and settled at three levels – practice, methodology, and goals – which are coupled and which mutually influence each other's evolution. Pickering (1984) has presented a detailed case for the symbiotic effect of experiment and theory in which choices, not uniquely constrained by experiment, are made by scientists to produce a theoretical framework (cf. Section 10.2 above). Shapere (1984) has consistently and repeatedly made the case for science as an enterprise whose methods and goals cannot be given *a priori*, but must be seen as learned and evolving under the constraints of the real world. Nickles (1987) emphasizes that discovery and justification are interdependent and the latter cannot be seen as a logical exercise separated from the former. There is an evolution of these methodologies toward a looser and less dogmatic characterization of science.

We list all of these examples just to remind the reader that there is no shortage of opinions about the nature of science and its methods. Who is correct? Well, they *all* are, in that each has identified an important aspect of scientific practice which can be (and often is) supported by case studies, such as the present one, of episodes from the history of science. Similarly, there is very much right in the elementary and standard representation of scientific method in terms of the series: observation, hypothesis, prediction, confirmation. The error is to argue for any *one* characteristic of theory choice as *the* characteristic of scientific practice. In Section 10.6 we shall illustrate (Cushing, 1984) some of the means that scientists use to produce a convergence of scientific opinion to a fairly localized position as that position itself evolves over time. The practice, methods and goals themselves are produced in this process. They are not completely arbitrary choices, nor are they fully determined by empirical considerations and none need remain fixed for all time. So, we are very much in the Fine–Laudan–Nickles–Pickering–Shapere camp (even though these philosophers may not consider themselves in each other's camps). They have all discussed representations of the scientific enterprise that emphasize an empirical approach to formulating models of scientific change and development. A common and central characteristic of their work is the recognition that single-component models of science are inadequate to represent reliably the richness of actual generation of scientific knowledge and that the methods used in the development and appraisal of scientific theories are themselves not fixed, but evolve over time. Hones (1987) has reached a similar conclusion in his case study of the weak-neutral-current experiments in high-energy physics. Fine (1986) has recently argued that we should view science as an historical entity with a contingent history that does not require science to have some fixed goal or aim, or even a fixed set of methods. Like any historical entity, science evolves over time and its characteristics at any given time are a product of its (contingent) past and of the environment it finds itself in at present. This aspect of Fine's characterization of science is certainly consonant with Shapere's.

We are supporting a multicomponent, contingent view of science by illustrating the role played by both empirical and nonempirical factors (including overarching, general, if somewhat elastic, principles such as causality) in the generation and epistemic justification of several influential research traditions in modern physics. In Section 9.3 theories used for this analysis included the old quantum theory, quantum field theories (especially gauge theories), the dispersion-theory and S-matrix theory programs, and current quantized-string theories. These examples were chosen because they are important episodes in the history of modern science. In addition to the essential support provided by a broad base of favourable empirical evidence, these theories often also appeal for justification and acceptance to general (non-empirically based) principles such as gauge invariance and various forms of the causality condition. The compound nucleus model of nuclear reactions was used as a paradigm to illustrate the coupling between empirical and nonempirical factors in theory development. In all of these cases the recycling of analogies and the key role played by dead-end theories are apparent, as is the prevalence of the emergence of the central physical concepts *from the mathematics employed* (examples of which are the bootstrap and string field theories from dual models). Our contention is that any methodology must be evaluated *holistically* by its ability to tell the best (overall) story about episodes in science. We must use judgment, not algorithms, to make this choice.

In considering specific examples, we must distinguish between 'big' theory choices and 'small' ones. The former are usually present in 'scientific revolutions'. Obvious examples are classical mechanics versus quantum mechanics and quantum field theory versus S-matrix theory. Then, within any one of these grand, overarching frameworks (or theories) there are many 'small' choices made among competing theories or models. Most scientific practice is concerned with the latter category of choice. For this type of choice the ground rules remain essentially fixed. But at certain crucial junctures ('bifurcation points') there are radically divergent possible paths available and the choice is critically underdetermined. That is why we (Cushing, 1982) have repeatedly stressed that the structure of science is very much that of a pyramid, with the major options of theory choice being presented to the community by the relatively few creators of significant theories. This brings up the important point that theories are not so much chosen between (or among) as they are created and evolved or developed. Only a rather small number of the best people (e.g., Chew, Goldberger, Gell-Mann) is able to engage convincingly in this game. Schweber (1984), in his history of modern quantum field theory, also stresses the pivotol role played by a select few.

> In fact, I perceive the history of quantum field theory from its inception as a history of the contributions of a group of exceptionally gifted *individuals*. They did form an elite community – socially and intellectually – who helped and criticized one another. Yet, it was the genius of the leading *individuals* of that community – in particular, Dirac, Heisenberg, Pauli, Weisskopf, Yukawa, Kramers – that was responsible for the decisive advances in the thirties. And the same is true for the advances in the forties: Tomonaga, Schwinger, Feynman, Bethe, and Dyson are the decisive *individuals*. (p. 43)

> All the major developments in quantum field theory in the thirties and forties have as their point of departure some work of Dirac's. (p. 51)

The rest of the practitioners of science must play the game (or one of the games) as already outlined. Their job is to cement a candidate theory into a stable matrix of accepted ideas, as was done with the compound nucleus model or dispersion relations once they had been brought under the umbrella of causality. They are the calculators and measurers who make up the bulk of the scientific community.

10.4   Changes in methodological rules

No one at all familiar with the history of science will find problematic the claim that accepted, successful scientific theories do change and are superseded. This replacement of one 'true' theory (or even worldview) by a later one is an obvious enough fact that one cannot seriously argue against it. This admission is independent of whether or not a person believes that science seeks (and finds) truth, approximate truth, etc. However, one might admit this change at the level of practice but still hold out for a fixed set of methods and goals in science at some metalevel. Changes in methods and goals may not be as apparent from an examination of this historical record as are changes in practice (e.g., theories), which are relatively commonplace. Even though we have been arguing in this chapter for changes at all three levels, we now turn to some specific examples of change in the methods and goals of science. Our purpose is to undermine any belief that these *must* remain (or, in fact, *have* remained) fixed in scientific practice. In a sense, examining changes at the metalevel may be analogous to looking for a 'second-order' effect against a 'first-order' background of relative stability at that level. But, it is an important exercise as a matter of principle.

Once we recognize the possibility that any of the components of a methodology of science are susceptible to change, we can appreciate that scientific practice itself becomes the *source* of philosophically (or perhaps better, of *methodologically*) interesting questions (rather than *vice versa*), a point already raised in the literature (Laudan, 1981; MacKinnon, 1982; Cushing, 1984). That is, there is something potentially very interesting *methodologically* about research in modern theoretical physics. As examples of significant changes from an earlier era, we mention just three here, before returning to current physics. Hacking (1975) has argued that probability as we now conceive it came into being only around 1660 with work by Huygens and by Pascal. Prior to the end of the Renaissance, there was no concept of inductive support for a theory on the basis of empirical evidence. The origin of this new mode of evidential support was in the 'low' sciences, such as alchemy and medicine, rather than in the 'high' sciences, such as mechanics or astronomy where it was believed that knowledge demonstrable from first principles was possible. The standards for valid knowledge had to be revised ('downward') before such evidential support became relevant to and accepted by science. Foucault (1966) has illustrated how, until the end of the sixteenth century, resemblance played a dominant and constructive role in the knowledge of Western culture. That is, reasoning by analogy, for example with a picture of man as a microcosm, was thereafter replaced by other styles of argument, with analysis becoming of central importance. Laudan (1981) has traced the emergence, during the last 300 years or so (and especially since about 1850), of the hypothetico-deductive method as *the* method of science.[2] MacKinnon (1982) has studied in detail the refinements of this method, as regards the requirements of adequacy and coherence, in the development of quantum mechanics in the first quarter of the present century. Some investigation of subsequent modifications in methodology found in more recent work in theoretical physics has been made (Cushing, 1982) – especially regarding the use of highly abstract models as the source or generator of new knowledge. Direct experience is no longer the preponderant origin of or guide for new theoretical entities (Dirac, 1931; Cushing, 1983a). We have mentioned this previously in the present monograph and shall return to it below. Nevertheless, the question remains whether or not there is anything qualitatively new or essentially different in the methodology of science to be found in studying contemporary basic research in modern theoretical physics. Or, does one simply generate *more* examples of the type of argument and reasoning already found in older, more familiar branches of physics? We can see in the *S*-matrix program one attempt in theoretical physics to modify the type of argument by which an 'explanation' is produced. The general scheme remains broadly hypo-thetico-deductive, but the coherence or uniqueness constraint becomes much tighter, as in the *S*-matrix program.

If we consider the general features of the evolution of methodology as used in physics (or even in science as a whole), the emergence and acceptance of the hypothetico-deductive method is clearly of major importance. The shift in recent times from direct experience to the use of abstract mathematical models (Cushing, 1982) as the source or generator of new hypotheses and theories is significant, but certainly not of the same proportion as the acceptance of the hypo-thetico-deductive method. The method of hypotheses was (reluctantly) resorted to by or forced upon the scientific community because of the necessity to admit ever more abstract or remote concepts into the circle of basic theoretical entities in theories (Laudan, 1981). Its use was, in a sense, precipitated by crisis (not by *one* specific crisis, though). Can we find a crisis *of method* in contemporary theoretical physics that might necessitate a major modification in the methodology used by practicing scientists? In part, the bootstrap conjecture (Freundlich, 1980; Red-head, 1980; Cushing, 1982) was just such a response. Since scientific methodology is basically conservative, it is not surprising that only difficulties perceived as profound would elicit such a radical move that was seriously considered for a time by a large part of the community of theoretical physicists. The bootstrap conjecture holds that a well defined but infinite set of self-consistency conditions determines *uniquely* the entities or particles which can exist. That is, once we are given any partial information about the actually existing world, nothing else about that world is contingent or arbitrarily adjustable. As we have seen, this almost Leibnizian idea is implemented in S-matrix theory through a unitarity equation, which is a statement of conservation of probability. The S-matrix program contrasts with quantum field theory which does have arbitrarily assignable quantities. Because of the highly constrained structure of S-matrix theory, that theory makes far fewer (but perhaps stronger) ontological and epistemological assumptions than does quantum field theory.

The anthropic principle of modern cosmology can also be cited as an attempt (perhaps of uncertain merit; Wilson, 1989) to make an apparently contingent fact – the actual occurrence of the extremely narrow range of initial conditions consistent with the eventual evolution of life as we know it – appear to be required by the presence of intelligent beings who observe the universe and its laws (Barrow, 1981; Barrow and Tipler, 1986; Gale, 1981, 1987). Indirectly, this principle had its (historical, even if not its logical) origin in a suggestion by Dicke (1957) to explain what seemed a remarkable coincidence of certain cosmological numbers. Subsequent developments of the anthropic principle have even tied the *existence* of the real world to the presence of intelligent life in that world (and hence to those initial conditions required for the subsequent appearance of such life). For example (Wheeler, 1977):

> Could the universe only then come into being, when it could guarantee to produce 'observership' in some locality and for some period of time in its history-to-be? (p. 5).

> Quantum mechanics has led us to take seriously and explore the directly opposite view that the observer is as essential to the creation of the universe as the universe is to the creation of the observer. (p. 27.)

This participatory or teleological version of the anthropic principle offers the actual existence of human life as an *explanation* for the conditions in the universe that led to that life.

Even in its more conservative forms, though, this principle represents an attempt to reduce the apparent arbitrariness or contingency of the world as we find it. 'Deviant' methodologies, such as the anthropic principle or the bootstrap conjecture, are resorted to in times of desperation. However, it is not clear that such attempts at an explanation are *a priori* any less plausible than the currently accepted, standard one. A key deciding factor is often empirical adequacy. The 'strong' versions of the anthropic principle do come up short on this score (Wilson, 1989). It is success in this that eventually largely *defines* what counts as an explanation (and, ultimately, as rational). After all, what could seem more bizarre (to one unenlightened by their successes) than action at a distance or the reduction of the wave packet (sometimes referred to in more sophisticated terminology as the projection postulate)? That which is of interest are often just those phenomena that can be explained (i.e., accounted for) by a theory (e.g., those relatively rare events successfully handled by the 'standard' Salam–Weinberg electroweak theory; Pickering, 1984).

An important characteristic of modern theoretical physics, exemplified in the episodes we have studied or outlined, is the use of abstract mathematics as the source of many new ideas in physics. This trend toward formalism as the *origin* of physical theories was presaged by Dirac in his famous 1931 paper on magnetic monopoles.

The steady progress of physics requires for its theoretical formulation a mathematics that gets continually more advanced. This is only natural and to be expected. What, however, was not expected by the scientific workers of the last century was the particular form that the line of advancement of the mathematics would take, namely, it was expected that the mathematics would get more and more complicated, but would rest on a permanent basis of axioms and definitions, while actually the modern physical developments have required a mathematics that continually shifts its foundations and gets more abstract. Non-euclidian geometry and non-commutative algebra, which were at one time considered to be purely fictions of the mind and pastimes for logical thinkers, have now been found to be very necessary for the description of general facts of the physical world. It seems likely that this process of increasing abstraction will continue in the future and that advance in physics is to be associated with a continual modification and generalisation of the axioms at the base of the mathematics rather than with a logical development of any one mathematical scheme on a fixed foundation.

There are at present fundamental problems in theoretical physics awaiting solution, e.g., the relativistic formulation of quantum mechanics and the nature of atomic nuclei (to be followed by more difficult ones such as the problem of life), the solution of which problems will presumably require a more drastic revision of our fundamental concepts than any that have gone before. Quite likely these changes will be so great that it will be beyond the power of human intelligence to get the necessary new ideas by direct attempts to formulate the experimental data in mathematical terms. The theoretical worker in the future will therefore have to proceed in a more indirect way. The most powerful method of advance that can be suggested at present is to employ all the resources of pure mathematics in attempts to perfect and generalise the mathematical formalism that forms the existing basis of theoretical physics, and *after* each success in this direction, to try to interpret the new mathematical features in terms of physical entities. (Dirac, 1931, p. 60).

Dirac then cited his own theory of the relativistic electron whose negative-energy states subsequently became identified with the positron. This need to go *from* mathematics and theory *to* experiment in current physics, rather than the reverse route which had been predominant in the past, has also been commented on by C. N. Yang (1980). We have seen several examples of this, such as quantum electrodynamics, renormalization, and non-abelian gauge fields, and could also cite the black holes of general relativity.

The use of abstract mathematical models is evident in all this as is the exploration of the surplus structure of these models – monopoles and solitons in gauge theories, topology in *S*-matrix theory. There is a predominance of mathematics as the source of new physical concepts – charm (from the suppression of unwanted terms in a theoretical model), the group-theory origin of quarks, the Higgs boson, and the topological 'quarks' of *S*-matrix theory (cf. Cushing, 1982). There has been a recycling of ideas as a result of individual workers recycling their expertise. This is especially evident in the way that most theoretical physicists are tied to their experience and technical skill in working with Lagrangian field theories (Chew, 1971; Pickering, 1984). It is also evident for theorists (such as Mandelstam, always an expert in the QFT formalism) who have shifted from the *S*-matrix program to the field theory one, as well as for those (such as Balázs) who have stayed with the *S*-matrix theory. The composite-model picture has gone from atoms, to nuclei, to quarks (and even, possibly, beyond). The analogy of quantum electrodynamics as a guide for gauge field theories, such as quantum chromodynamics, is apparent. Experiments and observed phenomena at times make sense and have meaning only within the framework of a theory which selects 'meaningful' events to study. That the order 'seen' is partly, but significantly, a function of the selection made has already been indicated, for example, by the fact that quantum chromodynamics and *S*-matrix theory are valid (in the sense of being calculationally adequate) in different domains of reactions (often characterized, respectively, as large versus small values of the transverse momentum). A calculationally successful theory, such as quantum chromodynamics, so *defines* the language of scientific discourse that even a radically different competitor, such as topological *S*-matrix theory, is cast in terms of that language.

These are some of the reasons why we suggest that great theorists are clever people who (with varying degrees of success and within rather broad limits set by data, which after all, are selected) *make* their theories work (as opposed to being fortunate enough to discover a 'law' already there). There are lots of pieces, *some* of which can be fit together into a workable theory to create a picture of the world as we see it. The details of continual fitting and stretching of theories and methods to obtain a faithful and coherent representation of the world raises questions about the uniqueness of the final product (or picture) generated by the exercise. This problem becomes especially acute for the increasingly abstract descriptions of nature created by modern theorists. It is ever more difficult to test these theoretical constructs (such as quarks, the Higgs boson or superstrings).[3] Let us emphasize again, though, as we did in Section 10.2, that we are *not* adopting the position that just *any* theory can be made to work. We now consider how plausible it is to claim that scientific theories give us a unique, accurate picture of our world.

10.5   A uniqueness in our theories?

There are actually two distinct but connected questions of the uniqueness, overwhelming tightness of fit, or reliability of the products of science. Are our theories in some sense unique and are the central entities in a theory to be taken realistically? Let us begin with the second issue. If, by creatively stretching and fitting models, theorists generate a network which they fit over the contours of the external world as they see it, the question naturally arises of the reality of the theoretical entities in such descriptions of nature. The *origin* of these concepts or theoretical entities is relevant to evaluating how seriously and literally they are to be taken since the reality question is not susceptible to proof or refutation either by direct logic (as for a mathematical theorem) or by direct observation. The issue is more one of plausibility than of proof. This is similar to deciding whether the stability of successful theories is better seen as resulting from their truth or from the difficulty of constructing them. While a realist could argue for his position on the basis of such stability, one can equally well see the explanation for that stability in very different terms. Which view one takes remains largely dependent upon one's predilections.

As one case in point, consider the picture that quantum field theory gives us of the 'vacuum' (aether?) as seething with particle-antiparticle pairs of every description and as responsible for spontaneously breaking symmetries initially present in the theory. Is nature *seriously* supposed to be like that? Perhaps we have a little more perspective today in questioning Newton's absolute space or the great certainty with which Maxwell argued for his aether. In a long article on the aether for the ninth edition of the *Encyclopaedia Britannica*, Maxwell concluded by discussing the vortex lines that would have to exist for an indefinitely long period of time in the neighborhood of a permanent magnet. Since any known fluid which can have vortices must also be viscous so that the vortices must ultimately dissipate as heat, he had to admit that the aether possessed another peculiar property. Nevertheless, he was adamant about its existence and even hinted at a function for the aether beyond the realm of mere physics.

> No theory of the constituion of the aether has yet been invented which will account for such a system of molecular vortices being maintained for an indefinite time without their energy being gradually dissipated into that irregular agitation of the medium which, in ordinary media, is called heat.
>
> Whatever difficulties we may have in forming a consistent idea of the constitution of the aether, there can be no doubt that the interplanetary and interstellar spaces are not empty, but are occupied by a material substance or body, which is certainly the largest, and probably the most uniform body of which we have any knowledge.
>
> Whether this vast homogeneous expanse of isotropic matter is fitted not only to be a medium of physical interaction between distant bodies, and to fulfill other physical functions of which, perhaps, we have as yet no conception, but also, ... to constitute the material organism of beings exercising functions of life and mind as high or higher than ours are at present, is a question far transcending the limits of physical speculation. (Maxwell, 1890, vol. 2, p. 775).

Maxwell's basic article of faith necessitating a material aether was that ultimately *all* forces must be forces generated by direct contact: a *vis a tergo* (a shove from behind), as he termed it.

The reality of theoretical entities – such as the aether – generated in the creation of theories seems accidental and implausible, except in some general sense of a quite lose and nonunique correspondence with the elements of the external world. This lends support to Laudan's (1984b) recent attack upon realism. Actually, Putnam's (1975, p. 73; 1978, p. 177) earlier 'miracle' (of the effectiveness of successful theories in the absence of a realist interpretation) would be better stated, not that successful theories can be created by trial and error, but rather that scientists are willing to work so long and hard on such an enterprise. One reason for the success (often partial, rather than total) of the theories of high-energy physics discussed in this book may lie in their being so complex and open-ended that they can never be fully explored and hence have the practical flexibility that allows them to be stretched to cope with new phenomena (but not, of course, completely arbitrarily and in the face of *any* eventuality).

Of course, we seek more from our theories than empirical and calculational adequacy alone. They give us a stable means of organizing and comprehending our world in terms of the vocabulary they define and through an interpretation of the central terms in those theories. However, we have no warrant for assigning truth or reality to these constructs with any meaning other than that they work (or, really, haven't failed us) in fulfilling these functions. They are *true* only in this very pragmatic sense.

So, a belief in the uniqueness of (even in the sense of equivalence of all various statements of) physical theories or of the view of the world they give us would seem implausible. Our successful theories and the worldview they give us bear (some) truth in the sense just indicated, but at a very foundational level (at least for quantum phenomena) essentially incompatible 'stories' are consistent with the data. We have discussed an example of this in Section 9.5 for the causal interpretation of quantum mechanics versus the standard, 'Copenhagen' one. A bad fit (or theory) is evident (Cushing, 1983a, p. 35; 1983b, p. 112), but a good fit is a poor, logical as well as pragmatic, exclusive argument for uniqueness. An image of theorists as *crafting* theories, as opposed to *discovering* them, may be closer to the reality of scientific practice. We have seen that mathematical models play an increasingly important role in helping one 'guess' hypotheses to try (Cushing, 1982, pp. 77–8). Chew's work, for example, can be cited as an example of this.

> I have often been motivated by theoretical models. The connection of the Mandelstam representation to the Schrödinger equation with a Yukawa potential has caused me frequently to use potential-scatter-

ing theory as a guide in my thinking. Also, Feynman-graph amplitudes have often been employed.

There remains the question of the relation between the mathematical formalism and a physical interpretation (in a sense, the relation of mathematics to reality). Again, Chew's view:

> [C]onsistency demands concurrent explanation of the measurement process that defines momentum and spin. Equivalently one may say that the meaning of a space-time continuum requires bootstrap elucidation.... The most general principle that I can presently identify is 'consistency'. A mathematical framework is appropriate if it facilitates a consistent view of nature. ... [I] do think of the central theoretical entities in mathematical terms.[4]

Are we to take *seriously* the pictures that accompany these mathematical models – strings, Kaluza–Klein compactification (as a means of making extra, unwanted dimensions in a theory 'curl up' and effectively disappear), quarks (just to mention a few here) – as representations of our physical world? (It is one thing to take (mathematical) string theories seriously as candidates for a theory and quite another to buy the string picture as literally true.) Or, is it that (relatively few of) the most creative scientists generate a workable network of ideas and ask all of the probing questions they can so that whatever theory survives must be accepted by the rest of the practitioners in the field, if they wish to play any game at all? Stability (not permanent, of course) is produced by default (Cushing, 1984, pp. 212–16), a theme we expand upon in the next section.

For example, consider the following. At the end of Chapter 2, we argued that questions arising directly out of Heisenberg's *S*-matrix program (especially Kronig's suggestion about imposing causality as a constraint on the *S*-matrix) produced a problem background out of which the relativistic dispersion-theory program of the 1950s emerged. Specifically, in late 1953, early 1954, Goldberger and Gell-Mann saw Rohrlich and Gluckstern's (1952) paper and became interested in dispersion relations. Their implementation, with Thirring, of the causality requirement through the vanishing of the commutator of field operators at spacelike-separated points began the enormously fruitful application of relativistic dispersion relations. The combination of dispersion relations with Chew's work on analytic reaction amplitudes led to the *S*-matrix program of the 1960s. Crossing symmetry was an essential ingredient in this program. It had been discovered in specific models and was then elevated to the status of an independent principle.

Chew felt that to have been the key missing in the old Heisenberg program.[5]

> Later I appreciated that what Heisenberg lacked was crossing.
>
> The pole–particle correspondence is imprecise without crossing, and bootstrap mechanisms are invisible.

That is, crossing is a special constraint on analyticity and is present only in a fully relativistic theory. The nonrelativistic Schrödinger amplitude can be unitary and analytic, but there is no crossing proper (because there is no crossed channel). Crossing gives dynamical content to the S-matrix program and makes the bootstrap possible.

From S-matrix theory came the duality models in the late 1960s and eventually the quantized string industry in the 1970s. Today, superstring theories offer a hope of providing a finite, consistent quantum theory of all the known interactions – electroweak, strong and gravitational. Without the S-matrix program, it is unlikely that quantized strings would ever have been discovered and studied. By inference, one can ask whether theoretical physics would have arrived at one of its promising constructs (superstring models) if it had not been for the existence, nearly four decades earlier, of a program (Heisenberg's S-matrix theory) which ultimately in its own right proved a dead end. It is impossible to *prove* that we would not have arrived at quantized string models by another route. However, the present detailed study of how these came to be seriously considered as a result of the S-matrix and duality programs makes it appear implausible that they would have been discovered had it not been for S-matrix theory. On the other hand, the road from quantum electrodynamics to modern gauge field theories owed nothing essential to the S-matrix program. So, while there might now still be *some* alternative proposed to overcome the difficulties of, say, quantum chromodynamics, that alternativle might not likely be superstring theories. Of course, *logically* quantized string theories *could* have been written down (in principle) anytime after the formulation of QFT. Locality and point particles seemed obviously essential for any successful quantum field theory. That is, even though a *point* electron had long been seen as a source of divergences (classically as well as quantum mechanically), an extended (say spherical) 'particle' produced difficulties for the Lorentz invariance of the theory. It was by no means obvious (and, in fact, has only been recognized recently as a plausibility after years of work on quantized strings) that a quantum string theory might avoid the divergences of point field theories.

Yet, if these quantized string theories do succeed, our 'picture' of physical reality will be one in which quantized strings underlie all matter. But, if our little scenario is at all plausible, this picture would not likely have arisen had it not been for an unsuccessful (wrong, etc.) idea which Heisenberg put forth and which might otherwise never have been followed up. One can at least question a belief in the uniqueness of our physical theories (and, by implication, their content). Certainly, there is an obvious response to this, one especially appealing if *the* theories we have are seen as being in principle uniquely determined by nature. For example, Richard Feynman's comment on this question was[6]:

> It seems to me that the answer is: if Heisenberg had not done it, someone else soon would have, as it became useful or necessary.

And, Richard Eden, one of the early contributors to the *S*-matrix program, takes a similar position[7]:

> ... a general view that I would support [is] that most people's research would have been discovered by someone else soon afterwards, if they had not done it. I think that this would have been true for Heisenberg's *S*-matrix work ...

However, even if one were to grant that *most* discoveries would still have been made by someone else, this leaves open the key question of whether, at critical junctures ('bifurcation points'), the *same* creative moves would nevertheless have been made by someone else.

Gell-Mann[8] has also expressed the belief that someone else would surely have made any given discovery. He believes that history should be used to help one see (retrospectively) the thread of ideas which eventually led to the correct theory and that what people thought at the time may be interesting, but it is unimportant. For him, those physicists who turn out to be correct are great, the others just not so good.[9] On the subject of this harsh judgement provided by *the* objectively correct theory, Gell-Mann (1987, p. 480) claims:

> I thought that it was not fair for a theorist to propose several contradictory theories at one time, that a theorist should save his money and then bet on one idea that he really thought was right. And I believed also that proposing a wrong theory counted as having written a wrong paper. ... My feeling was that a theorist should be judged by the correctness of his guesses about nature, that his reputation should be gauged by the number of right minus the number wrong, or even the number right minus twice the number wrong.

This shows a curious mixture: *guessing* about the theories of nature, yet imputing right versus wrong to the professional standing of the individual on the basis of these guesses. In such a view, the (true) laws of nature are discovered (versus created), although they may be express-ible in various forms.[10] Objective truth, the rationality and inevitability of science and the serendipity of the enterprise all coexist harmoniously here. Gell-Mann's ability to countenance simultaneously ideas that appeared incompatible to others made an early impression on David Olive:[11]

> My first contact with Gell-Mann was when he visitd Cambridge for a few months in 67. I was particularly impressed by his dictum (expressed then, or possibly later) that apparently contradictory theories such as the quark model, and S-matrix theory could indeed be different aspects of the same, more comprehensive theory. This open mindedness was in striking contrast to the more entrenched attitudes then current.
>
> I remember thinking that the dual theory vindicated this point of view ....

(Recall (Chapter 8) that dual models could be formulated either in terms of an S-matrix theory or in terms of a quantized field framework.)

Another illustration of this eclectic attitude is Gell-Mann's belief that renormalization is a necessary guide along the path to the correct theory (i.e., to reject bad theories), but that *the* correct theory must be *finite* and *all* of its parameters (except possibly one) should be calculable from the theory.[12] Here we have a true bootstrapper who is able to live with renormalization as a necessary interim tool. This is reminiscent of Einstein's dictum that the theoretical physicist may appear to be an unscrupulous opportunist who borrows whatever ideas are useful, without always worrying about ultimate consistency (Einstein, 1970, p. 684):

> The scientist, however, cannot afford to carry his striving for epistemological systematic that far. He accepts gratefully the epi-stemological conceptual analysis; but the external conditions, which are set for him by the facts of experience, do not permit him to let himself be too much restricted in the construction of his conceptual world by the adherence to an epistemological system. He therefore must appear to the systematic epistemologist as a type of unscrupu-lous opportunist: he appears as a *realist* insofar as he seeks to describe a world independent of the acts of perception; as *idealist* insofar as he looks upon the concepts and theories as the free inventions of the human spirit (not logically derivable from what is empirically given); as *positivist* insofar as he considers his concepts and theories justified *only* to the extent to which they furnish a logical representation of relations among sensory experiences.

So, practicing scientists do often believe in the inevitability of the position that science has arrived at at any given time. Yet in the events described in this monograph, a cycle or oscillation between a secure, conservative approach (basically QFT) and a more free-wheeling, 'desperate' one (SMT) in times of crisis is apparent (Cushing, 1985; Gale, 1986):

'old' QFT→SMT$_1$(~1945)
→QED→SMT$_2$(1954–65) & dispersion theory
→QCD & strings

There is little sense of *inevitability* in the final outcome or position. It is not implausible that the (*unrefuted*) *S*-matrix program *could* have been pressed on with in spite of its mathematical complexity. (After all, who would claim that gauge QFT or superstrings in their present forms are mathematically *simple*?) Rather, another approach (gauge QFT) offered greater hope of more progress once SMT bogged down. But, as we have seen in Section 10.2, even this 'progress' depended upon experimentalists and theorists consciously choosing to study a class of very rare scattering events (which gauge QFT could handle) to the exclusion of the much more common class of events (which it could not). In large measure, SMT and QFT had *different* domains of success (Pickering, 1984). We are not particularly uncomfortable with a lack of inevitability in other areas of history. That point is nicely made in the review of a book that examines whether it was *inevitable* that the Confederacy should lose the United States Civil War in the nineteenth century (vann Woodward, 1986, p. 3).

> Inevitability is an attribute that historical events take on after the passage of sufficient time. Once the event has happened and enough time has passed for anxieties and doubts about how it was all going to turn out to have faded from memory, the event is seen to have been inevitable. Different outcomes become less and less plausible, and before long what did happen appears to be pretty much what had to happen. To argue about what might have happened or whether and why the presumably inevitable turned out to be thought so strikes many people as a waste of time.

Any convincing power our examples or possible scenarios here have comes not from the general logical point that things might *in principle* have been different, but from the plausibility they have based on a detailed examination of the actual historical record of the construction and justification of real scientific theories. The method of this monograph in the philosophy of science is largely a philosophically sensitive (but hopefully not prejudicially biased) examination of the actual historical record of certain major episodes in current physics. One must look in detail at the means by which theories are constructed and judged to see whether 'uniqueness' claims and the like are compelling.

As a relevant example that undermines the belief of many scientists in any type of uniqueness, necessity or (approach to) *truth* of a currently-accepted theory, let us consider (Cushing, 1987a) gauge field theories and the analysis of effective Lagrangian theories given recently by Schweber (1987). The formalism of QFT has shown a great degree of adaptability in this century. This can be seen as an indication that it provides a general framework or a language for relating empirical data, rather than a fundamental *ab initio* theory. This is the essence of Weinberg's (1985) folk theorem to which Schweber refers and which we discussed in Section 8.6. Let us elaborate briefly on this insight of Weinberg's. William Frazer (1985) labels as a 'harmonic oscillator theorem' Weinberg's claim that the general S-matrix postulates can be implemented in a given regime by an effective Lagrangian in QFT. Presumably his reason for calling it a harmonic oscillator theorem is by analogy with the elementary result in classical mechanics that 'any' mechanical system subjected to a 'small' displacement from a stable equilibrium configuration (and then released) will execute simple harmonic motion about that equilibrium configuration. Just as this result would not motivate us to say that simple harmonic (i.e., Hooke's law spring) forces are fundamental in nature, so effective Lagrangian QFT need not be seen as a fundamental theory. In fact, in the late 1950s, Lehmann, Symanzik and Zimmermann (LSZ) constructed a formalism of causal asymptotic field theory as a means of implementing the general S-matrix principles (such as conservation of probability) using as input operators corresponding to scattering results, but avoiding use of either a Lagrangian or perturbation theory. Schweber's emphasis on effective Lagrangian field theories is an important one for an evaluation of the epistemological and ontological implications of the quantum field theory program generally. Rather than being fundamental theories, these may be simply a convenient phenomenological scheme for fitting data.

The use of a series (or 'tower') of effective Lagrangians in nested domains is possible because the physical phenomena themselves are nested in domains demarcated by (length) scales that are set *intrinsically* by the masses of the particles involved in or used to characterize those phenomena. Recall that in the arcane system of units often employed by particle theorists – namely, $\hbar = c = 1$ – the unit of mass becomes the inverse of a unit of length: $m \Leftrightarrow l^{-1}$. Given, as a brute fact of nature, that this layered structure of elementary-particle phenomena exists, it becomes understandable why a succession of effective Lagrangians is so empirically useful. We are able essentially to neglect those phenomena involving higher mass particles (i.e., smaller length scales) than those present at the energy scale of the experiments under consideration. Nature presents us with a layered structure of phenomena and we can tell the story of peeling the onion in terms of a sequence of effective Lagrangian theories. We end up telling the same type of story over and over. If nature did not have this structure in terms of largely nonoverlapping domains of phenomena, science as we know it might not be possible. Still, our success in telling a repetitive story does not argue for the literal truth of that story, any more than does the effectiveness of the small oscillations theorem (of classical mechanics) establish the literal truth of 'spring' forces as fundamental in nature.

10.6 Convergence of scientific opinion

We have argued at some length for a model of science in which the ground rules are discovered in actual scientific practice and which evolve along with that practice. The emphasis has been on change rather than on absolutes. So, the problem we are now faced with is one of understanding how scientific opinion is made to converge to certain laws and theories. Associated with that is the question of the status of the knowledge these laws and theories provide us with. Theories of science usually attempt to present science as a rational enterprise. The spectrum of arguments runs from the nearly *a priori* in which philosophically satisfying canons of rationality are first agreed upon and then conformity of scientific practice to these is sought, to the other extreme of accepting that scientific practice *must* be a paradigm of rationality and then examining scientific practice to learn what rational knowledge generation is. These two approaches produce, respectively, a study of many 'in-principle' problems (concerning the convergence and content of scientific knowledge) that might arise in the practice of science versus a careful study and description of actual scientific practice. The current trends to make the philosophy of science itself an empirical, scientific activity stress in varying degrees the need for the latter approach. However, this approach does not, and probably cannot, answer the other school's in-principle objections, which are serious and potentially disastrous for the rationality of the scientific enterprise if rationality is characterized too absolutely. A tight, precise, rule-governed process of evidence-theory evaluation may be a chimera, an interesting 'game' to toy with but an unattainable goal (see, for example, Earman, 1983). In principle, once this elasticity or even partial arbitrariness in the rules is admitted, the foundation of the entire scientific enterprise *could* fall apart, but it doesn't in practice. It's all a matter of *degree*. By agreement, some questions are left alone.

We must not err by attempting to present science as being better than it is. We are reduced to doing the best we can in constructing and evaluating scientific theories and then we must suitably enlarge our definition of and criteria for scientific rationality. This enlarging of the class of what we are willing to label 'rational' is analogous to a move often made in mathematics. When a class, such as continuous functions, turns out to be too narrow for the problems at hand, one enlarges the definition in such a way as to keep the desired essential properties of the original class and yet to include more cases of interest. Thus, for functions one generates the sequence: continuous, piecewise continuous, integrable, generalized (i.e., distributions). Similarly, we ought not be more stringent in our criteria of what counts for rational activity than can actually be accommodated in practice. In a sense, to be rational may be to do the best one is able (at a given time under a given set of circumstances). So, although we do the *best* we can, that best turns out to be considerably less than we might at first have hoped for. While we must avoid the extremes of demanding too much or too little from science, we must surely begin with an adequate *description* of scientific practice.

We have already advocated Laudan's (1984a) triadic model as providing the broad outline or general matrix within which to view scientific practice. Laudan's basic position is that scientific debates are conducted and settled at three levels: that of fact or practice, that of methodology and that of goals or aims. This is reminiscent of Buchdahl's earlier (1970) triadic methodological structure. Laudan does not see this as a unidirectional hierarchical scheme, but one coupled at all three levels so that there is a network of linked causes that produces the effect of convergence of scientific debate to a common opinion. There is no single, overriding factor that is dominant in all cases and no algorithm for settling debates, any more than there is just one scientific method. As we have already seen, Galison (1987) has made a similar case for the multi-level, interlocking arguments used to arrive at a consensus that a new class of phenomena has been discovered in the laboratory. This stability of opinion is not to be equated to any logically necessary uniqueness.

Basically, through trial and error scientists find rules by which they make choices to arrive at empirically adequate theories (or to agree that an experimental 'fact' has been observed). We demand such theories and in some fields we are able to create what we demand. This empirical adequacy is relative to problems that are taken as being important at a given time. Precisely what this criterion means need not remain the same for all time.

The question is: How is convergence of scientific opinion produced and maintained around a fairly localized center as that center itself shifts? Let us use as the framework within which scientific debate takes place Laudan's coupled, triadic model of practice, methodology and goals. One of his claims is that significant shifts occur at all three levels. This does not mean that such shifts occur at *all* levels in *every* or even in most cases. In fact, that would be a disaster. Different factors or levels are more important at different times. Certainly the most common, and therefore the most important practically, although perhaps the least interesting philosophically, is debate carried on at the level of practice. We take this level to include not only specific experiments and their interpretation, but the actual testing of theories and the choice between them based on empirical adequacy. While each reader could generate his or her own (long) list of examples at this level of debate, let us cite just a few from fairly modern physics (see Brown and Hoddeson, 1983, for others): the discovery of the particle nature of cosmic rays, the conclusion that there are two types of mesons (the strongly interacting π and the weakly interacting μ), the breakdown of Dirac single-electron theory *via* the Lamb shift and the anomalous magnetic moment of the electron and the explanation of these effects by quantum electrodynamics. Another case that we have discussed is the shift from quantum field theory in the late 1950s and early 1960s to S-matrix theory for most of the 1960s and the subsequent shift in the mid 1970s back to quantum field theories (in the form of gauge field theories) (Cushing, 1982, 1983a). In this last case, the movement of concentration of effort among the community of physicists from one theory to another took place at the level of *practice*, but it had other potential implications as well, as we have seen (e.g., the criterion for a fundamental explanation in SMT).

As we have discussed in the previous section, Laudan (1981) has given his own example of a major shift at the level of methodology. However, as an illustration of a significant shift, perhaps bordering on the level of methodology as an allowable resource for explanation, we can cite the postulation of theoretical entities (such as quarks) that cannot, even in principle, exist as individual or free objects. It turns out, interestingly enough as we saw in Section 6.7, that one of the motivations for postulating quarks to be unobservable in field theory was the bootstrap conjecture which is usually associated with the rival *S*-matrix program (Gell-Mann, 1987).

Finally, at the level of the goals or aims of science, a requirement of long standing has been that our theories give us an intelligible explanation, one which by and large accords with our common sense. Certainly, in pre-quantum physics, locality and determinism would have been among the unquestioned canons of an acceptable explanation. However, quantum mechanics (and all the associated implications of the Bell theorems) does not permit such a constraint (on microprocesses, at the very least) (cf., Cushing and McMullin, 1989; Cushing, 1989c). There is today a great struggle in the philosophy of science to attempt to salvage some form of (local) realism from all this debate. Another possible route in this choice between locality and determinism is Bohm's (1980; pp. 77, 80 ff.) nonlocal quantum potential, which was discussed in Section 9.5. There one buys determininism at the price of nonlocality. Either choice, however, requires a profound change in our 'standards' of intelligibility. Similarly, *S*-matrix theory, had it become the dominant theory of high-energy physics, could have required a profound shift in emphasis away from ontology and toward epistemology alone as the reasonable (almost instrumentalist) level at which to ask scientific questions (Cushing, 1985), a point we developed in Section 7.3.

These examples give further support to two points we have made previously. First, shifts *do* occur at all three levels (practice, methodology and goals) of scientific debate. Second, these three levels are interconnected and changes effected at one level can produce changes at another (higher or lower). The time evolution is *continuous* (if examined on the proper scale) and different sets of criteria are the most important at different times. This fine-grained continuity of concepts, even through periods of rapid change, is important if we are to understand scientific activity as rational (i.e., as taking place in accord with some rules or standards). Thus, the empirical success of quantum mechanics (level of pracrice) produced profound conceptual changes at the level of goals regarding the requirement of intelligibility. Again, effectiveness of the *S*-matrix program at the level of phenomena implied major revisions at the conceptual level (ontology versus epistemology). These are both instances of our common sense being revised or defined by successful theories. Action at a distance could be taken to be another from an earlier era. Virtually any conceptual concession will be made for an *empirically* successful theory. What could be more remarkable (and obviously counter-commonsensical) than, say, the Pauli exclusion principle; the Aharonov–Bohm (1959) effect (on the status of potentials in quantum mechanics); Fine's (1982) suggestion of a possible 'harmony of random choices' as one attempt to save locality; or Bohm's (1980) implicate order to save determinism? The basic issue is: How much could science change and still remain science? *In principle*, it probably could by almost any amount over time, a point that has also been made by Shapere (1984).

Now that changes at all three levels have been established, how do we account for the resolution of these debates and for the convergence of scientific opinion? We must (and this is one of Laudan's central points as well) examine several factors which, together, produce convergence. Even more important, we must avoid certain 'philosophers' games' that confront us with problems that do not really exist in practice and that can appear to make any type of meaningful convergence seem illusory. Let us give two examples of these. One ploy is to take the statement of a general insight extremely literally and produce a *reductio* argument to show that the statement is nonsense. In this spirit, we must surely see the lasting significance of Hempel's famous Ravens' Paradox not as refuting as totally incorrect the intuitively appealing idea that more confirming instances are better for a theory – for there is certainly something correct about that – but rather as indicating that the process of theory evaluation is not as simple as we might like it to be. Or, typically, a representation given is that there is a set of data or of empirical facts E and a set of theories $T_1, T_2, \ldots T_n$ which cover that data. Riddle: How do we choose *one* theory from among this class of undetermined theories? Well, the Duhem–Quine underdetermination thesis notwithstanding in principle, one is not in actual practice presented with that type of choice. Usually there are a few (say, one or two) empirically adequate theories around at any given time. We need not look far to find the reason for this. Workable theories are sufficiently difficult to construct, and so few scientists have the ability to create them, that the scientific community is faced with a choice *between* theories rather than *among* (a plethora of) theories (Cushing, 1982, p. 72; 1983a, p. 35.)

So, once we have avoided such captious moves that might in principle make convergence appear either remarkable or ungrounded, we can see one evident factor tending to produce convergence. Anyone who talks with working scientists will be aware that most (if not nearly *all*) practicing scientists *believe* both in the *truth* of a successful theory and that, moreover, in principle there exists *one true theory* which could explain everything. We are not claiming that all the creators of successful theories hold this view, but only that the overwhelming majority of the members of the scientific community do. At the moment we do not care whether or not such a belief is well-grounded or justified (by, say, a realist position, etc.). It is the *existence* of the belief that is important for the present discussion, not the correctness of that belief. Connected with this belief and with the painful difficulty of getting things to work (be they theories or experiments) is the primacy, for practicing scientists, of empirical adequacy and the ability to control phenomena as a test for an acceptable theory or law. Part of the reason for this is, again a belief, that such tests constitute an impartial court of appeal (although we reiterate some questions about this below). We take this to be part of what Laudan (1984a) means by his Leibnizian ideal.

Now just as an actual choice must be made between theories rather than among all possible theories, so the criterion of empirical adequacy is applied to some selected set of important problems, not to all possible problems. It becomes essential to realize how much theory and experiment feed on one another and that different theories often work for different sets of phenomena. Andy Pickering (1984, especially Chapter 14) has recently termed this incommensurability at the level of practice. He uses as an example the 'old' (pre-mid 1970s) physics versus the 'new' physics. His point is that new classes of phenomena were studied for the new gauge field theories. To make contact with something we mentioned before (at the cost of an oversimplification, admittedly), let us associate a class of high-energy experiments, termed soft-scattering events, with S-matrix theory (SMT) in the sense that that theory produced a coherent account of those results and was able to suggest new experiments to be performed. As gauge field theories (QFT) came to the fore, a new class of hard-scattering experiments proved more fruitful for mutually advancing (the new) theory and for empirical confirmation. The important point is that the new theory (QFT) does not do well at handling the old (soft-scattering) data, while the old theory (SMT) is not effective at accounting for the new (hard-scattering) experiments. A new class of phenomena is studied to support the new theory and the old phenomena are forgotten about (or at least moved to the back burner for a while). The relevance of this for the present discussion is twofold: the court of appeal (phenomena) is not wholly observer-independent since we choose, and at times even create (Hacking, 1983), what we look at and, more important, this strong coupling between theory and experiment keeps the center of mass of activity fairly localized, thus contributing to convergence of activity. The latter effect is enhanced, at least in current high-energy physics, by the need for experimentalists to *do* something and to exhibit *progress* toward something. The experiments are large, only relatively few can be done, and they had better produce a significant result. The practical character of the ongoing scientific enterprise can get lost in abstract philosophical discussions of methodology. One of the reasons for the recent success of gauge field thories is that they have been able to interact fruitfully with experimental practice to produce the convergence necessary for (at least apparently) progressive scientific activity (cf., Galison, 1987). That is, reality has its (largely objective) constraints, but the choices of what to look at are ours. The experimental phenomena are objective in the sense of being stubborn in spite of a preconceived theoretical framework that may exist. Still, the class of these objective phenomena we choose to study can shape our picture of the world.

Perhaps this is a plausible and reasonably faithful picture of how convergence of scientific opinion is *produced* at various levels. Certainly the central feature of this activity is the requirement of empirical adequacy. A theory sufficiently successful at the level of practice will eventually produce the necessary changes at the level of methodology or of goals. Still, though, what is taken as significant for empirical adequacy is often determined or selected by these higher-level criteria. It is truly a coupled or bootstrapping system.

Even if we accept in broad outline this picture of how scientific opinion becomes and remains fairly localized around a central position as that position itself evolves in time, we can further ask why it is that the final theories thus obtained remain so stable. These theories provide the base upon or the background against which this ongoing evolution takes place. We *fit* our theories (Cushing, 1983a) to the phenomena we have selected. Once these theories have broken down (or been falsified) so that we know the limits of their validity, we then have theories that have been variously termed mature (Putnam, 1975, p. 73; 1978, p. 21), mythical or archival (Misner, 1977), and established (Rohrlich and Hardin, 1983). Misner has gone so far as to claim that immutable truth in science is to be found *only* in these falsified or archival theories. That is a reasonable position since the regime of validity of such theories has been sharply delimited and they have been extensively tested there. In a sense, they are (merely) equivalent to, or faithful representations of, the phenomena they cover. They have been exhaustively tested for nonfalsity (Cushing, 1983a; p. 35). We can be confident that they give us true predictions in their restricted domains (although the matter of true explanations remains open to debate; Laudan, 1984b). Looked at in this way, the explanation of their success does not appear remarkable or miraculous (even though man's ability and struggle in actually discovering or creating them remains so). The goal has been empirical adequacy and we have obtained that through a process of accommodation. Necessary epistemological and ontological accommodations have been made in the interpretation of successful theories to produce coherence. We have seen an example of this with the adjustment of the pragmatic meaning of the term *causality* in nucler and high-energy physics (cf. Section 9.4). That is, goals and practice have been mutually adjusted so, of course, there is success. This is not to claim that just *any* set of practice, methods and goals can be forced to coexist harmoniously. Laudan (1984b) has made similar observations about the symbiotic effect between goals and practice in his discussion of scientific realism versus relativism in the interpretation of scientific theories.

We can summarize our argument with a 'story' about three areas or groups of phenomena to be studied. Let us call these groups A, B and C. Group A consists of questions for which many theories or explanations can be found. Here we have an obvious case of underdetermination so that there is no clear choice and hence no convergence of opinion. Group B consists of problems for which even the brightest people are able to create just one or two theories that work, so (by default) there is convergence. Finally, group C has questions so difficult that people cannot find *any* theories that handle all the problems deemed important, so again there is no convergence of opinion to any one of these partial theories. This explanation looks for the source of success in some areas in the subject matter (i.e., set of phenomena) studied. Some phenomena lend themselves to an empirical approach. This is not a matter of science versus nonscience, but a difference among disciplines.

It may be that in science (say, in physics) we are able to create only one theory (or *very* few theories) that covers a large set of phenomena because of the relatively simple nature of the questions raised. If there were too many 'good' theories (i.e., they were too easy to make), there would be divergence; if *none* (e.g., philosophy, psychology), there would again be (as there is) divergence. We have finally found *some* (few) areas where this empirical approach works and has important, useful, etc. results and applications. If this is so, then any *internal* demarcation problem (about the boundary between science and other forms of rational knowledge generation) 'evaporates' or becomes a nonproblem (as does the null result of the Michelson–Morley experiment for special relativity or the equivalence of inertial and gravitational masses for the equivalence principle). So, once science works, it seeks out those phenomena and problems for which it continues to work. There is no compelling reason to believe such an approach will necessarily be extendible to *all* fields.

We have argued for a representation of the scientific enterprise in which the stability of successful theories is a reflection of the *difficulty* of constructing them, rather than of their objective and unique truth. An idea or a theorist must enter the field and hold it against all comers simply because that is the best way we have discovered of producing *stable* knowledge. That is, the creators of this knowledge and the most able critics in the field have asked all of the probing questions they have been able to think of and the theory (if successful) has survived. So, of course, such a theory is stable (for a longer or shorter period of time), until some more clever person finds the crucial chink. This stability is related to the fact that *very* few people have the ability *and* good fortune to create a theory that can cover a set of data and keep them nailed down while the theory is adjustd to cover some new phenomenon. A case can be made (Cushing 1982, p. 78) for scientists as clever people who make their theories work, rather than as discovering laws of nature that pre-exist outside their own minds. Of course, not just any theory can be made to work. Theorists generate a network that covers reality. This is in some sense analogous to a suit of clothes that fits a person. There is certainly no unique suit which is the only one that fits. However, a bad fit is evident. Perhaps the success of a theory resides as much in the cleverness and continued ingenuity of its creator as in its external, objective status as a 'law' of nature. A similar view of successful philosophical systems or explanations put forth by pugnacious and brilliant philosophers who argue well, write extensively, and bludgeon their opponents into submission (or at least silence) would possibly not raise great objection in some quarters.

We have described a convergence of scientific *activity*, which must be distinguished from productivity and progress toward a fundamental theory or a 'true' view of the world, a point already made in the previous section. That is, this case study, as well as the extensive field theory enterprise that was outlined at the end of Chapter 1, has been a story of much *activity* in the field of strong-interaction physics, but perhaps of relatively little *progress*. The strong interaction remains an unsolved problem. Michael Moravcsik puts this very succinctly as:[13]

[A]s far as strong interactions are concerned, we have not made any substantial physical progress since Yukawa in 1935. His suggestion that forces are created by particle exchange, and that there is a relationship between the range of the force and the mass of the exchanged particle, remains the only physical idea strong interaction physics has been existing on, and the history since 1935 has been simply a series of mathematical reformulations of this physical idea: ... [O]ld-fashioned field theory, then dispersion theory, then Regge poles, and now QCD are simply reincarnations of the same Yukawa idea.

The point is that everything bogs down after one-'particle' exchange calculations (such as those associated with Figures 1.3, 1.4, 3.1(b), 4.1, 4.3, 4.4, 4.5, 4.7, 4.8, 4.9, 4.10(b), 5.1(c), 6.7, 6.9 or 6.11 of the text). It is basically Yukawa's 'force' term (the pole term of QFT or SMT) which drives Heisenberg's S-matrix equations for calculations. Attempts to go significantly beyond this are so complex as to be at best ambiguous in their success. In nuclear physics even today, the 'Bonn' (Machleidt, Holinde and Elster, 1987) and the 'Paris' potentials (Vinh Mau, 1979, 1985) are QFT and SMT attempts, respectively, to handle particle exchange phenomena in generating the basic nucleon–nucleon force. These are really semiphenomenological schemes, rather than calculations solely from first principles.

Although the preponderance of opinion today is overwhelmingly in favor of the quark-gauge-field-theory model, it is perhaps worthwhile, even if a bit sobering, to emphasize just how far the pendulum had swung the other way by the mid 1960s. George Zweig (1980, p. 36), an independent creator of the quark (or aces, in his terminology) model usually associated with Gell-Mann, recalls what happened when he sought an academic position:

> When the physics department of a leading University was considering an appointment for me, their senior theorist, one of the most respected spokesmen for all of theoretical physics, blocked the appointment at a faculty meeting by passionately arguing that the ace model was the work of a 'charlatan'. The idea that hadrons, citizens of a nuclear democracy, were made of elementary particles with fractional quantum numbers did seem a bit rich. This idea, however, is apparently correct.

Finally, there is the very interesting question of judgment for a scientist in deciding just when to stop working on a theory which appears 'stalled'. To cite a 'bandwagon effect' is no explanation. That simply reflects the fact that most cannot create new theories and, because of the premium science puts on solving problems (as opposed to merely working on solving problems), the majority must join an active and productive program. Nevertheless, one finds that some successful creators of scientific theories continue to stay with a program long after most workers have deserted it. Einstein's quest for a unified field theory is an obvious example from an earlier era. Chew's dedication to topological *S*-matrix theory could also be cited, as could Schwarz's (and Green's) research on (super) string models. Such people enter a desert – some manage to come out and some don't. Mandelstam and Olive made a transition from the *S*-matrix program to the string one and continued to work on it in the lean years. That was certainly not a bandwagon switch. So, there is a whole spectrum of judgment thresholds for motivating change from one program to another. Such judgment is an essential part of the scientific enterprise. But, what are the criteria of good versus bad judgments? If these can be assessed only retrospectively (*á la* Gell-Mann; cf. Section 10.5) *after* the eventual success (or failure) of the program worked on, then 'bad' choices are not necessarily irrational ones. Certain schemes just can't be fit to nature and one's fortunes there are partly luck of the draw.

10.7   *A* view of science

Much of the philosophy of science in the present century, including the work of the logical positivists and of many of the 'postpositivists', has been concerned with attempting to identify and articulate a special, even unique, hallmark of scientific knowledge. This has usually focused on the methods of science as providing a warrant for the rationality of the scientific enterprise. Although the most desirable situation might be one in which absolutely fixed and true laws or theories would be produced, it was long ago recognized that the 'true' scientific theories of one age were often overthrown, sometimes with accompanying radical conceptual revisions. That is, it was obvious that the actual historical record is blatantly incompatible with a claim that science produces final, unchangeable theories (although the hope of a much earlier age had been necessary, and hence unique, laws and theories). Nevertheless, rationalist or foundationist postpositivist schools in the philosophy of science (for example, Popper, 1963, and Lakatos, 1976, retain much of this spirit) continued to maintain that the rationality of science could still be secured by the fixed methods peculiar to science (perhaps by a logic of justification or by the testing of its proposed theories) and that the goal of science remained truth (even if only in some sense of an approximate correspondence with reality). An apparently reasonable corollary to this view of science is (one form or another of) scientific realism which takes as actually existing the theoretical entities postulated by successful scientific theories.

A foundationist approach attempts to lay down certain principles that *must* be adhered to. These may be based upon *a priori* arguments or upon requirements necessary if one is to avoid the possibility of contradiction. An example of this type of maneuver is Watkins' (1984) attempt to establish that there must be *one* goal for science in order to prevent (p. 123) two different theories from being judged equally good, each satisfying equally well two (possibly) quite different goals. An appeal to coherence is made in order to block such an occurrence. However, in fact, rival (even incompatible) goals *could* be adopted by different groups of scientists. That might be undesirable and we might not like it, but that does not mean it could not occur. Among the requirements that Watkins sets for any goal of science is that it must be coherent, feasible and impartial (p. 124). It is not obvious that the first two will be mutually consistent in the real world, a point we return to when we discuss the explanations that quantum mechanics allows. Furthermore, the claim of impartiality is problematic when case studies are considered in detail. Although philosophy of science may once have set truth as a goal to which science aspires (Watkins, 1984, p. 155), in fact actual scientific practice must often be content with a noncontradiction, arranged through a process of accommodating our initially hoped-for demands with what nature will sustain. Watkins (1984), for example, begins with his Bacon–Descartes ideal and then is forced realistically to cut back drastically to a much lesser goal, but one that he hopes can still defeat the rationality skeptic. But even these reduced demands remain more in the nature of a wish, rather than a necessity that logic demands.

Although the simplicity and epistemic security provided by a foundationist model of science does give it a certain initial appeal, a closer examination of the historical record of actual scientific practice has shown that things are not as simple as we might hope them to be. The logical positivist school not only provided a foundationist reconstruction of what science ought to be – one which failed to accord at all with real scientific practice – but it also was unsuccessful in elucidating a logic of induction. So too, the rationalist postpositivists have been unable, in spite of many proposed methodologies, to identify *the* hallmark of scientific rationality. Again, it has been the complexity of, and the variation in, actual scientific practice that has confounded such attempts, even though these methodologies have often identified some important (but not universal) aspects of scientific practice. As an alternative to the failure of rationalist programs in the philosophy of science, naturalized and socialized descriptions and explanations of science have been proposed. The best known example of this approach is Kuhn's (1970), with the Edinburgh 'strong programme' school (Bloor, 1976) providing a radical example of the sociology of knowledge that would account for not only the form but even the very content of scientific knowledge in terms of social factors.

Actually, in the Continental tradition, Bachelard, as long ago as the 1930s, had developed a naturalized philosophy of science. In his *The New Scientific Spirit* (1934), we find what could almost be taken as a slogan for a central theme of the present monograph: 'Science in effect creates philosophy' (1934, p. 3). Bachelard takes psychology and epistemology to be equally important for the dialectic of science (p. 27) and appreciates that syntax and the meaning of terms 'bootstrap' each other (p. 53). He not only questions any (necessary) permanence of the forms of rationality (p. 54), but even sees mathematics as producing essential changes in the scientific spirit (i.e., in the practice, methods and goals of science) (p. 55). The latter is consistent with his view of science as substituting equations for descriptions and quantities for qualities (p. 68). Our inherent belief in simplicity is a reason that we use the concept of determinism in an attempt to rationalize reality (pp. 102–3). Causality is seen as '... still a fundamental category of objective thought' (p. 111). Bachelard asks (p. 137) '[A]re there methods [of science] that are exempt from obsolescence? The answer would seem to be no.' He (1947) sees science progressing by changing its *methods*. This has been precisely our contention in arguing for an evolving and essentially changing framework of practice, methods and goals.

A relativist or irrationalist (perhaps better, *a*rationalist in contrast to the rationalist) school sees scientific knowledge as contingent, being determined by social and historical factors, so that the specific laws of science become largely arbitrary. However, just as the purely logical or rationalist program in the philosophy of science foundered on the harsh reality of historical scientific practice, so the relativist or arationalist one fails to be supportable in the face of a detailed examination of specific episodes in the history of science. While the logicists have made important contributions to the philosophy of science by identifying and elucidating the often compelling internal structure of scientific practice, the sociologists of knowledge have quite correctly provided important insights into the role played by external factors, not only in the formation of the scientific community, but even in the production of scientific knowledge itself. Each of these facets of science is central to an understanding of science, but neither alone is sufficient for providing an accurate picture of science.

The model of scientific change we advocate occupies a middle ground between the overly rational reconstructions of the logicists and the overly arational reconstructions of the sociologists of knowledge. It is not fruitful to argue whether it is possible to compromise between rationalism and arationalism. The point to be appreciated is that each approach has provided important and valid insights into the scientific enterprise and these insights should be incorporated into any model of change in science. The model we have discussed is *naturalized*, in that it is based upon a detailed examination of the actual historical record of (current) real scientific practice to provide the 'data' to which the methods of science are applied in studying the scientific enterprise itself; *socialized*, in that it takes cognizance of 'external' sociological and individual factors in the construction and justification of scientific theories; *highly constrained*, in that it recognizes and examines in detail the (not uniquely determining) 'internal', often strongly compelling, logic of a theoretical framework once that framework has been (even tentatively) accepted for development and testing. The process of constructing scientific knowledge is a bootstrapping one and there is no given, fixed foundation upon which to base it. The same is true for constructing a methodology of science.

A priori approaches to methodologies of science (and to scientific knowledge itself) have not only failed, but they even seem implausible as a reliable means of underpinning epistemic claims given the new, virtually unimaginable (in advance of actual discovery) ideas and images of our world that have been produced by science (an observation Shapere, 1984, has often made): non-Euclidian spaces used in physics; Bohr's principle of complementarity; the non-separability of quantum mechanical systems; the possibility, both in the QFT and the SMT frameworks, that at a fundamental level the concept of space-time may not be meaningful (Kaplunovsky and Weinstein, 1985; Chew and Stapp, 1988); and the existence of fractals, just to cite a few of the more dramatic examples from fairly recent theoretical physics. In fact, the record of scientific practice provides a source of fruitful analogies for new ideas in the philosophy of science, rather than new methodological principles flowing *from* the philosophy of science itself, as in an earlier era. This is again a bootstrap strategy, since case studies are both the source of and the test of methodology. Laudan (1984a) has suggested a dynamic scheme in which the theory, methods and goals of science mutually influence and bootstrap one another, so that not only the theory (or laws and practice) of science, but also its methods and goals, evolve over time. Just as Laudan has argued that the method of hypothesis represented a major change in the methodology of science, we have made a case that the bootstrap conjecture of modern S-matrix theory could have produced a significant modification in the methods (and even in the goals) of science. The cosmological anthropic principle (Barrow and Tipler, 1986), in some of its more radical (teleological or participatory) forms, is another illustration of a possible revision at these metalevels of science. Forman (1971), Pickering (1984), Schweber (1984) and Galison (1987) – each in quite different ways – have shown the importance of social and sociological factors in the formulation and acceptance of scientific theories and of the world views they foster. The present extensive case study of a major episode in modern theoretical physics has indicated the effect of the pyramid structure of scientific community in which only relatively few of the ablest people can create and convincingly sustain new concepts and theoretical frameworks which the vast majority of scientists must then work on. This extreme difficulty of producing a viable theory and the even greater ingenuity required to accommodate a nascent research program to new challenges are central factors in producing a convergence of scientific opinion to a stable and fairly localized framework as that framework itself evolves (Cushing, 1984). It is just not so that *any* theory can be made to work. But, once one has bought into a particular (mathematical or theoretical) set of assumptions, the internal logic of that framework can be compelling in leading one to specific conclusions or predictions. This is evident in modern theoretical physics where the mathematical formalism often guides physical intuition in providing new insights (or suggestions for the postulation of new physical entities or processes), rather than the physical intuition indicating developments in mathematics as in previous times. Once the theory or model produces a 'successful' result, there is a tendency to look retrospectively, from the vantage point of this accepted formalism, back at the developments that led to the success and to see that chain of events as 'inevitable'.

These same case studies, though, make it implausible that the theoretical framework accepted by the scientific community at a given time is in any sense unique or alone compellingly demanded by the facts available. But for accidents of history, not only might the 'relevant' empirical data be based on a very different set of chosen phenomena, but even the interpretation or representation of a given set of phenomena could be quite different.

An acceptable methodology, or perhaps really only a good *description* of scientific practice, must stand or fall largely on having a proper spirit or emphasis in its representation of science. Of course, it must be faithful to the historical facts, but that alone is not enough since many, often conflicting methodologies can be 'fit' to various episodes in the history of science. We should recognize that science is what scientists have done[14], not what a philosopher tells us the scientists meant to do, were *really* doing, or should have done. Successful theories are *made* to work; they just don't work (on their own or because nature demands it). Once we are inside a formalism (e.g., a mathematical one), we may feel that its own internal logic seems compelling. For example, Raine and Heller (1981) in their retrospective reconstruction of mechanics from Aristotle to Einstein, as seen from the vantage point of general relativity, claim (p. 57):

> The *natural history* of scientific ideas is determined by its own internal logic to a much greater extent than a superficial glance at the historical record might indicate. The logic becomes more evident once one translates the history into a technical language of the present.

However, this statement is made within the framework of the present, 'correct' theory. An essential aspect overlooked by such an approach is how one buys into the starting assumptions of the formalism. Of course, not just *any* theory can be made to work. The external world has its constraints, but within these limits there is latitude and the need for choices for which a (never completely or uniquely compelling) argument must be made and sustained (much as in the philosophy of science).

From case studies and methodologies, we can draw up a set of *desiderata* for a scientific theory to satisfy. The more the better, but these criteria are neither necessary nor sufficient for acceptance of a theory at a given time. Today in science, the (eventual) requirement of (a not wholly theory-independent) empirical adequacy has attained the status of an almost necessary condition for theory acceptance. Fertility, or generative potential, is another fairly common characteristic of successful scientific theories. No theory satisfies *all* of the desirable requirements we might wish for. Scientists make the best arguments they can in a given set of circumstances. Some carry the day and some do not. There is no algorithm for success in this enterprise. Detailed, philosophically sensitive examination of specific episodes in the history of science can reveal what successful stratagems have been, but not universal characteristics. In fact, focusing on such criteria is too narrow an exercise to reveal the many-faceted and holistic nature of the scientific enterprise.

In summary, then, and to return to the scheme we outlined in the preface, we have argued for a particular framework within which to organize and view science. As a starting point we suggest a matrix within which practice, methods and goals are mutually coupled and evolving. Under practice, with its facts, laws and theories, we include the overwhelming bulk of scientific activity which encompasses experimental work and the construction of laws and theories. For many this may seem to be just about the whole of science. Shifts at this level are familiar enough and few, if any, would argue for an absolute stability there. A long list of significant changes in theories and constructs is readily generated and we have discussed several examples. The next level, or layer (depending upon your point of view), is that of methodology or rules by which one judges theories and constructs. The method of hypotheses (or hypothetico-deductive method), induction, generally accepted modes of explanation and rules of inference are found there. There is a component of formal logical structure that is important, but it is not everything. Canons of rationality are included here. However, rationality, as giving 'good' reasons, consists largely of fitting new developments and conjectures into a presently accepted conceptual framework. This level is relatively more stable than that of practice, but it is not immune to change. The method of hypotheses was not always the accepted mode of argument for the existence of (at-present) unobservables. We have also offered the bootstrap conjecture from the S-matrix program and the cosmological anthropic principle(s) as plausible candidates for changes in methodology. At a 'higher' level (or perhaps 'innermost' layer) are the goals or aims of science. Truth (approximate, etc.) is often given as (*the*?) goal of science. However, it is by no means evident that the goal of the dispersion-theory program was truth in the sense of any fundamental understanding of nature. Rather, one could represent that activity as an attempt to organize and correlate empirical results. Furthermore, science has arguably traditionally sought intelligible explanations, ones that largely accord with our common sense. Quantum mechanics, as a reflection of the phenomena of the real world, appears to have changed all of that.

The point we wish to make here is that *none* of these levels or components is immune to change, either in principle or in fact.[15] This triad is mutually coupled and evolving. The outer layers of practice are subject to more rapid variation, but these changes also affect the inner core. This coupled network of practice–method–goals evolves over time under the pressure of constraints from the real world (e.g., experiment), as well as from external (e.g., sociological) factors, an example of which we have illustrated in the dispersion-theory program as the need to *do* something. We cite Medawar (1967) on the premium placed on scientists' *solving* problems, as opposed to merely *working* on solving problems. This extremely pragmatic bent of physics since the Second World War has been emphasized by Schweber (1989) and by Cini (1980). In a very different context, Stephen Gould has made a similar point in writing about theory choices in 'the great Devonian controversy' in geology during the 1830s in England. In commenting upon a reaction to excess speculation, Gould (1986, p. 9) observes:

> But they understood the cardinal principle of all science – that the profession, as an art, dedicates itself above all to fruitful doing, not clever thinking; to claims that can be tested by actual research, not to exciting throughts that inspire no activity.

This coupling of 'internal' (to science) and 'external' factors is an essential feature of scientific practice.

In this evolution of the coupled components of our triad, practice is typically the overriding and ultimate determiner of this evolution. That is, we claim that new innovations in methodology and in goals are ultimately rooted in successful scientific practice (i.e., what proves to be *doable*). In a time of extreme difficulty or crisis, there is a willingness to countenance a change in the rules (*any* rules, if necessary) of the game. This evolution is not an erractic or chaotic process, however. Language (of current discussion and practice) provides continuity. As we have already emphasized, much of scientific rationality consists of fitting new ideas into a matrix or network of accepted concepts. Convergence to a fairly localized position as that position itself evolves over time is produced in large part by the difficulty of fitting theories to the constraints provided by the real world and by the fact that viable theories can be produced by relatively few people. Schweber (1984) and MacKinnon (1982) have also stressed the importance of the few in the development of quantum mechanics and of quantum field theory. This 'convergence by default' may be local rather than global. That is, once a not absolutely necessary or not uniquely required choice has been made at a critical juncture, a set of options or a range of possibilities is delimited 'internally' by the logic of theory and by the empirical data. However, detailed examination of the actual route of historical development at least makes it plausible that things might have gone quite differently. Among the examples we have considered already – most especially the SMT versus QFT competition – we can cite (Section 9.5) the Bohm–Vigier (1954; Bohm, 1952; Bohm, Hiley and Kaloyerou, 1987) version of quantum mechanics versus the standard Heisenberg–Schrödinger version (today codified as the 'Copenhagen' interpretation). This case also illustrates that the methodological phenomena we discuss are not peculiar to current high-energy physics alone. More details of this example will be forthcoming in an extensive study (Cushing, 1989b).

We have seen that the Bohm–Vigier formulation does allow one to maintain an essentially classical view of determininism, at the price of some apparently innocuous violations of locality (c.f. Cushing and McMullin, 1989, for some comment on the nonlocality issue). Perhaps less plausibly, there is also the point that a 'superdeterminism' (i.e., no freedom of choice for the experimenters themselves) could in principle remove any conflict between locality and empirical results (e.g., Peres and Zurek, 1982). These examples show that, but for historical accident or extrascientific disinclinations toward certain schemes, our currently accepted worldview of fundamental phenomena could be very different from what it is. Who gets to the top of the hill first holds the high ground and must be dislodged by 'newcomers'. The mere fact that there is later a logically consistent alternative theory is not in itself sufficient motivation for a community of practitioners to change to a new expertise. At certain critical junctures theory choices can be badly underdetermined by existing constraints. Such 'bifurcation points' represent times of great instability for the historical evolution of a theory.

We have emphasized repeatedly that not just *anything* can be made to go in science. At the same time, though, the theoretical framework that science settles on is neither uniquely required by empirical considerations nor by logic. Methodologies of science should not reconstruct scientific activity as a more rational enterprise than it is. Nancy Cartwright (1983), in her *How the Laws of Physics Lie*, has made a case, really, that the general laws and theories of physics do not tell the *whole* truth. That is, *simple*, general, underlying laws of physics often give a deceptive unity to an explanation of actually extremely complex phenomena. They gain their unifying explanatory power at the price of belying the complexity of the real world. Similarly, too rational a reconstruction of scientific practice represents a poor caricature of actual science. One must balance rational, sociological and historical factors in any description of science.

The developments in modern physics that we have discussed in this monograph cannot *force* one to abandon a belief in a fixed set of (meta) principles. While each of the characteristic features of these episodes (e.g., mathematics as the *source* of new ideas in physics, the essential use of mathematical models, abstraction from specific models, the origin of new 'metaphysical' ideas rooted in highly technical, quite abstract practice) may be found in some form and to some degree in the science of previous eras, the *degree* to which modern theoretical physics relies essentially on these modes of operation is a significant change. Such a quantitative difference can produce a qualitatively distinct enterprise. Of course, if one is willing to contort the (meta) framework enough or make it sufficiently vacuous, then one can continue to maintain the position of an invariant set of meta-principles.[16] To make an analogy, which is not meant to be either captious nor irreverent, continued belief in a fixed set of methodological principles is in many ways similar to a belief in the existence of a personal, loving God. That is, given the recognized circularity of arguments from design and the scale of evil in the world, some people continue to believe in God in spite of these, while others simply can no longer. It is not a matter of what we would *like* to be the case, but rather of what we finally find plausible or believable, given our present information. We seem to have a felt need to find or construct ultimate explanations and these may arguably have some survival value (a theme developed long ago by Meyerson, 1908). Nevertheless, it does not follow that the evidence will necessarily support beliefs that satisfy such needs. My argument has been that an examination of practice in some areas of modern physics does not require or warrant a belief in a fixed set of methodological principles.[17] However, it cannot categorically rule out the possibility of such a belief either. The matter is reduced to a question of plausibility.

We have focused on potential and actual *change* in the scientific enterprise. This is not to deny the aura of stability of much of science. At the metalevels, stability is often more apparent than is change so that stability could seem the first-order (or dominant) effect and change a second-order one. In different contexts, one characteristic may be more important than the other. For everyday scientific practice, methodological stability predominates. For questions of *principle* in methodology, change is of central importance. In our story, philosophy of science has been reduced largely to an analysis of the history of (recent) science (to detect actual or potential changes in methodology and goals) and of the insights provided into methodology by science.

Appendix

In order to be able to illustrate the central concepts in subsequent developments in the main body of this case study, we establish a certain technical background against which we can discuss the origins of Wheeler's and of Heisenberg's scattering matrix. This outline will provide a brief and selective summary of nonrelativistic Schrödinger scattering theory, classical electromagnetic wave theory and the theory of the interaction of a quantized atomic system with the radiation field. The material presented here I take to be given and unproblematic by the mid to late 1930s. As a justification for this assumption let me point out that the essentials of all these topics can be found in the 1933 edition of Mott and Massey's *The Theory of Atomic Collisions*, in the 1933 edition of Slater and Frank's *Introduction to Theoretical Physics* and in the 1936 edition of Heitler's *The Quantum Theory of Radiation*.[1] These were all standard reference works at the time so that their contents can be assumed to have been known to practicing theoretical physicists of that era. The order of presentation here is not necessarily the historical one and very few specific references will be given for this background material. Also, a unified notation and a fairly modern presentation will be used to facilitate reading and subsequent reference for illustrations. The technical details given for this introductory background will be more complete than for most of the historical presentation in the main body of the text. The reason for this is that the mathematics is sufficiently simple here that the interested general reader should be able to follow the arguments leading to specific important results. Many later developments can be understood by analogy with these simpler examples, since the underlying ideas are often quite similar. The reader who is already familiar with this background material or who is not interested in it can move on to the historical development of the S matrix in Chapter 2.

The wave function $\Psi(\mathbf{r}, t)$ for a single particle interacting with a fixed force center represented by a potential $V(\mathbf{r})$ is determined by the Schrödinger equation[2]

$$i\frac{\partial \Psi}{\partial t} = H\Psi \tag{A.1}$$

where the Hamiltonian H is given as

$$H = -\nabla^2 + V(\mathbf{r}). \tag{A.2}$$

For steady-state solutions we take

$$\Psi(\mathbf{r}, t) = e^{-iEt}\psi(\mathbf{r}). \tag{A.3}$$

Then the wave function $\psi(\mathbf{k}_0; \mathbf{r})$ for a single particle moving in the field of a spherically symmetric potential $V(r)$ is the solution to the time-independent Schrödinger equation

$$[-\nabla^2 + V(r)]\psi(\mathbf{k}_0; \mathbf{r}) = E\psi(\mathbf{k}_0; \mathbf{r}) \tag{A.4}$$

with the energy E given by

$$E = k^2. \tag{A.5}$$

Here $\mathbf{k}_0$ is the momentum vector specifying the incident beam direction and $\mathbf{k}$ specifies the direction to the detector. Since we consider here *elastic* scattering, we have set $|\mathbf{k}_0| = k$. If a beam of free particles, represented by the incident plane wave

$$\phi_i = e^{i\mathbf{k}_0 \cdot \mathbf{r}}, \tag{A.6}$$

interacts with a scattering center $[V(r)]$ as illustrated in Figure A.1, then the asymptotic form of the scattering state is given as

$$\psi^+(\mathbf{k}_0; \mathbf{r}) \xrightarrow[r \to \infty]{} e^{i\mathbf{k}_0 \cdot \mathbf{r}} + f(k, \theta)\frac{e^{ikr}}{r} \equiv \phi_i + \psi_{sc}. \tag{A.7}$$

Here $f(k, \theta) = f(\mathbf{k}_0, \mathbf{k})$ is the scattering amplitude and ψ_{sc} the scattered wave. We use θ to denote the angle between $\mathbf{k}$ and $\mathbf{k}_0$.

An integral equation, equivalent to Eq. (A.4) and incorporating the boundary conditions of the incident plane wave of Eq. (A.6) and the outgoing spherical wave of Eq. (A.7), can be obtained (Cushing, 1975, pp. 509–10) in terms of the Green's functions $G_0(\mathbf{r}; \mathbf{r}')$ which satisfies

$$(\nabla^2 + k^2)G_0(\mathbf{r}; \mathbf{r}') = \delta(\mathbf{r} - \mathbf{r}'). \tag{A.8}$$

The required solution is

$$G_0(\mathbf{r};\mathbf{r}') = -\frac{e^{ik|\mathbf{r}-\mathbf{r}'|}}{4\pi|\mathbf{r}-\mathbf{r}'|}. \tag{A.9}$$

Equations (A.4), (A.8) and (A.9) then imply that $\psi^+(\mathbf{k}_0;\mathbf{r})$ is the solution to

$$\psi^+(\mathbf{k}_0;\mathbf{r}) = e^{i\mathbf{k}_0\cdot\mathbf{r}} - \frac{1}{4\pi}\int\frac{e^{ik|\mathbf{r}-\mathbf{r}'|}}{|\mathbf{r}-\mathbf{r}'|}V(r')\psi^+(\mathbf{k}_0;\mathbf{r}')dV' \tag{A.10a}$$

$$\xrightarrow[r\to\infty]{} e^{i\mathbf{k}_0\cdot\mathbf{r}} - \frac{e^{ikr}}{4\pi r}\int e^{-i\mathbf{k}\cdot\mathbf{r}'}V(r')\psi^+(\mathbf{k}_0;\mathbf{r}')dV'. \tag{A.10b}$$

Comparison of Eqs. (A.7) and (A.10b) yields a formal expression for the scattering amplitude

$$f(k,\theta) = -\frac{1}{4\pi}\int e^{-i\mathbf{k}\cdot\mathbf{r}'}V(r')\psi^+(\mathbf{k}_0;\mathbf{r}')dV'. \tag{A.11a}$$

$$= -\frac{1}{4\pi}\langle\phi_f|V|\psi_i^+\rangle. \tag{A.11b}$$

Here ϕ_f is a plane wave state with the final-state momentum vector $\mathbf{k}$ (i.e., Eq. (A.6) with $\mathbf{k}_0\to\mathbf{k}$) and the index i on ψ_i^+ labels the initial-state momentum vector $\mathbf{k}_0$ (cf. Eq. (A.10a)).

The flux density $\mathbf{j}$, representing the number of particles crossing unit

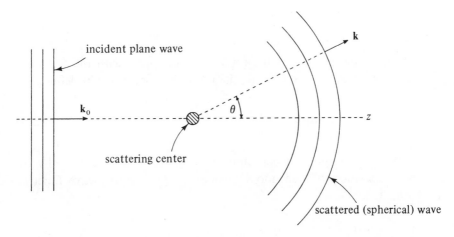

Figure A.1   Scattering of a plane wave by a center of finite extent.

area in unit time, is

$$\mathbf{j} = \frac{1}{i} (\psi^* \nabla \psi - \psi \nabla \psi^*). \tag{A.12}$$

Since a cross section σ is defined (for a single scattering center) as

$$\sigma = \frac{\text{number of events/unit time}}{\text{number of incident particles/unit area/unit time}}, \tag{A.13}$$

the differential scattering cross section per unit solid angle $d\Omega$ is

$$\frac{d\sigma_{sc}}{d\Omega} = \lim_{r \to \infty} \frac{1}{2k} \mathbf{j}_{sc} \cdot \frac{d\mathbf{A}}{d\Omega} = |f(k,\theta)|^2. \tag{A.14}$$

Here $\mathbf{j}_{sc}$ is the flux density of Eq. (A.12) computed for the scattered wave ψ_{sc} only.[3] The denominator $2k$ comes from the incident flux computed with Eqs. (A.12) and (A.6). The element of surface area $d\mathbf{A} = r^2 \, d\Omega \hat{r}$ is located a large distance r from the scattering center. This elastic scattering (i.e., $|\mathbf{k}| = |\mathbf{k}_0|$ or simply $k = k_0$) cross section is a measure of that part of the incident flux diverted (or scattered) into the solid angle $d\Omega$. The total elastic cross section is defined as

$$\sigma_{sc} = \int \frac{d\sigma_{sc}}{d\Omega} \, d\Omega \tag{A.15}$$

where the integral is carried out over a complete solid angle (4π). Similarly, the absorption or reaction cross section σ_r measures the net influx over a large sphere surrounding the scattering center

$$\sigma_r = \lim_{r \to \infty} \left\{ -\frac{1}{2k} \int \mathbf{j} \cdot d\mathbf{A} \right\}. \tag{A.16}$$

The $\mathbf{j}$ in this integral is the total flux, computed with the full wave function of Eq. (A.7). Finally, the total cross section σ_t is defined as

$$\sigma_t = \sigma_{sc} + \sigma_r \equiv \sum_{l=0}^{\infty} \sigma_{t,l}. \tag{A.17}$$

It will be important here and for use later to have a partial-wave decomposition (Faxén and Holtsmark, 1927) of the wave function $\psi(\mathbf{k}_0; \mathbf{r})$ as

$$\psi(\mathbf{k}_0; \mathbf{r}) = \sum_{l=0}^{\infty} \psi_l(k,r) P_l(\cos \theta) \tag{A.18}$$

where the radial wave functions $\psi_l(k, r)$ satisfy the equation

$$\frac{1}{r}\frac{d^2}{d^2 r}(r\psi_l) + \left[k^2 - V(r) - \frac{l(l+1)}{r^2}\right]\psi_l = 0. \tag{A.19}$$

The asymptotic form of the wave function for our scattering problem can then be written

$$\psi^+(\mathbf{k}_0; \mathbf{r}) \xrightarrow[r \to \infty]{} \sum_{l=0}^{\infty} \frac{(2l+1)}{2ikr}[(-1)^{l+1} e^{-ikr} + e^{2i\delta_l} e^{ikr}] P_l(\cos\theta), \tag{A.20a}$$

$$\psi_{sc} \xrightarrow[r \to \infty]{} \sum_{l=0}^{\infty} \frac{(2l+1)}{2ikr}(e^{2i\delta_l} - 1) e^{ikr} P_l(\cos\theta), \tag{A.20b}$$

where $\delta_l(k)$ is the phase shift and is a measure of the strength of the potential $V(r)$. The phase shift can be expressed, through Eqs. (A.10b) and (A.20b), in terms of the spherical Bessel functions $j_l(kr)$ as

$$(2l+1)e^{i\delta_l}\sin\delta_l = -(-1)^l k \int_0^{\infty} r^2\,dr\,j_l(kr)V(r)\psi_l^+(k, r). \tag{A.21}$$

Since $\delta_l(k) \xrightarrow[k \to 0]{} k^{2l+1}$, one can obtain a low-energy expansion (stated only for the $l = 0$, or s, wave here) (cf. Messiah, 1965, pp. 392 and 409)

$$k\cot\delta = -\frac{1}{a_0} + \frac{1}{2}r_{eff}k^2, \tag{A.22}$$

where a_0 is the scattering length and r_{eff} the effective range. Equation (A.22) serves as the *definition* of the parameters a_0 and r_{eff} in the Taylor expansion of $k\cot\delta(k)$. Comparison of Eqs. (A.21) and (A.7) shows that

$$f(k, \theta) = \frac{1}{k}\sum_{l=0}^{\infty}(2l+1)e^{i\delta_l}\sin\delta_l P_l(\cos\theta) \tag{A.23a}$$

$$= \frac{1}{2ik}\sum_{l=0}^{\infty}(2l+1)(S_l - 1)P_l(\cos\theta) \tag{A.23b}$$

with the definition

$$S_l(k) = e^{2i\delta_l(k)}. \tag{A.24}$$

If Eqs. (A.20b) and (A.20a), respectively, are used in Eqs. (A.15) and (A.16), then the orthogonality of the Legendre polynomials $P_l(\cos\theta)$ yields

$$\sigma_{sc}(k) = \frac{\pi}{k^2}\sum_{l=0}^{\infty}(2l+1)|S_l - 1|^2, \tag{A.25}$$

$$\sigma_r(k) = \frac{\pi}{k^2} \sum_{l=0}^{\infty} (2l+1)(1-|S_l|^2).$$ (A.26)

From Eqs. (A.17). (A.25), (A.26), and (A.23b) we obtain an important result

$$\sigma_t(k) = \sigma_{sc}(k) + \sigma_r(k)$$

$$= \frac{2\pi}{k^2} \sum_{l=0}^{\infty} (2l+1)[1 - \operatorname{Re} S_l(k)] = \frac{4\pi}{k} \operatorname{Im} f(k, \theta = 0).$$

The relation

$$\sigma_t(k) = \frac{4\pi}{k} \operatorname{Im} f(k, \theta = 0)$$ (A.27)

is known as the optical theorem (Feenberg, 1932). It states that the imaginary part of the forward (i.e., $\theta = 0$ or $\mathbf{k} = \mathbf{k}_0$) elastic scattering amplitude gives the total cross section (which includes *both* scattering *and* reaction processes). This is a very general result and it will appear several times in the subsequent discussion.

Since the cross section can never be negative in any partial wave l (recall the definition of Eq. (A.13)), Eq. (A.26) implies that

$$|S_l(k)|^2 \leqslant 1.$$ (A.28)

In fact, when the scattering is purely elastic, so that $\sigma_r = 0$, we obtain the constraint

$$|S_l(k)|^2 = 1$$ (A.29)

or

$$S_l(k)S_l^*(k) = 1.$$ (A.30)

This implies that the phase shift $\delta_l(k)$ of Eq. (A.24) must be real for purely elastic scattering. In terms of Eq. (A.20a), we see that Eq. (A.29) is a statement that the flux of the incoming spherical waves (e^{-ikr}/r) is equal to the flux of the outgoing spherical wave (e^{ikr}/r). In this simple example we see that conservation of flux, or of probability, is associated with the unimodular value for $S_l(k)$. That is, if purely elastic scattering is the only process possible (or, the only open channel, the term *channel* being used to denote an allowed reaction), then Eq. (A.29) follows. We can also relate the reality of the phase shifts (or, equivalently, Eq. (A.29)) to a property of the interaction potential $V(r)$ of Eq. (A.2). If V is a real (as opposed to a complex) potential, then the Hamiltonian H of Eq. (A.2) is hermitian ($H^\dagger = H$). It is well-known that the hermiticity of

H implies that the total probability integral

$$\int |\Psi(\mathbf{r}, t))|^2 \, dV = 1 \tag{A.31}$$

remains a constant in time. If we prepare an incoming packet of finite spatial extent and demand that the incident flux in this open channel equal the outgoing scattered flux in that same channel, then we obtain Eq. (A.29).

We now apply this simple single-channel formalism to a specific example that will serve to illustrate several key points in subsequent developments. Let us take the interaction potential $V(r)$ to be a square well of the form

$$V(r) = \begin{cases} -V_0, & r < a \\ 0, & r > a \end{cases} \tag{A.32}$$

where V_0 is a constant. To simplify the discussion to the essentials, we consider only the s-wave (that is, the $l = 0$ partial wave) equations. We denote the wave function for $l = 0$ as $\psi^+(r)$, rather than as $\psi_0^+(k, r)$. According to Eq. (A.18) we are taking $\psi(\mathbf{k}_0; \mathbf{r}) \simeq \psi_0^+(k, r) = \psi^+(r)$. Equation (A.20a) requires the asymptotic form

$$\psi^+(r) \xrightarrow[r \to \infty]{} \frac{i}{2kr} [e^{-ikr} - S e^{ikr}] = \frac{e^{i\delta} \sin(kr + \delta)}{kr}. \tag{A.33}$$

The boundary condition that $\psi^+(r)$ be finite at $r = 0$ is met by the solution

$$\psi^+(r) = \begin{cases} \dfrac{A \sin(Kr)}{Kr}, & r < a & \text{(A.34a)} \\[2ex] \dfrac{i}{2kr} [e^{-ikr} - S(k) e^{ikr}], & r > a & \text{(A.34b)} \\[2ex] = \dfrac{e^{i\delta} \sin(kr + \delta)}{kr} & & \text{(A.34c)} \end{cases}$$

where K is defined as

$$K = \sqrt{V_0 + k^2} \tag{A.35}$$

and

$$S(k) = e^{2i\delta(k)}. \tag{A.36}$$

We have again dropped the subscript 0 on S and on δ. The constant A of Eq. (A.34) can be determined by demanding the continuity of $\psi(r)$ and of its derivative at $r = a$. However, since we are interested now in $S(k)$, we simply demand continuity of

$$R(k) \equiv [r\psi / d(r\psi)/dr]_{r=a}, \tag{A.37}$$

where $R(k)$ is known as the derivative function. (It is just the reciprocal of the logarithmic derivative.) Continuity of $R(k)$ as $r \to a_+$ and $r \to a_-$ becomes

$$R(k) \equiv \frac{\tan(ka+\delta)}{k} \equiv \frac{e^{-ika} - S(k)e^{ika}}{-ik[e^{-ika} + S(k)e^{ika}]} = \frac{\tan(Ka)}{K} \tag{A.38}$$

so that $S(k)$ is

$$S(k) = e^{2i\delta} = e^{-ika} \left[\frac{1 + i\sqrt{k}\,R(k)\sqrt{k}}{1 - i\sqrt{k}\,R(k)\sqrt{k}} \right] e^{-ika} \tag{A.39a}$$

$$= \frac{[1 - i\tan(ka)]\left[1 + \dfrac{ik}{K}\tan(Ka)\right]}{[1 + i\tan(ka)]\left[1 - \dfrac{ik}{K}\tan(Ka)\right]}. \tag{A.39b}$$

Notice that the Eq. (A.39a) is quite general and holds for *any* potential of finite range, whereas Eq. (A.39b) is valid only for the square well of Eq. (A.32).

Now, if we were to require a bound state for the potential of Eq. (A.32), so that $|\psi(r)|^2$ would have to have a finite volume integral, then we would demand that

$$r\psi(r) \to \text{const.}\, e^{-\kappa a} \tag{A.40}$$

for $r > a$. In this case, continuity of R implies that

$$\frac{\tan(Ka)}{K} = -\frac{1}{\kappa}. \tag{A.41}$$

If, for reference later, we set $k = i\kappa$, $\kappa > 0$, then Eq. (A.41) can be written as

$$\frac{ik}{K}\tan(Ka) = 1. \tag{A.42}$$

Therefore, at least in this simple example, the allowed bound-state energies $E_n = -\kappa_n^2 < 0$ given by the solutions to Eq. (A.41) are poles of $S(k)$ (Eq. (A.39b)) on the positive imaginary axis in the complex k plane. (Notice that these bound states also correspond to zeros of $S(k)$ on the negative imaginary k axis.)

If there are many processes possible (i.e., many channels open), then the time-independent Schrödinger equation, Eq. (A.4), becomes a matrix equation

$$H\psi_\alpha = E\psi_\alpha \tag{A.43}$$

where the operators H and E have the form

$$H_{\beta\gamma} = -\delta_{\beta\gamma}\nabla_\gamma^2 + V_{\beta\gamma}, \tag{A.44a}$$

$$E_{\beta\gamma} = \delta_{\beta\gamma}k_\gamma^2, \tag{A.44b}$$

and the wave function ψ_α can be considered a column vector with components $\psi_{\alpha\beta}$. We take α (fixed) to label the entrance channel and β the exit channel. For N open channels we have $\alpha, \beta = 1, 2, \ldots, N$. As a generalization of Eq. (A.7), the full wave function has the asymptotic form

$$\psi_{\alpha\beta}^+(\mathbf{k}_\alpha; \mathbf{r}_\beta) \xrightarrow[r_\beta \to \infty]{} \delta_{\alpha\beta}\, e^{i\mathbf{k}_\alpha \cdot \mathbf{r}_\alpha} + \frac{e^{ik_\beta r_\beta}}{r_\beta} f_{\alpha\beta}(\mathbf{k}_\beta, \mathbf{k}_\alpha). \tag{A.45}$$

By the same type of argument that led from Eq. (A.7) to Eq. (A.14), one can show that the differential transfer cross sections are

$$\frac{d\sigma_{\alpha\beta}}{d\Omega} = \frac{v_\beta}{v_\alpha} |f_{\alpha\beta}(\mathbf{k}_\beta, \mathbf{k}_\alpha)|^2, \tag{A.46}$$

where v_α and v_β are the velocities of the incoming and outgoing particles, respectively. The elastic differential cross section would be $d\sigma_{\alpha\alpha}/d\Omega$. The proof of the optical theorem (Feenberg, 1932), using conservation of flux and the orthogonality of the ψ_α, still goes through and Eq. (A.27) becomes

$$\sigma_t(k_\alpha) = \frac{4\pi}{k_\alpha} \operatorname{Im} f_{\alpha\alpha}(\theta = 0). \tag{A.47}$$

We also need for reference some elementary results from the classical theory of the scattering of electromagnetic waves. However, in order to avoid the additional complexities that the vector nature of electromagnetic waves introduces, we use the scalar wave equation to illustrate the results we want. The arguments for the electromagnetic case are very similar, but technically more involved.[4] If $\psi(\mathbf{r}, t)$ represents, say, the displacement of a medium of density ρ due to the propagation of a pulse at a speed v through the medium, then ψ satisfies the scalar wave equation

$$\left(\nabla^2 - \frac{1}{v^2}\frac{\partial^2}{\partial t^2}\right)\psi(\mathbf{r}, t) = 0. \tag{A.48}$$

The energy density u per unit volume

$$u = \tfrac{1}{2}\rho\left[\left(\frac{\partial\psi}{\partial t}\right)^2 + v^2\nabla\psi\cdot\nabla\psi\right] \tag{A.49}$$

and the vector $\mathbf{S}$ representing the energy per unit time transported across unit area by the wave

$$\mathbf{S} = -\rho v^2 \frac{\partial\psi}{\partial t}\nabla\psi \tag{A.50}$$

satisfy, by virtue of Eq. (A.48), the continuity equation

$$\nabla\cdot\mathbf{S} + \frac{\partial u}{\partial t} = 0. \tag{A.51}$$

To obtain solutions to Eq. (A.48), we can Fourier decompose the $\psi(\mathbf{r}, t)$ as

$$\psi(\mathbf{r}, t) = \int_{-\infty}^{\infty}\psi(\mathbf{r}, \omega)e^{-i\omega t}\,d\omega. \tag{A.52}$$

The wave equation reduces to the Helmholtz equation

$$(\nabla^2 + k^2)\psi(\mathbf{r}, \omega) = 0. \tag{A.53}$$

The wave number k is defined as

$$k = \frac{\omega}{v} = n(\omega)\frac{\omega}{c} \tag{A.54}$$

where $n(\omega)$ is an index of refraction and c the wave speed in some reference medium.[5] As previously, we write

$$\psi^+(\mathbf{r}, \omega) \equiv \psi_i(\mathbf{r}, \omega) + \psi_{sc}(\mathbf{r}, \omega) \xrightarrow[r\to\infty]{} \psi_i + f(\mathbf{k}_0, \mathbf{k})\frac{e^{ikr}}{r} \tag{A.55}$$

with $\psi_i(\mathbf{r}, \omega)$ being a plane wave

$$\psi_i(\mathbf{r}, \omega) = e^{i\mathbf{k}_0\cdot\mathbf{r}}. \tag{A.56}$$

(The $1/r$ fall off for the intensity of the outgoing spherical wave from a localized source in Eq. (A.55) is a direct result of energy conservation.) That is, we are interested in the scattering of an incident plane wave by an obstacle to produce an outgoing scattered wave. Figure A.1 will again do well enough to illustrate this situation, if we denote by Σ the surface of the obstacle (or scatterer). Since ψ_{sc} (as well as ψ_i) is a solution to Eq. (A.53), we can use the Green's function $G_0(\mathbf{r};\mathbf{r}')$ of Eqs. (A.8) and (A.9) and the same type of manipulations which led to Eq. (A.10a) to obtain a formal expression for the scattering amplitude

$$f(\mathbf{k}_0, \mathbf{k}) = -\frac{1}{4\pi} \int_\Sigma e^{-i\mathbf{k}\cdot\mathbf{r}'} (\nabla'\psi_{sc} + ik\psi_{sc}) \cdot d\mathbf{A}'. \tag{A.57}$$

Here the integral is evaluated on Σ, the surface of the scattering obstacle. The power scattered and the power absorbed, respectively, are given as

$$P_{sc} = \int_\Sigma \mathbf{S}_{sc} \cdot d\mathbf{A}', \tag{A.58a}$$

$$P_{abs} = \int_\Sigma \mathbf{S}_{tot} \cdot d\mathbf{A}'. \tag{A.58b}$$

The total cross section is just the total power P_t divided by the incident (plane wave) flux $2\omega k\rho v^2$. Direct calculation yields

$$\sigma_t(k) = \frac{1}{k} \text{Re} \left[\int_\Sigma e^{-i\mathbf{k}_0\cdot\mathbf{r}'} (i\nabla'\psi_{sc} - \mathbf{k}_0\psi_{sc}) \cdot d\mathbf{A}' \right]. \tag{A.59}$$

Comparison of Eqs. (A.57) and (A.59) again yields an optical theorem

$$\sigma_t(k) = \frac{4\pi}{k} \text{Im}\, f(\mathbf{k} = \mathbf{k}_0). \tag{A.60}$$

Let us build from the plane waves of Eq. (A.56) an incident wave packet that has a sharp front and travels with a speed c,

$$\psi_i(\mathbf{r}, t) = \int_{-\infty}^{\infty} g(\omega) e^{i(\mathbf{k}_0\cdot\mathbf{r} - \omega t)}\, d\omega. \tag{A.61}$$

We take $g(\omega)$ to be a function sharply peaked around some value $\omega = \omega_0$. An exact solution to Eq. (A.48) (with $v = c$) can be written in terms of the $\psi^+(\mathbf{r}, \omega)$ of Eq. (A.55) as

$$\psi(\mathbf{r}, t) = \int_{-\infty}^{\infty} g(\omega)\psi^+(\mathbf{r}, \omega) e^{-i\omega t}\, d\omega \tag{A.62a}$$

$$\xrightarrow[r\to\infty]{} \int_{-\infty}^{\infty} g(\omega) \left[e^{i\mathbf{k}_0\cdot\mathbf{r}} + f(\mathbf{k}_0, \mathbf{k})\frac{e^{ikr}}{r} \right] e^{-i\omega t}\, d\omega \tag{A.62b}$$

$$= \psi_i(\mathbf{r}, t) + \frac{1}{r} \int_{-\infty}^{\infty} g(\omega) f(\mathbf{k}_0, \mathbf{k}) \, e^{i(kr - \omega t)} \, d\omega. \tag{A.62c}$$

Due to rapid oscillations of the exponential factor, the integral in Eq. (A.62c) has appreciable values only when $t \approx \dfrac{k_0 r}{\omega_0} = \dfrac{r}{c}$ and does, indeed, represent an outgoing spherical wave.

We now restrict Eq. (A.62c) to forward scattering and write $f(\omega) = f(\mathbf{k} = \mathbf{k}_0)$ for the forward scattering wave amplitude. The amplitude $\psi_{sc}(\omega)$ of the Fourier component of the scattered wave

$$\psi_{sc}(r, t) \equiv \frac{1}{r} \int_{-\infty}^{\infty} \psi_{sc}(\omega) \, e^{i(kr - \omega t)} \, d\omega \tag{A.63}$$

is given as

$$\psi_{sc}(\omega) = f(\omega) g(\omega) \tag{A.64}$$

where $g(\omega)$ is the amplitude of the incident wave (cf. Eq. (A.61)). If $\psi_i(\mathbf{r}, t)$ of Eq. (A.61) does represent a sharp wave front that would arrive at the scattering center (located at the origin) of Figure A.1 at $t = 0$, then the fact that the wave travels with a *finite* velocity c tells us two things. First, in the *forward* direction, $\psi_i(\mathbf{r}, t)$ cannot reach a point a distance r to the right of the obstacle before a time $t \geqslant t_0 \equiv r/c$. Second, at this same *forward* position, the scattered wave $\psi_{sc}(\mathbf{r}, t)$ must also vanish for times prior to t_0.

In order to see the implications of these observations[6], let us write the Fourier transform of Eq. (A.61) (again, for the forward direction).

$$g(\omega) = \frac{1}{2\pi} \int_{-\infty}^{\infty} \psi_i(\mathbf{r}, t) \, e^{-i(kr - \omega t)} \, dt \tag{A.65a}$$

$$= \frac{1}{2\pi} \int_{t_0}^{\infty} \psi_i(\mathbf{r}, t) \, e^{i\omega(t - t_0)} \, dt. \tag{A.65b}$$

The second form follows because $\psi_i(\mathbf{r}, t) \equiv 0$, $t < t_0$. To begin with the simplest case, we first assume that $g(\omega)$ (or, equivalently, that ψ) is square integrable for real $\omega \in (-\infty, \infty)$. Since $(t - t_0) > 0$ in the integral of Eq. (A.65b), it follows that $g(\omega)$ exists everywhere in the upper half complex plane (i.e., $\operatorname{Im} \omega > 0$ gives an exponential damping factor), as does $dg/d\omega$, and that both vanish as $\omega \to \infty$ there. This implies that $g(\omega)$ is an analytic function in the upper half complex ω plane. From the inverse of Eq. (A.63) we can make a similar argument that $\psi_{sc}(\omega)$ is analytic in the same region. Hence, the $f(\omega)$ of Eq. (A.64) must also be analytic (aside from possible zeros of $g(\omega)$, which we do not consider explicitly here) for Im $\omega > 0$. The important point is that this *analyticity* of the forward scattering amplitude is a result of *causality*.

If this analytic $f(\omega)$ vanishes sufficiently rapidly as $\omega \to \infty$, Im $\omega > 0$, then Cauchy's integral formula can be applied around a semicircular contour C along the real axis (cf. Figure 2.2) and closed at infinity in the upper half ω-plane to yield

$$f(\omega) = \frac{1}{2\pi i} \int_C \frac{f(\omega')\,d\omega'}{(\omega' - \omega)} = \frac{1}{2\pi i} \int_{-\infty}^{\infty} \frac{f(\omega')\,d\omega'}{(\omega' - \omega)} \tag{A.66}$$

for ω in the upper half complex plane. When ω approaches the real axis from above, Eq. (A.66) reduces to

$$f(\omega) = \frac{1}{2\pi i} \mathbf{P} \int \frac{f(\omega')\,d\omega'}{(\omega' - \omega)} + \tfrac{1}{2} f(\omega) \tag{A.67a}$$

or to

$$f(\omega) \equiv \operatorname{Re} f(\omega) + i \operatorname{Im} f(\omega) = \frac{1}{\pi i} \mathbf{P} \int_{-\infty}^{\infty} \frac{f(\omega')\,d\omega'}{(\omega' - \omega)}. \tag{A.67b}$$

Since (in our *present* example, but not *always*), the physical fields $\psi_i(r, t)$ and $\psi_{sc}(r, t)$ of Eqs. (A.61) and (A.63) must be real, we find that

$$f^*(\omega) = f(-\omega) \tag{A.68}$$

for real values of ω. This in turn implies

$$\operatorname{Re} f(-\omega) = \operatorname{Re} f(\omega), \tag{A.69a}$$

$$\operatorname{Im} f(-\omega) = -\operatorname{Im} f(\omega). \tag{A.69b}$$

The dispersion relation for $\operatorname{Re} f(\omega)$ finally becomes

$$\operatorname{Re} f(\omega) = \frac{2}{\pi} \mathbf{P} \int_0^{\infty} \frac{\omega' \operatorname{Im} f(\omega')\,d\omega'}{(\omega'^2 - \omega^2)} \tag{A.70a}$$

$$= \frac{2}{\pi^2} \mathbf{P} \int_0^{\infty} \frac{\omega'^2 \sigma_t(\omega')\,d\omega'}{(\omega'^2 - \omega^2)}. \tag{A.70b}$$

The last form requires use of the optical theorem, Eq. (A.60). Therefore, in principle, the total cross section $\sigma_t(\omega)$ determines both Im $f(\omega)$ and Re $f(\omega)$. That is, $\sigma_t(\omega)$ fixes the complete forward scattering amplitude $f(\omega)$ through dispersion relations. If it should happen that $f(\omega)$ does not vanish sufficiently rapidly as $\omega \to \infty$, then we write a dispersion relation for $f(\omega)/\omega$ and we obtain

$$\mathrm{Re}\,(f(\omega)-f(0))=\frac{2\omega^2}{\pi}\,\mathrm{P}\!\int_0^\infty\frac{\mathrm{Im}\,f(\omega')\,\mathrm{d}\omega'}{\omega'(\omega'^2-\omega^2)} \qquad (A.70c)$$

$$=\frac{\omega^2}{2\pi^2}\,\mathrm{P}\!\int_0^\infty\frac{\sigma_t(\omega')\,\mathrm{d}\omega'}{(\omega'^2-\omega^2)}. \qquad (A.70d)$$

An extension of this procedure will work as long as $f(\omega)$ behaves no worse than ω^n as $\omega\to\infty$. Such dispersion relations, but for the optical index of refraction $n(\omega)$ (which can be related to $f(\omega)$) were written down by Kramers (1927, 1929) and by Kronig (1926). They actually began with a specific model for the index of refraction (based on a classical or quantum-mechanical damped harmonic oscillator), examined the analytic properties of that $n(\omega)$ and demonstrated the existence of a dispersion relation for this complex index of refraction. Kramers observed that these analyticity properties were related to the finite speed of propagation of a signal.

Finally, we summarize the interaction of atomic charges with the electromagnetic field. Taking account of the classical radiation damping of a charged harmonic oscillator (of natural frequency ω_0), we have for the equation of motion of such a charged oscillator[7]

$$\ddot{x}-\tau\dddot{x}+\omega_0^2x=0 \qquad (A.71)$$

where $\tau=\frac{4}{3}e^2$ is the characteristic time over which radiative effects are important. Since the radiation damping term is small, we take $\dddot{x}\simeq-\omega_0^2\dot{x}$, the value which would be valid if $\tau=0$. With $\gamma\equiv\tau\omega_0^2=(\tau\omega_0)\omega_0\ll\omega_0$, Eq. (A.71) then becomes approximately equivalent to

$$\ddot{x}+\gamma\dot{x}+\omega_0^2x=0. \qquad (A.72)$$

This is the same equation that governs a dampened oscillating mechanical system. To the level of approximation already made, the solution to Eq. (A.72) is

$$x(t)=x_0\,\mathrm{e}^{-\gamma t/2}\,\mathrm{e}^{-\mathrm{i}\omega_0 t}. \qquad (A.73)$$

The motion is damped and the total mechanical energy of the system decreases as $\mathrm{e}^{-\gamma t}$ so that $1/\gamma$ is the lifetime of this oscillator. Since the radiated electric field is proportional to the acceleration of the charge, this field has the time dependence

$$E(t)=E_0\,\mathrm{e}^{-\gamma t/2}\,\mathrm{e}^{-\mathrm{i}\omega_0 t}. \qquad (A.74)$$

We obtained the spectrum of the emitted electromagnetic radiation by Fourier analyzing $E(t)$ as

$$E(\omega)=\frac{1}{2\pi}\int_0^\infty E(t)\,e^{i\omega t}\,dt=\frac{1}{2\pi}E_0\frac{1}{i(\omega-\omega_0)+\gamma/2}. \tag{A.75}$$

The intensity of this radiation is proportional to $|E(\omega)|^2$ so that

$$I(\omega)=I_0\frac{\gamma}{2\pi}\frac{1}{(\omega-\omega_0)^2+\gamma^2/4} \tag{A.76}$$

where I_0 is the total intensity (i.e., $I(\omega)$ integrated over all frequencies). This distribution is peaked around $\omega=\omega_0$ and has a full breadth (or width) γ at half maximum.

Equation (A.71) and the subsequent discussion applied to a freely oscillating charge which has a certain amount of mechanical energy at $t=0$ and which then radiates away that energy. If we instead consider such a charged oscillator driven by an electric field $E_0\,e^{-i\omega t}$, then Eq. (A.72) is replaced by

$$\ddot{x}+\gamma\dot{x}+\omega_0^2x=2\,eE_0\,e^{-i\omega t}. \tag{A.77}$$

The steady-state solution is

$$x(t)=x_0\,e^{-i\omega t}=\frac{2\,eE_0}{\omega_0^2-\omega^2-i\omega\gamma}\,e^{-i\omega t}. \tag{A.78}$$

The power radiated by an accelerated charge is proportional to the square of the acceleration so that the scattering cross section becomes

$$\sigma_{\rm sc}(\omega)=\frac{32\pi\,e^4}{3}\frac{\omega^4}{(\omega^2-\omega_0^2)^2+\omega^2\gamma^2} \tag{A.79a}$$

$$\approx\frac{3}{2}\frac{\pi}{\omega_0^2}\frac{\gamma^2}{(\omega-\omega_0)^2+\gamma^2/4}, \tag{A.79b}$$

where the last form holds for ω close to ω_0.

Furthermore, if one considers a (not too dense) collection (N per unit volume) of charged oscillators whose motion is given by Eq. (A.78), then the dielectric constant ε for such a medium is just

$$n^2(\omega)=\varepsilon(\omega)=1+\frac{8\pi N\,e^2}{\omega_0^2-\omega^2-i\omega\gamma} \tag{A.80}$$

Here $n(\omega)$ is the optical index of refraction, which is a complex quantity. It was essentially for $n(\omega)$ and $\varepsilon(\omega)$ that Kramers (1927, 1929) and Kronig (1926) wrote dispersion relations. For a dilute collection of scatterers $\varepsilon(\omega)$ can be related to the forward scattering amplitude as

$$\varepsilon(\omega) = 1 + \frac{4\pi N}{\omega^2} f(\omega). \tag{A.81}$$

If we take $\varepsilon(\omega)$ to be finite (on physical grounds), then we would expect $f(0) = 0$ in this classical framework. (This is consistent with scattering by a tightly bound electron (i.e., Rayleigh scattering) at low ω. On the other hand, for the scattering from a free electron (i.e., Thomson scattering), $f(0) = -2 e^2$.) As $\omega \to \infty$ (i.e., $\omega \gg \omega_0$), $\varepsilon(\omega)$ should approach its value for a free electron, or according to Eq. (A.80),

$$\varepsilon(\omega) \underset{\omega \to \infty}{\longrightarrow} 1 - \frac{8\pi N e^2}{\omega^2}, \tag{A.82}$$

or

$$f(\omega) \underset{\omega \to \infty}{\longrightarrow} -2 e^2. \tag{A.83}$$

Equation (A.70d) can be written as

$$\text{Re} f(\omega) = -\frac{1}{2\pi^2} \mathbf{P} \int_0^\infty \frac{\sigma_{sc}(\omega') \, d\omega'}{[1 - (\omega'/\omega)^2]}. \tag{A.84}$$

In the limit that $\omega \to \infty$, we obtain

$$\int_0^\infty \sigma_t(\omega) \, d\omega = 4\pi^2 e^2. \tag{A.85}$$

This is an example of a sum rule that the total cross section must satisfy. Such sum rules are an important test for dispersion relations.

We have indicated how Eqs. (A.78) and (A.80) are obtained in *classical* theory, when the charges are treated as classical particles. Even as early as 1924, Kramers and Heisenberg (1925) had used the old semiclassical quantum theory (prior to the invention of quantum mechanics) to obtain a result of the form of Eq. (A.80) for the scattering of light from an atom. We now outline the quantum-mechanical treatment of the interaction of an atom with a classical radiation field. Particular examples are the light emitted when an unstable state decays and also the scattering of light by an atom. These problems were studied by Weisskopf and Wigner (1930), by Weisskopf (1931), and by Breit (1932, 1933).

Let the states of the atomic system be denoted by u_n. If H_0 is the Hamiltonian for the unperturbed atomic system, then the u_n are the solutions to

$$H_0 u_n = E_n u_n. \tag{A.86}$$

We take the u_n to be a complete, orthonormal set of functions. If H' is the Hamiltonian for the perturbation (say an external radiation field) and $\Psi(t)$ the state vector for the system, then the Schrödinger equation is

$$i\frac{\partial \Psi}{\partial t} = H\Psi \equiv (H_0 + H')\Psi. \tag{A.87}$$

We can expand this $\Psi(t)$ in terms of the unperturbed solutions $u_n e^{-i\omega_n t}$, where $\omega_n = E_n$ in our system of units,

$$\Psi(t) = \sum_n a_n(t) u_n e^{-i\omega_n t}. \tag{A.88}$$

Here $|a_n(t)|^2$ is the probability of finding the atom in the state u_n at time t. From Eqs. (A.87) and (A.88) there follows a set of equations for the expansion coefficients $a_n(t)$

$$i\frac{d}{dt}a_k(t) = \sum_n \langle u_k | H' | u_n \rangle a_n(t) e^{i\omega_{kn}t} \tag{A.89}$$

with

$$\omega_{kn} \equiv \omega_k - \omega_n. \tag{A.90}$$

Equation (A.89) is exact and is completely equivalent to Eq. (A.87). Suppose that at $t=0$ the atom is in level E_m (i.e., state u_m), where m is a *fixed* index. The initial condition on the $a_k(t)$ is then

$$a_k(t=0) = \delta_{km}. \tag{A.91}$$

The essential approximation consists of assuming that H' is sufficiently weak that only those levels near E_m will be populated and that just *one* photon will be exchanged in this process. That is, we take m to label a state with *no* photons present and n ($\neq m$) to be a state with *one* photon present. Because H' is essentially the electromagnetic vector potential, it can connect only states differing by one photon. Furthermore, we now restrict ourselves to just *two* atomic energy levels, E_i and E_f with

$$E_i - E_f = \omega_0 > 0. \tag{A.92}$$

The index n hereafter labels the energy ω_n of the one photon present, where $a_0(t)$ will be the amplitude for the original no photon state (i.e., level E_i). To this level of approximation, Eq. (A.89) reduces to

$$\frac{d}{dt}a_0(t) = \sum_n H'_{0n}a_n(t) e^{i(\omega_0 - \omega_n)t}, \tag{A.93a}$$

$$\frac{d}{dt}a_n(t) = H'_{n0}a_0(t) e^{i(\omega_0 - \omega_n)t}. \tag{A.93b}$$

Here we have written $\langle u_m|H'|u_n\rangle$ as H'_{0n}. These equations have the approximate solution for $\omega_0 t \gg 1$[8]

$$a_0(t) = e^{-\gamma t/2}, \tag{A.94a}$$

$$-a_n(t) = H'_{n0} \frac{[e^{i(\omega_n - \omega_0)t - \gamma t/2} - 1]}{(\omega_n - \omega_0 + i\gamma/2)}, \tag{A.94b}$$

$$\gamma \approx 2\pi \sum_n |H'_{n0}|^2, \tag{A.94c}$$

where the width γ is equal to the total transition probability per unit time from E_i to E_f. The long-time net probability of emitting a photon of frequency ω is $|a_n(t\to\infty)|^2$, so that the intensity for the photon of energy ω is

$$I(\omega) = \omega \sum_n |a_n(t\to\infty)|^2 = \frac{\gamma}{2\pi} \frac{\omega}{(\omega - \omega_0)^2 + \gamma^2/4}. \tag{A.95}$$

Realizing that we have assumed $\omega \approx \omega_0$ in this discussion, we can identify $I_0 \approx \omega \approx \omega_0$ as the total intensity so that Eq. (A.95) is approximately the same as the classical result of Eq. (A.76). (We can also directly integrate Eq. (A.95) over all frequencies. This too yields I_0.)

Weisskopf (1931) made the analogous calculation for resonance fluorescence, which is the scattering of light by an atom when the frequency of the incident light includes the resonant frequency of the atom (i.e., the frequency corresponding to a transition from, say, the ground state to an excited state). We have just seen the nature of the solution and of the emission spectrum generated when an atom decays from an excited state to its ground state. Physically, we picture the scattering of light by an atom as the absorption of a photon by the atom, causing excitation of the atom to a higher energy level, and the eventual transition back to the ground state accompanied by the emission of a photon. On this basis, we would expect the steady-state situation (representing the scattering of light by an atom) to be a superposition of these processes so that the final result for the scattered intensity or cross section σ_{sc} should be similar to Eq. (A.95). The details of calculation produce a result like Eq. (A.95) and one similar to the classical result of Eq. (A.79b). When the frequency spectrum of the incident light does not include the resonant frequencies of the atom, direct application of perturbation theory leads to a dispersion relation of the Kramers–Heisenberg type, Eq. (A.80) (Schrödinger, 1926; Dirac, 1927; Klein, 1927; Waller, 1927, 1928). Such dispersion formulas were important in the development of quantum mechanics.

Therefore, we see that, both classically and quantum mechanically, a damped system has a response that exhibits a resonance (or bump) phenomenon in the neighborhood of certain frequencies characteristic of the scatterer. When the cross sections for the scattering of neutrons by nuclei were found to have a resonance structure, it was natural to associate these peaks with nuclear resonances. The basic analogy made was that the incident neutrons were like the photons and that there were quasi-stationary energy levels of the system (nucleus plus neutron). On the basis of this analogy, Breit and Wigner (1936) calculated the amplitude and cross section for the transfer reaction from channel α to channel β. In modern notation, the Breit–Wigner one-level resonance formula, for a reaction initiated in the lth partial wave of channel α, is (cf. Breit, 1940a)

$$\sigma_{\alpha\beta}(E_\alpha) \approx (2l+1)\frac{\pi}{k_\alpha^2}\Gamma_\alpha^s\Gamma_\beta^s/[(E_\alpha - E_\alpha^s)^2 + (\Gamma^s/2)^2]. \tag{A.96}$$

Here Γ_α^s is the width or probability of decay of the compound nucleus (or quasiexcited state) into channel α, Γ_β^s that for the decay into channel β, and Γ^s is the total probability of decay by all possible processes. The entrance channel energy is E_α, while E_α^s is the entrance channel energy which corresponds to the resonances of the compound nucleus.

Even though a formula such as Eq. (A.96) proved useful for fitting experimental data, the basis of its derivation was certainly open to question since perturbation theory, based on the assumption of a weak coupling between the incident beam and the scatterer, had been used. However, the interaction between the neutron (or proton) and the nucleus was known to be strong, so that a perturbation approach, in which higher-order interactions were neglected, was not valid. Kapur and Peierls (1938) and Siegert (1939) provided a derivation of the resonance formula that did not require any assumption about the interactions being weak. The argument was also given in terms of the one-particle Schrödinger equation and clarified the meaning of the resonant or compound nuclear states. For the s-wave, we replace the $r\psi^+(r)$ of Eq. (A.33) by $\phi(k,r)$ and write

$$\phi(k,r) = \frac{I\sin(kr)}{k} + \frac{If(k)}{k}e^{ikr} \tag{A.97}$$

where $f(k)$ is the s-wave scattering amplitude and I an arbitrary multiplicative factor. Equation (A.97) is the solution to the Schrödinger equation valid for $r \geq a$ where a is the range of the potential (i.e.,

$V(r) \equiv 0$, $r \geqslant a$). A simple calculation shows that

$$f(k) = \frac{\phi(k,a)\cos(ka) - \phi'(ka)\sin(ka)/k}{\phi'(k,a) - ik\phi(k,a)} e^{-ika}. \tag{A.98}$$

Let the regular solutions $\phi_n(r)$ to the Schrödinger equation be defined by the boundary conditions

$$\phi_n(r=0) = 0, \tag{A.99a}$$

$$(\phi'_n - ik_n\phi_n)|_{r=a} = 0. \tag{A.99b}$$

Here the k_n will be complex and we write $k_n^2 \equiv E_n - i\Gamma_n/2$ where E_n and Γ_n are real. With these functions, the scattering amplitude $f(E)$ can be shown to consist of a sum of resonance terms (or 'bumps') of the form $[i(E_n - E) - (1/2)\Gamma_n]^{-1}$ plus a smooth function of E for all real values of the energy $E = k^2$. In the neighbourhood of one of these resonant energies E_n, $f(E)$ reduces to

$$f(E) = -\frac{e^{i\delta_n}}{2k_n}\frac{\Gamma_n}{(E - E_n) + i\Gamma_n/2} + f_{\text{pot}}(E) \tag{A.100a}$$

$$\equiv f_{\text{res}}(E) + f_{\text{pot}}(E). \tag{A.100b}$$

Here $f_{\text{pot}}(E)$ (the so-called potential scattering amplitude) is a smoothly varying function of E near $E = E_n$. Since, for $l = 0$ scattering, $\sigma_{\text{sc}} = 4\pi|f(E)|^2$, the scattering cross section reduces to a Breit–Wigner form of Eq. (A.96) in the neighborhood of a resonance.

There is also a simple but important relation between the width Γ of a state with complex energy $E - i\Gamma/2$ and the lifetime τ of the state. Since the time-dependent solution to the Schrödinger equation has a factor $e^{-i(E - i\Gamma/2)t}$ and since the probability for that state at time t goes as the square of the modulus of the wave function, the probability decays as $e^{-\Gamma t}$. That is, the lifetime τ is given as

$$\tau = \frac{1}{\Gamma}. \tag{A.101}$$

In other words, compound nuclear (or resonance) states with a very long lifetime must have a very small width Γ (or, equivalently, a small probability of decay).

Furthermore, directly from the Schrödinger equation satisfied by $\phi_n(r)$ and the boundary conditions (A.99), Siegert (1939) showed that

$$\Gamma_n \int_0^a |\phi(r)|^2 \, dr = (k_n + k_n^*)|\phi(a)|^2. \tag{A.102}$$

Therefore, for very narrow resonances ($\Gamma_n \approx 0$), in which the wave function is largely confined to the interior of the nucleus, Eqs. (A.102) and (A.99b) imply that

$$\phi'_n(a) \approx 0. \tag{A.103}$$

In this case the approximate resonant energies are those corresponding to the wave functions such that $\phi'(r = a) = 0$.

The term 'resonance' can be used in two different senses in discussing scattering phenomena. A phase shift $\delta_l(k)$ (cf. Eq. (A.23a)) is said to resonate when it increases rapidly through an odd multiple of $\pi/2$ (i.e., $\delta_l(k) \to (2n + 1)\pi/2$, $n = 0, 1, 2, \ldots$), as E passes through a resonance energy $E_r = k_r^2$. If we consider *only* s-waves ($l = 0$) in Eqs. (A.23a) and (A.25), the scattering cross section is

$$\sigma_{sc}(k) \approx \frac{4\pi}{k^2} \sin^2[\delta(k)]. \tag{A.104}$$

At a phase-shift resonance, this cross section becomes as large as possible,

$$\sigma_{sc}(k) \underset{k \to k_r}{\longrightarrow} \frac{4\pi}{k^2}. \tag{A.105}$$

Equation (A.105) is known as the unitarity limit and represents an absolute upper bound to this cross section (for the s-wave). Since, from Eq. (A.23a) for s-wave,

$$f(k) \approx \frac{1}{k} e^{i\delta} \sin \delta = \frac{\tan \delta}{k(1 - i \tan \delta)} \tag{A.106}$$

a simple Breit–Wigner form for $f(k)$ will result when $\tan \delta$ is given as

$$\tan \delta(k) \approx \frac{\Gamma/2}{(E_r - E)}. \tag{A.107}$$

So, at a phase-shift resonance, when only *one* partial wave contributes appreciably to the scattering amplitude, the approximate form of $f(k)$ is

$$f(k) \approx -\frac{\Gamma}{2k_r(E - E_r + i\Gamma/2)}, \tag{A.108}$$

which in turn reduces Eq. (A.104) to

$$\sigma(E) \underset{E \to E_r}{\longrightarrow} \frac{4\pi}{k_r^2} \frac{(\Gamma/2)^2}{[(E - E_r)^2 + (\Gamma/2)^2]}. \tag{A.109}$$

This is a single-channel version of Eq. (A.96). A phase-shift resonance shows up as a peak (or 'bump') in the cross section and this peak saturates (or reaches) the unitarity limit of Eq. (A.105). In such cases the background term is negligible.

However, the term 'resonance' can also be used more generally to denote any peak in the cross section, even when that peak does not equal the unitarity limit of Eq. (A.105). That is, such a resonance stands out as a peak (large or small) against some background, as in Eq. (A.100a). It is this type of resonance that is usually being referred to when one speaks of a Breit–Wigner (BW) resonance. Of course, a phase-shift resonance is simply a particular (i.e., 'strong') type of resonance in this more general sense of the term. The discussion of resonances given in the text accompanying Eqs. (2.40)–(2.49) refers to resonances in this broad sense.

Let us illustrate some of these points with a finite-range potential, and in particular, with the example of the square well treated at the beginning of this appendix. A resonance in the BW sense occurs when (cf. Eqs. (A.37) and (A.38))

$$R(k) \equiv \frac{r\psi}{(r\psi)'}\bigg|_{r=a} \approx \frac{1}{k_r'} \frac{\Gamma'/2}{(E_r'-E)}. \tag{A.110}$$

As Eq. (2.43) of the text indicates, $R(k)$ has an *infinite* sequence of such resonances. (In terms of the γ of Eq. (2.43), we have $\gamma^2 = \Gamma/2k_r'$.) The scattering amplitude $f(k)$ of Eq. (A.23a) for the s-wave becomes,

$$f(k) \equiv \frac{e^{2i\delta}-1}{2ik} = \frac{e^{2i(ka+\delta)}e^{-2ika}-1}{2ik}$$

$$= \frac{i\,e^{-2ika}}{2k}\left[(e^{2ika}-1) - \frac{2ikR}{1-ikR}\right]. \tag{A.111}$$

The first term is the potential, or hard core, amplitude since, for an impenetrable sphere (i.e., $V_0 \to -\infty$ in Eq. (A.32)), the wave function of Eq. (A.33) must vanish at $r=a$. Then $S(\text{hard sphere}) = e^{-2ika}$ and the corresponding scattering amplitude becomes

$$f(\text{hard sphere}) = \frac{S(\text{hard sphere})-1}{2ik} = \frac{i}{2k}(1 - e^{-2ika}). \tag{A.112}$$

This is just the first term in Eq. (A.111). It corresponds to the limit $R \to 0$ in Eq. (A.39a) (or in Eq. (A.111)). If we rewrite Eq. (A.111) and take k to be near the k_r' of Eq. (A.110), we find

$$f(k) = f_{pot}(k) + f_{res}(k) \approx \frac{i}{2k_r'} (1 - e^{-2ik_r'a}) - \frac{e^{-2ik_r'a}\Gamma'}{2k_r'[(E - E_r') + i\Gamma'/2]}.$$

(A.113)

We can see the origin of the expected behavior of Eq. (A.100a). It also becomes clear that all of the effects of the interior region of the nucleus (i.e., $r < a$) on the general scattering amplitude $f(k)$ of Eq. (A.111) can be parameterized or characterized by the value of the wave function and its derivative at the nuclear surface ($r = a$) through the R of Eq. (A.37). The nucleus is a 'black box' characterized by $R(k)$.

For the finite square-well potential (cf. Eqs. (A.32)–(A.38)), we find from the continuity of $\psi^+(r)$ and its derivative at $r = a$ (cf. Eq. (A.38))

$$R(k) \equiv \frac{\tan(ka + \delta)}{k} = \frac{\tan(Ka)}{K} \approx \frac{\Gamma'/2}{k_r'(E_r' - E)},$$

(A.114)

where the last form holds in the neighborhood of any of the infinite set of resonances defined by

$$K_r'a \equiv \sqrt{V_0 + k_r'^2}\, a = (2n + 1)\frac{\pi}{2}$$

(A.115a)

or

$$k_r'a + \delta_r = (2n + 1)\frac{\pi}{2}.$$

(A.115b)

Notice that, as verification of Eq. (2.47) of the text in this particular case, we find

$$\frac{dR(k)}{dk} = \frac{k}{K}\frac{dR}{dK} = \frac{k}{K^3\cos^2(Ka)}[Ka - (1/2)\sin(2Ka)] > 0.$$

(A.116)

This in turn implies that

$$\frac{dR}{dk} = \frac{1}{k\cos^2(ka + \delta)}\left\{a + \frac{d\delta}{dk} - \frac{1}{2k}\sin[2(ka + \delta)]\right\} > 0 \quad \text{(A.117a)}$$

or

$$\frac{d\delta}{dk} > -a - \frac{1}{2k},$$

(A.117b)

which is the Wigner inequality of Eq. (2.60) of the text.

The normalization constant of the interior wave function of Eq. (A.34a) is just

$$A = e^{i\delta} \frac{K}{k} \frac{\sin(ka+\delta)}{\sin(Ka)} \tag{A.118}$$

so that near one of the resonances of Eq. (A.114) we obtain

$$|A|^2 \approx \frac{k_r'^2}{k_r'^2 \sin^2(K_r'a)} \frac{(\Gamma'/2)^2}{[(E-E_r')^2 + (\Gamma'/2)^2]}. \tag{A.119}$$

Therefore, the probability of finding the projectile inside the nucleus (or interaction region) is

$$\int_0^a |A|^2 \frac{\sin^2(Kr)\,dr}{K^2} = \frac{|A|^2}{K^2}\left[\frac{a}{2} - \frac{\sin(2Ka)}{4K}\right] \approx \frac{a}{2} \frac{|A|^2}{K^2}$$

$$\approx \frac{a}{2} \frac{1}{k_r'^2 \sin^2(K_r'a)}\left[\frac{\Gamma'^2}{4(E-E')^2 + \Gamma'^2}\right]. \tag{A.120}$$

This probability exhibits peaks in the resonance regions.

From Eq. (A.114) we can obtain an expression for the phase shift $\delta(k)$ as

$$\tan\delta(k) = \frac{\dfrac{k}{K}\tan(Ka) - \tan(ka)}{1 + \tan(ka)\dfrac{k}{K}\tan(Ka)}. \tag{A.121}$$

A phase-shift resonance, in the sense of Eq. (A.107), will occur at those energies $E_r = k_r^2$ such that the denominator of Eq. (A.121) vanishes,

$$\tan\delta(k) \approx \frac{\Gamma/2}{E_r - E}. \tag{A.122}$$

(Notice that we have used primes to denote the general Breit–Wigner resonance energies $E_r' = k_r'^2$, as in Eq. (A.114), and unprimed energies $E_r = k_r^2$ for phase-shift resonances, as in Eqs. (A.109) and (A.122).) Of course, the set $\{E_r'\}$ and the set $\{E_r\}$ do not coincide, since there are infinitely many members in $\{E_r'\}$ (cf. (A.114)), whereas there is only a finite number of members (in some cases none) in $\{E_r\}$. However, one of the E_r' may become an E_r. If we compare Eqs. (A.108) and (A.113) we see that the smoothly varying background term, $f_{\text{pot}}(k)$, will vanish when

$$2k_r'a = m\pi, \quad m = 0, 1, 2\ldots. \tag{A.123}$$

If $f_{res}(k)$ resonates at one of these energies, then Eq. (A.113) reduces to Eq. (A.108) at that resonant energy and the unitarity limit is reached, as must be the case for a phase-shift resonance. But this is a *very* particular set of circumstances on k'_r and on K'_r simultaneously and amounts to a constraint on the value of $V_0 a^2$ (i.e., on the potential). So, the BW resonance formula and the arguments used to obtain it allow one to understand how a large number of peaks in the cross section *may* arise. It does not *demand* any pronounced peaks.

Notes*

Preface

1. This material is based upon work supported by the National Science Foundation. Any opinions, findings, and conclusions or recommendations are mine and do not necessarily reflect the views of the Foundation.

Chapter 1 (Introduction and background)

1. I wish to thank Professor Res Jost for raising this issue (in a letter of 1/22/86) and for the reference to Jacob Burckhardt, the great historian.
2. On this point let me state that Professor Murray Gell-Mann is most emphatic in insisting upon a clean distinction between dispersion theory as on-mass-shell field theory and (an autonomous) S-matrix theory as a program opposed to field theory. This issue will arise again throughout our case study. Professor Gell-Mann has been extremely generous in contributing recollections to this study and, most especially, in criticizing successive drafts of this work (in a lengthy discussion on 5/4/85). Although I am cognizant of his strong feelings on this point and have taken account of them, I am fairly certain that he may still not be happy with my handling of that issue. Professor Gell-Mann prefers to see the basic theme of this episode as another in a series of problems encountered with standard field theory methods and of successful solutions within a field theory framework.
3. It will also become evident that we must distinguish between the axiomatic SMT and the autonomous SMT programs. These have at time overlapped, but they are distinct.

---

* When it is necessary to refer to a note from a previous chapter, we give the chapter number first. Thus, note 2.4 is note 4 of Chapter 2.

316

Chapter 2 (Origin of the S matrix: Heisenberg's program as a background to dispersion theory)

1. The material from a preliminary version of this chapter was presented at the International Congress for the History of Science in Berkeley on August 5, 1985 and has since appeared in an expanded version (Cushing, 1986a).
2. Interview at Notre Dame (Spring, 1981) and private correspondence (11/10/81).
3. Private correspondence (6/20/84) and telephone conversation (9/21/84).
4. Telephone conversation (9/28/84) and interviw in Chicago (10/10/84).
5. Wheeler (1937b) uses the term 'scattering matrix' (pp. 1116 and 1117) and the notation c_{mn} (pp. 1115–17) for the elements of this matrix. His S_J (p. 1114 ff) should *not* be mistaken for the modern notation for a partial-wave S-matrix element. Wheeler's S_J is the kernel of the Fredholm integral equation that determines the scattering wave functions. However, for uniformity of notation we refer to his scattering matrix as the S matrix.
6. Wheeler (1937b), Eq. (30) and pp. 1116–18.
7. I thank Professor Ole Knudsen for raising this question.
8. Brown and Rechenberg (1987) have discussed the friendship between Dirac and Heisenberg. Interestingly, they point out that Dirac sent Heisenberg a copy of the 1935 edition of his (Dirac's) text on quantum mechanics.
9. In Eq. (2.10) the indices i and k on the state vectors represent the total momentum vectors in the initial and final states (and, in this sense, $i=k$ there), whereas, on the matrix S, i and k are compound indices standing for each of the individual momentum vectors in each state.
10. In terms of a more complete notation, the time-dependent scattering states $\Psi_i(t)$ would be

$$\Psi_i(t) = e^{-iH(t-t_0)} e^{-iH_0 t_0} \phi_i = e^{-iE_i t} \psi_i^+$$

with Eq. (2.14) being

$$\psi_i^+ = \lim_{t \to +\infty} e^{iE_i t} \Psi_i(t).$$

Heisenberg has taken $t_0 = 0$ here. In modern treatments it is common to take $t_0 \to -\infty$ so that

$$\Psi_i^+(t) = e^{-iHt} \psi_i^+ \xrightarrow[t \to -\infty]{} \Phi_i(t) = e^{-E_i t} \phi_i.$$

The S-matrix elements then become expressible as

$$S_{fi} = \lim_{t \to +\infty} \langle \Phi_f(t) | \Psi_i^+(t) \rangle = \langle \psi_f^- | \psi_i^+ \rangle.$$

This definition of S makes it clear that S_{fi} is the probability amplitude for beginning in the distant past in a free-particle state Φ_i (corresponding to a plane wave ϕ_i) and ending in the distant future in a free-particle state Φ_f.

11. Letter of 7/10/85. I thank Dr. Rechenberg for this correspondence.
12. I thank Professor Wheeler for a brief conversation on this point (Crystal City American Physical Society Meeting, 4/26/85) and for subsequent correspondence (9/24/85).
13. I thank Professor Ole Knudsen for a question on this point in correspondence (10/11/85).
14. Heisenberg actually writes $S = 1 + R$, rather than Eq. (2.17).
15. I thank Professor Wouthuysen for this correspondence (10/29/85).
16. This paper is also referrenced in Møller (1946b, p. 406) and in Ma (1947, p. 195). On 4/4/85 Dr. F. Coester of Argonne National Laboratories provided me with the typed manuscript of a paper by Heisenberg titled 'Die Behandlung von Mehrkörperproblemen mit Hilfe der η-Matrix.' On 5/4/85 Dr. H. Rechenberg gave me a copy of a handwritten manuscript of one of Heisenberg's papers, which Rechenberg considers to be the fourth S-matrix paper. Coester's and Rechenberg's manuscripts are identical. Coester recalls that he obtained the manuscript in Switzerland around 1945, possibly from Wentzel. The manuscript has appeared in Heisenberg's *Collected Works* (Blum, Dürr, and Rechenberg, 1989).
17. I thank Professor Jost for his letter of 10/16/85.
18. As in several other places in this book, the notation here and in the immediately following equations is often symbolic in the sense that only discrete indices are used and Dirac delta functions are sometimes omitted. The purpose of the equations is just to indicate the structure of the arguments.
19. A summary of Stückelberg's work can be found in the obituary by Rivier (1984).
20. In terms of Note 10 above, we see that a formal expression for $U(t, t_0)$ is

$$U(t, t_0) = e^{iH_0 t} e^{-iH(t - t_0)} e^{-iH_0 t_0}$$

so that the physical S-matrix elements become

$$S_{fi} = \langle \phi_f | S | \phi_i \rangle = \langle \phi_f | U(\infty, -\infty) | \phi_i \rangle.$$

Here ϕ_f and ϕ_i are free-particle (plane wave) states.
21. A formal relation like Eq. (2.32) was also discussed by Stückelberg (1944).
22. I thank Professor Dyson for this letter of 8/15/85.
23. Letter from Dyson to J. R. Oppenheimer (10/17/48). I thank Professor Dyson for informing me of this (letter of 8/15/85) and for sending me a copy of it.
24. Ma (1947) acknowledges Pauli for advice on this work on the false zeros. Grythe (1982) has presented an interesting commentary on some of the early work which followed Heisenberg's initial papers on the S matrix. The private correspondence cited provides insights into the reactions of Kramers, Møller, and Pauli to this project.
25. The question of just what to make of these false or redundant zeros is a somewhat complicated one. Grythe and Haug (1983) have examined this in

detail and shown that it is not possible in all cases to rule out the false zeros on the basis of Møller's condition, Eq. (2.21). The attitude today would be to see these redundant zeros as unphysical remnants of a nonrelativistic theory which ought not be present in a correct, relativistic theory. However, these considerations were not possible in 1946–47 and so played no role in the historical episode being related here.

26. The $f(k,r)$ and $\phi(k,r)$ used here correspond to $r\psi(k,r)$ of the Appendix. Also, we follow fairly closely Jost's original conventions in defining $f(k,r)$ and $f(k)$. These differ from those for the Jost function found in some modern works (e.g., Newton, 1966, Chapter 12).

27. As a cautionary note, we warn the reader not to confuse the scattering amplitude $f(k,\theta)$ of Eq. (A.7) with the Jost function $f(k)$ of Eq. (2.35)

28. It is unclear that Grythe (1982) is correct in stating that 'The early S-matrix theory had come to an end with the discovery of Ma's redundant zeros' (p. 201). Theorists I have talked with who were active at that time did not see Ma's work as decisive (in the negative sense) for Heisenberg's program. It was a difficulty, but one to be thought about. (Møller (1946a, p. 45) referenced Ma's work.) Dr. Rechenberg in a discussion (5/1/85) concurred with this position of mine. I wish to thank Inge Grythe for his correspondence on the subject of the redundant zeros.

29. I thank Professor Jost for this letter of 10/16/85.

30. Subsequently, Wigner (1951) and Wigner and von Neumann (1954) gave an elegant mathematical formulation of a theory for $R(E)$.

31. Equation (2.48) is actually a matrix equation, with

$$\delta_{\alpha\beta}\,e^{-ik_\alpha a_\alpha} \text{ and } \delta_{\alpha\beta}\sqrt{k_\alpha}$$

being the diagonal elements of the corresponding matrices.

32. I thank Professor Eden for correspondence (10/21/85) on this point.

33. This use of the term 'causality' (in physics) is very different from the formulation of causality given by philosophers.

34. Letter to me from Professor Jost (10/16/85).

35. I thank Professor Rohrlich for an interview on this (10/19/84).

36. I thank Professor van Kampen for this correspondence (11/17/85).

37. Equation (2.60) can easily be verified directly for the square well phase shift of Eq. (A.39b) by starting from

$$\tan(ka+\delta) = \frac{k}{K}\tan(Ka).$$

Chapter 3 (Dispersion relations)

1. I thank Professor Coester for an interview (9/25/84) on this.

2. Telephone conversation with Goldberger (9/21/84).

3. I thank Professor Eden for this correspondence (10/21/85).

4. Professor Rohrlich (10/19/84) and Professor Coester (9/25/84) each stressed this continuity.
5. Our convention for four vectors will be

$$x = (x_0, \mathbf{x}) = (t, \mathbf{r}), \quad k = (k_0, \mathbf{k}) \text{ with } kx \equiv k_\mu x_\mu \equiv k_0 x_0 - \mathbf{k} \cdot \mathbf{x}$$

$$\text{and } \Box^2 = \frac{\partial^2}{\partial t^2} - \nabla^2.$$

6. For fermion fields, which satisfy anticommutation relations, we require the anticommutator to vanish at spacelike distances.
7. In Eq. (3.17) P is the Dyson chronological product and this expression for S has been obtained in the interaction (or Dirac) representation. We are intentionally not explicitly making a distinction between the Heisenberg and the Dirac representations because our main interest is to indicate the structure of the arguments used, not the calculational details themselves.
8. Goldberger (1955a) essentially used (3.20) to compute $[\phi(x), \phi(y)]$ and then related this to the forward scattering amplitude.
9. We again warn the reader that we do not display all the kinematical factors for $T_{\beta\alpha}$ here. For the details of this reduction technique, see Jackson (1961, pp. 14–19).
10. In the previous chapter and in the Appendix, where we were concerned mainly with nonrelativistic situations, we set $2m = 1$. For relativistic problems it will often be more convenient to set $m = 1$, as we do in this chapter.
11. In fact, $f(\infty) \neq -e^2$ (Goldberger and Low, 1968; Goldberger, 1970, p.687).
12. That is, assuming isospin invariance, one can analyze the $\pi^- p$ and $\pi^+ p$ scattering data to obtain the isospin phase shifts and then, from these, reconstruct the $\pi^0 p$ amplitude.
13. Goldberger, letter of 6/20/84.
14. Letter of 1/22/86 from Jost.
15. Letter of 12/18/85 from Polkinghorne.
16. The axioms of general quantum field theory are discussed, for example, in the books by Jost (1965) and by Streater and Wightman (1964).
17. A good summary of much of the later work in the axiomatic QFT–SMT program can be found in Lasalle (1974) and in Bros and Lasalle (1975).
18. We simply suppress the complications due to the spin of the nucleon.

Chapter 4 (Another route to a theory based on analytic reaction amplitudes)

1. In addition to an interview at Notre Dame (Spring, 1981), there are letters (11/10/81, 6/8/82, 6/10/82, 12/8/82) in which Chew attempts to recall and reconstruct the major episodes in the development of the S-matrix program. Much of this reconstruction subsequently appeared in Capra's (1985) interview with Chew.

2. Letter of 11/10/81 from Chew.
3. Here our states ϕ_α are normalized to a Dirac delta function, whereas those in the Appendix are unnormalized plane waves.
4. Letter of 12/8/82 from Chew.
5. For a discussion of this model, see Mandl (1959, pp. 18–21).
6. Figure 4.2 is a sketch based on Jackson (1958, p. 13 and p. 15). A discussion of the spin and isospin complications for the phase shifts can be found on pp. 16–18 of Jackson.
7. A very similar principle, the substitution law (Jauch and Rohrlich, 1955, pp. 161–3), is also familiar in quantum electrodynamics. Specific cases of this law have long been known (Bethe and Heitler, 1934; Kockel, 1949; Baumann, 1953).
8. This observation belongs to the historical development of Chapter 5, but for pedagogical reasons it is natural to make it here. However, Figure 4.6 and its interpretation *via* Eqs. (4.23) will not be used in this chapter.
9. Letter of 12/8/82 from Chew.
10. Letter of 12/6/85 from Moravcsik.
11. Here $\sum\limits_{n}$ stands for both a sum and an integration. A detailed exposition of the Chew–Low model can be found in Schweber's (1961) QFT textbook.
12. The details of this solution can be found, for example, in Cushing (1975, pp. 394–6). Here z is (up to an overall multiplicative factor) the center of momentum energy squared. We have taken the mass of the particle to be unity.
13. Actually, depending on the particular δ_{2T2J} chosen, there are known (Chew and Low, 1956a, p. 1574) numerical factors multiplying the term $1/f^2$.
14. Goldberger, Miyazawa and Oehme (1955, p. 988) pointed out that their relativistic dispersion relations (Eqs. (3.39) above) provided a rigorous basis for an effective range expansion. Notice that in our example
$$\text{Re } G(x) = 2\sqrt{x(x-4)} \cot \delta(x).$$
15. Interview with Goldberger on 9/21/84.
16. The coupling constant g here is that for the *relativistic* interaction Hamiltonian $H_1 = -ig\bar{\psi}\gamma_5 \sum\limits_{\lambda} \tau_\lambda \phi_\lambda \psi$, which, in the nonrelativistic limit, reduces to the static interaction Hamiltonian of Eq. (4.14).
17. For the relations among the definitions of various coupling constants see Jackson (1958, pp. 6, 7, 21 and 25).
18. We have omitted known kinematic factors. A modern derivation of Eqs. (4.39) and (4.40) can be found, for example, in Källen (1964, pp. 167–77).

Chapter 5 (The analytic S matrix)

1. Letter of 6/20/84 from Goldberger.
2. Telephone conversation (9/28/84) with Gell-Mann.

3. At the 1985 Fermilab Conference on Particle Physics in the 1950s, Nambu recalled that Symanzik had proposed some double dispersion relations to him (around 1956), but that he (Nambu) had discouraged Symanzik from publishing them.
4. Letter of 10/21/85 from Eden.
5. Letter of 11/12/84 from Mandelstam.
6. We have given only the simplest case here. In general there are pole terms, as well as other possible subtraction terms. The curves of Figure 5.2 show the *largest* possible regions of support for $\rho(s,t)$.
7. This $A(s,t,u)$ is related to the invariant differential cross section as

$$\frac{d\sigma}{dt} = \frac{\pi}{q^2} \frac{d\sigma}{d\Omega} = \frac{16\pi}{s(s-4m^2)} |A(s,t,u)|^2.$$

8. Chew recognized the importance of Mandelstam's work for his own program. He essentially 'kidnapped' Mandelstam from the 1958 American Physical Society Meeting in Washington after hearing his paper (Frazer, 1985).
9. We have omitted some possible subtractions here.
10. For purposes of illustration we have used in Figure 5.4 the boundaries for the equal-mass relativistic case. The asymptotes are somewhat different, but qualitatively the same, for nonrelativistic kinematics.
11. We have glossed over how bound-state poles and substractions are handled.
12. Demonstration of the actual *existence* of such iterative solutions was mucn later established in a mathematical *tour de force* by Atkinson (1968a, 1968b, 1969, 1970) using fixed-point theorems. It is also interesting that there is no (nonzero) solution for $\rho(s,t)$ if elastic unitarity holds exactly for *all* three channels *simultaneously* (Aks, 1965). That is, the dispersion-theory program *requires* production processes for self-consistency just as does quantum field theory.
13. I thank Professor Frazer for providing his comments (letter of 1/6/86) on this work done by him and Fulco. A fuller account of this electromagnetic structure problem for nucleons is given in Sec. 7.7 of Frazer's (1966) book.
14. Letter from Chew on 11/10/81.
15. Letter of 1/22/86 from Jost.
16. Written comments from Polkinghorne (7/1/88).
17. As an interesting aside on this paper of Cutkosky's, which was widely used and referenced since it was published in 1960, it was only in 1966 that Rolnik published a note showing that Cutkosky's result was off by a factor of 2! Had all those people *really* read the paper?
18. Chew's recollection (Capra, 1985, p. 259); letter from Frautschi (8/20/84).
19. Letter of 11/12/84 from Mandelstam.

Chapter 6 (The bootstrap and Regge poles)

1. For the details of obtaining these Pomeranchuk theorems, see Frautschi (1963, pp. 154–8). The Pomeranchuk theorem on constant asymptotic cross sections follows rigorously from the assumption that the scattering amplitude approaches the *same* limit as $s \to +\infty$ and as $s \to -\infty$ (along the real axis). However, the argument we give here is probably easier to understand in physical terms.
2. We neglect the possibility of subtractions here, although these are important in the s-wave.
3. Actually, there are separate bounds for $t = 0$ and $t < 0$, but, for our discussion here, the essential point is made by Eq. (6.18).
4. Letter from Frazer (1/6/86).
5. Letter from Chew (11/10/81).
6. In this approximation, it is assumed that the elastic double spectral functions $\rho^{(el)}(s, t)$ (which come from the narrow strips $4m^2 \leqslant s(t, u) \leqslant 16m^2$ in the diagram of Figure 5.2)) provide the dominant contributions to the scattering amplitudes.
7. Telephone conversation of 9/28/84 with Gell-Mann.
8. Telephone conversation of 9/21/84 with Goldberger.
9. Letter of 9/10/84 from Mandelstam.
10. Letter of 9/6/84 from Low.
11. Letter from Chew (11/10/81).
12. The strip approximation work was left for now since Regge poles were more interesting (Frautschi, letter of 8/20/84).
13. As we have been previously, there are many indications that it was Mandelstam who first appreciated the importance of Regge asymptotic behavior for the relativistic case (Frautschi, 1985, p. 44; Polkinghorne, 1985, p. 23; Chew, 1962a, p. 396). Perhaps one of the best recollections of this point in a publication from this period is Frautschi's (1963, p. 144):

   > It was Mandelstam who first urged, in private discussion, the importance of Regge poles for high-energy scattering.

   A similar statement appears in Frautschi, Gell-Mann and Zachariasen (1962, p. 2204).
14. Letter of 12/8/82 from Chew.
15. Blankenbecler and Goldberger (1962, p. 766) had made a similar suggestion at the La Jolla Conference:

   > ... we suggested the possibility of associating Regge poles with 'elementary' particles as well as with dynamical resonances.

   Figures 5 and 6 of their paper are plots of the nucleon family. They also pointed out the likelihood of Regge cuts, in addition to Regge poles, in the relativistic case.

16. A good sense of the phenomenology in vogue during the 1960s can be gotten from Barger and Cline's (1969) book.
17. Schwimmer and Zachariasen (1975) much later showed that it was still possible in some models for a simple Regge pole phenomenology to be correct at asymptotically high energies.
18. Capra (1985, p. 259); Frautschi (letter of 8/20/84).
19. Interview on 10/10/84 with Gell-Mann.
20. Letter from Chew (12/8/82).
21. Interview with Gell-Mann (10/10/84).
22. Gell-Mann, Goldberger, Low, Marx and Zachariasen (1964, p. B145).
23. However, Chew himself found this merely a simplified version of the Chew–Mandelstam mechanism, and here a cutoff had been introduced (letter from Chew on 11/10/81).
24. Interview on 10/10/84 with Gell-Mann.
25. Again, though, it is only fair to point out that extreme complexity of Regge theory in SMT need not necessarily exclude that approach from becoming a fundamental theory (through, say, the Reggeon calculus (Gribov, 1968)). After all, Reggeon calculus is not obviously more horrendous than gauge QFT or superstring calculations. The history of some attempts to reconcile SMT and QFT through a Reggeization program has been studied by Cao (1988).

Chapter 7 (An autonomous *S*-matrix program)

1. Letter of 10/4/85 from Dyson.
2. Letter of 9/6/84 from Low.
3. Letter of 12/27/85 from Low.
4. Letter of 9/10/84 from Manedlstam.
5. Conversation of 9/28/84 with Gell-Mann.
6. Bogoliubov and Shirkov's (1959) authoritative book on quantum field theory derisively dismissed such arguments.
7. Chou and Dresden (1967) gave a formal *S*-matrix discussion of electromagnetic interactions, but the theory did not cope with the modifications in analyticity produced by the exchange of massless particles.
8. The *S*-matrix program of my 1985 paper is by and large what I here refer to as the *autonomous* *S*-matrix program.
9. Letter from Stapp (6/22/84).
10. My translation.
11. Letter of 2/3/86 from Iagolnitzer.
12. Bloxham (1967), in his Cambridge Ph.D. dissertation, avoided using global analytic properties in inductively establishing a certain minimum set of singularities through application of the unitarity equations. (See also, Bloxham, Olive and Polkinghorne, 1969a, 1969b, 1969c).

13. In fairness, though, it is certainly arguable that the great complexity that SMT faced was not markedly worse than that of gauge QFT and of superstring theory today.
14. Letter 6/22/84 from Stapp.

Chapter 8 (The duality program)

1. Schwarz, a Ph.D. student of Chew's, was heavily involved in the Regge duality program. He would later become a central figure in the quantized string program, as we show below.
2. Of course, $F(v, t)$ depends upon both v and t, but, since t is held constant, we suppress the t dependence. This t dependence does appear in $\alpha(t)$. It is precisely because Eq. (8.5) holds for *all* values of t that Eq. (8.5) is termed a *super*convergence relation.
3. Naturally, the Stirling approximation ceases to be valid when $z \to -\infty$ along the negative *real* axis.
4. Although we have not discussed signature at length previously, it was introduced in order to allow analytic continuation in l of suitable combinations of partial-wave amplitudes (Froissart, 1961; Gribov, 1962).
5. A simple formal way to see this is with the relativistic expression for the energy of a particle

$$E = \frac{m_0}{\sqrt{1 - (v/c)^2}}.$$

Since E must be real, m_0 being purely imaginary (i.e., $m_0^2 < 0$) requires that $v > c$.
6. More formally this is stated as there being only nonexotic states.
7. Dr. Tian-Yu Cao of Cambridge and Harvard Universities has done a study of some aspects of attempts to Reggeize field theories that are related to this program (Cao, 1988).
8. Things are actually more complicated than this, since all possible orderings of (a, b, c, d) must be considered (cf., Chew and Rosenzweig, 1978, pp. 267–70, especially their Figure 2.8). Still, our illustration indicates the nature of the actual argument.
9. Interview with Rosenzweig on 1/22/86.
10. Letter of 1/28/86 from Nicolescu.
11. Letter of 2/1/86 from Nicolescu.
12. An indication of why the residue of the pole term must be positive can be gotten from the identity

$$\frac{1}{z - \alpha} = P \frac{1}{z - \alpha} + \pi i \, \delta(z - \alpha)$$

and the fact that the unitarity condition, Eq. (2.20), requires Im $A > 0$.

13. The actual case is more complicated with

$$[a_\mu^{(m)}, \quad a_\nu^{\dagger(n)}] = -g_{\mu\nu}\delta_{mn} = 1, 2, 3, \ldots,$$

where $g_{\mu\nu}$ is a metric tensor.

14. A review of the quantized string program of this period can be found in Scherk (1975).

15. Interestingly, the work on supersymmetries, begun by Wess and Zumino (1974), was an outgrowth of supergauge transformations first used in field theories associated with the (Nevu-Schwarz, 1971a) string model. Things had in a sense come full circle with Brink, Schwarz and Scherk (1977).

16. This possibility of having just *one* underlying entity is reminiscent of Heisenberg's nonlinear spinor quantum field theory.

17. Recall that a field theory is renormalizable if it contains only a finite number of *different* types of infinities (i.e., diagrams which lead to undefined terms). If new types of infinities arise in successively higher orders of a perturbation series, then no *finite* number of redefinitions of physically observable quantities (such as charge, mass, etc.) can produce a theory that can predict (finite) measurable quantities.

18. Letter of 1/13/86 from Capps.

Chapter 9 ('Data' for a methodological study)

1. Telephone interview of 9/21/84 with Goldberger.
2. Telephone interview of 9/21/84 with Goldberger.
3. A detailed history of early QED by Schweber (1984) has appeared since my own 1982 outline and a further extension of his work is in press (1990).
4. In this section a few formulas from Chapter 2 are repeated, because that makes the section easier to read and because some readers may not have read Chapter 2 in detail.
5. I do not mean to imply that there could not be problems for this reconstructed story. Rather, it is a scenario worth further study. I (1989b) am currently involved in an extensive case study of the causal quantum theory.

Chapter 10 (Methodological lessons)

1. The recognition of the role of extrascientific reasons in theories of high generality is not recent, having already been discussed by Frank (1957, pp. 354–60).

2. Jardine (1986) has attempted to downplay the importance of such diversity of styles of scientific inquiry in earlier periods by distinguishing between the 'original reasons' for considering a theory and the 'consensual reasons' for finally accepting it. The impact of such 'serendipity' (p. 123) in the discovery of knowledge is muted, in Jardine's opinion, because the 'good' results, however uncovered, can be accounted for today by our superior methods of modern science. This defense has a somewhat 'Whiggish' air about it. However, we are concerned here only with acknowledgment of the fact that methodological rules *have* changed. McMullin (1988) has argued for the rationality of science largely in terms of a continuing commitment to empirical adequacy even though he allows that reason and practice do interact (p. 10), that, by a process of internalization, previously transcendent values become part of science and, thereby, subject to revision (p. 20), that the rationality of science may change to reach a given goal (p. 22), and that new goals may be set for/by science (p. 31). Given that these significant changes have taken place at the metalevel, it becomes unclear just what does characterize science in an essential or invariant manner.

3. Oldershaw (1988) has recently given a useful nontechnical discussion, illustrated with several examples from particle physics and cosmology, of the effective nontestability of much of current physical theory.

4. Letter of 6/10/82 from Chew.

5. Letter of 11/10/81 from Chew.

6. Letter of 10/21/85 from Feynman.

7. Letter of 10/21/85 from Eden.

8. Conversation on 11/5/85 with Gell-Mann.

9. Interview with Gell-Mann on 5/4/85.

10. Interview with Gell-Mann on 10/10/84.

11. Letter of 2/13/86 from Olive.

12. Conversation with Gell-Mann on 11/1/84.

13. Letter of 12/16/85 from Moravcsik.

14. A slogan such as 'Science is what scientists do' need not be circular because we characterize scientists by the activities they engage in, not by merely saying they do science.

15. We are excluding as uninteresting constraints on scientific practice and reasoning those already present in the normal, common-sense demands placed on everyday argument (such as the usual rules of deductive reasoning, say, or an acknowledgment of some relevance of facts). In setting aside such general constraints as uninteresting (but not as unimportant), we claim that there are no *invariant* (ahistorical, atemporal) characteristics of science that distinguish it from nonscience. Rorty (1988, p. 60) has pointed out that the use of hypotheses is not peculiar to science and has even claimed that there is no *special* scientific method. Toulmin (1961) claimed there is no *one* aim or method for science (pp. 13–17) and pictures scientific explanation and understanding in terms of *evolving* ideals

of natural order. On the question of whether or not scientific methodology is atemporal in any interesting sense, see Cushing (1989d).

16. See the comments in Note 15 above on McMullin's attempt to find a unifying theme amidst the change of practice and method in science, and Cushing (1989d).

17. Rorty (1988, p. 70) has offered a *psychological* basis for our need to give science a special status.

Appendix

1. We also make a few (historically) retrospective observations that put the results in a form useful for illustrations in the main body of the text.

2. For convenience, here and in the early part of the main body of the book, we usually employ units such that $\hbar$ (Planck's constant divided by 2π) $=1$, c (speed of light in vacuum) $=1$, m (mass of particle) $=\frac{1}{2}$. More details of scattering theory in modern notation can be found in Blatt and Weisskopf (1952), Messiah (1965) and Newton (1966).

3. A *rigorous* justification of this use of ψ_{sc} only (rather than the full ψ of Eq. (A.7)) to compute σ_{sc} requires that wave packets (rather than plane-wave states) be employed. The basic idea is that once one gets out of the dead forward (i.e., $\theta = 0$) direction one will, for a wave packet, be out of the incident beam and see only ψ_{sc}. For details see Messiah (1965, pp. 372–80) and Low (1959).

4. In fact, one can make a fairly complete analogy between the scalar wave field $\psi(\mathbf{r}, t)$ and the $\mathbf{E}$ and $\mathbf{B}$ fields of electromagnetic theory with the following correspondences:

$$\mathbf{D} = \varepsilon \mathbf{E} \qquad\qquad\qquad \mathbf{B} = \mu \mathbf{H}$$

$$u = \frac{1}{8\pi}(\mathbf{E} \cdot \mathbf{D} + \mathbf{B} \cdot \mathbf{H}) \qquad\qquad \mathbf{S} = \frac{c}{4\pi} \mathbf{E} \times \mathbf{H}$$

$$\mathbf{E} \to 2\sqrt{\pi}\sqrt{\rho_0}\frac{-\partial\psi}{\partial t} \qquad\qquad \mathbf{B} \to -\sqrt{2\pi}\sqrt{\rho_0}\,c\nabla\psi$$

$$\varepsilon \to \rho/\rho_0 \qquad\qquad\qquad \mu \to \frac{\rho}{\rho_0}\frac{c^2}{v^2}$$

$$u \to \frac{1}{2}\rho\left[\left(\frac{\partial\psi}{\partial t}\right)^2 + v^2\nabla\psi \cdot \nabla\psi\right] \qquad \mathbf{S} \to -\rho v^2 \frac{\partial\psi}{\partial t}\nabla\psi$$

5. For the electromagnetic case, of course, c is the speed of light in vacuum.

6. These and the immediately following statements are not intended to be mathematically precise and rigorous. For the necessary mathematical details, see Titchmarch (1937, pp. 128–9).

7. Again, for simplicity, we consider only one-dimensional motion, $x(t)$, rather than three-dimensional motion, $r(t)$. The results are the same in either case. A modern reference having more details of classical radition theory is Jackson (1975).

8. We have left out several technical steps here, among which is the level shift (see Heitler (1936), pp. 110–13).

References

Abers, E., Zachariasen, F., and Zemach, C. (1963). Origin of internal symmetries, *Physical Review*, **132**, 1831–6.

Aharonov, Y., and Bohm, D. (1959). Significance of electromagnetic potentials in quantum theory. *Physical Review*, **115**, 485–91.

Aks, S. Ø. (1965). Proof that scattering implies production in quantum field theory. *Journal of Mathematical Physics*, **6**, 516–32.

Alessandrini, V., Amati, D., Le Bellac, M., and Olive, D. (1971). The operator approach to dual multiparticle theory. *Physics Reports*, **1**, 269–346.

Allen, A. D. (1973). The bootstrap from the perspective of formal logic. *Foundations of Physics*, **2**, 473–5.

Amati, D., Fubini, S., and Stanghellini, A. (1962a). Asymptotic properties of scattering and multiple production. *Physics Letters*, **1**, 29–32.

Amati, D., Stanghellini, A., and Fubini, S. (1962b). Theory of high-energy scattering and multiple production. *Il Nuovo Cimento*, **26**, 896–954.

Anderson, H. L., Davidon, W. C., and Kruse, U. E. (1955). Causality in the pion–proton scattering. *Physical Review*, **100**, 339–43.

Arenhrövel, H., and Sauris, A. M. (eds.) (1981). *Lecture Notes in Physics*, **137**. Berlin: Springer–Verlag.

Ascoli, R., and Heisenberg, W. (1957). Zur Quantentheorie nichtlinearer Wellengleichungen. *Zeitschrift für Naturforschung*, **12a**, 177–87.

Asquith, P. D., and Kitcher, P. (eds.) (1984). *PSA 1984, Proceedings of the 1984 Biennial Meeting of the Philosophy of Science Association*, vol. 1. East Lansing, MI: Philosophy of Science Association.

Asquith, P. D., and Nickles, T. (eds.) (1983). *PSA 1982, Proceedings of the 1982 Biennial Meeting of the Philosophy of Science Association*, vol. 2. East Lansing, MI: Philosophy of Science Association.

Atkinson, D. (1968a). A proof of the existence of functions that satisfy exactly both crossing and unitarity (I). Neutral pion–pion scattering. No subtractions. *Nuclear Physics*, **B7**, 375–408.

Atkinson, D. (1968b). A proof of the existence of functions that satisfy exactly both crossing and unitarity (II). Charged pions. No Subtractions. *Nuclear Physics*, **B8**, 377–90.

Atkinson, D. (1969). A proof of the existence of functions that satisfy exactly both crossing and unitarity (III). Subtractions. *Nuclear Physics*, **B13**, 415–36.

Atkinson, D. (1970). A proof of the existence of functions that satisfy exactly both crossing and unitarity (IV). Nearly constant asymptotic cross sections. *Nuclear Physics*, **B23**, 397–412.

Aubert, J. J. *et al.* (1974). Experimental observation of a heavy particle J. *Physical Review Letters*, **33**, 1404–6.

Augustin, J.-E. *et al.* (1974). Discovery of a narrow resonance in e^+e^- annihilation. *Physical Review Letters*, **33**, 1406–8.

Bachelard, G. (1934). *Le nouvel esprit scientifique*. Paris: Presses Universitaires de France. (Appeared in translation (by A. Goldhammer, 1984) as *The New Scientific Sprit*, Boston; Beacon Press.) Any page references we give are to the 1984 (English) edition.

Bachelard, G. (1947). *La formation de l'esprit scientifique*. Paris: Librairie Philosophique J. Vrin.

Bacon, F. (1620). *The New Organon*, (ed. F. H. Anderson, 1960). Indianapolis, IN: The Bobbs-Merrill Co. Inc.

Balázs, L. A. P. (1962a). Dispersion-theoretical fitting of nucleon–nucleon scattering. *Physical Review*, **125**, 2179–84.

Balázs, L. A. P. (1962b). $I = \frac{1}{2}, J = \frac{1}{2}$, state in πN scattering with the nucleon as a bound state. *Physical Review*, **128**, 1935–9.

Balázs, L. A. P. (1962c). Low-energy pion–pion scattering. *Physical Review*, **128**, 1939–44.

Balázs, L. A. P. (1963). Low-energy pion–pion scattering. II *Physical Review*, **129**, 872–6.

Balázs, L. A. P. (1976). Thresholds of the pomeron in a dual multiperipheral model. *Physics Letters*, **61B**, 187–90.

Balázs, L. A. P. (1977a). Vacuum singularities in a dual multiperipheral model. *Physical Review*, **D15**, 309–18.

Balázs, L. A. P. (1977b). Planar bootstrap based on a Padé approximation to the dual multiperipheral model. *Physical Review*, **D15**, 319–26.

Balázs, L. A. P. (1977c). Planar bootstrap without the dual-tree approximation. *Physical Review*, **D16**, 885–95.

Balázs, L. A. P. (1977d). Generation of a Pomeron and an f Regge pole in the dual-unitarization program. *Physical Review*, **D16**, 2905–7.

Balázs, L. A. P. (1977e). Meson spectrum in a single planar boostrap model. *Physics Letters*, **71B**, 216–8.

Balázs, L. A. P. (1986). Could there be a Planck-scale bootstrap underlying the superstring? *Physical Review Letters*, **56**, 1759–62.

Balázs, L. A. P., Gauron, P., and Nicolescu, B. (1984). Self-consistent lowest-order dual topological-unitarization Regge-tragectory and coupling calculation. *Physical Review*, **D29**, 533–5.

Balázs, L. A. P., and Nicolescu, B. (1980). Spherical boostrap calculation of qqq-baryon and multiquark-hadron masses. *Zeitschrift für Physik*, **C6**, 269–82.

Balázs, L. A. P., and Nicolescu, B. (1983). Self-consistent light-quark qqq and qqq̄q̄ mass spectrum. *Physical Review*, **D28**, 2818–22.

Ballam, J., Fitch, V. L., Fulton, T., Huang, K., Rau, R. R., and Treiman, S. B. (1956). *High Energy Nuclear Physics, Proceedings of the Sixth Annual Rochester Conference*. New York: Interscience Publishers.

Bardakci, K., and Ruegg, H. (1968). Reggeized resonance model for the production amplitude. *Physics Letters*, **28B**, 342–7.

Barger, V. D., and Cline, D. E. (1969). *Phenomenological Theories of High Energy Scattering*. New York: W. A. Benjamin.

Bargmann, V. (1949a). Remarks on the determination of a central field of force from the elastic scattering phase shifts. *Physical Review*, **75**, 301–3.

Bargmann, V. (1949b). On the connection between phase shifts and scattering potential. *Reviews of Modern Physics*, **21**, 488–93.

Barrow, J. D. (1981). The lore of large numbers: some historical background to the anthropic principle. *Quarterly Journal of the Royal Astronomical Society*, **22**, 388–420.

Barrow, J. D., and Tipler, F. J. (1986). *The Anthropic Cosmological Principle*. Oxford: Clarendon Press.

Baumann, K. (1953). Eine einfache Herleitung der Streuformel von Bhabba. *Acta Physica Austriaca*, **7**, 96–7.

Belinfante, J. G., and Cutkosky, R. E. (1965). Baryon Supermultiplet. *Physical Review Letters*, **14**, 33–5.

Bethe, H., and Heitler, W. (1934). On the stopping of fast particles and on the creation of positive electrons. *Proceedings of the Royal Society*, **A146**, 83–112.

Bethe, H. A., and Rohrlich, F. (1952). Small angle scattering of light by a Coulomb field. *Physical Review*, **86**, 10–16.

Blankenbecler, R., Cook, L. F., and Goldberger, M. L. (1962). Is the photon an elementary particle? *Physical Review Letters*, **8**, 463–5.

Blankenbecler, R., Coon, D. D., and Roy, S. M. (1967). S-matrix approach to internal symmetries. *Physical Review*, **156**, 1624–36.

Blenkenbecler, R., and Goldberger, M. L. (1962). Behavior of scattering amplitudes at high energies, bound states, and resonances. *Physical Review*, **126**, 766–86.

Blankenbecler, R., Goldberger, M. L., Khuri, N. N., and Treiman, S. B. (1960). Mandelstam representation for potential scattering. *Annals of Physics*, **10**, 62–93.

Blatt, J. M., and Weisskopf, V. F. (1952). *Theoretical Nuclear Physics*. New York: John Wiley & Sons.

Bloor, D. (1976). *Knowledge and Social Imagery*. London: Routledge & Kegan Paul.

Bloxham, M. J. W. (1967). Physical region singularities of the S-matrix, unpublished Ph.D. dissertation, Department of Applied Mathematics and Theoretical Physics, University of Cambridge.

Bloxham, M. J. W., Olive, D. I., and Polkinghorne, J. C. (1969a). S-matrix singularity structure in the physical region. I. Properties of multiple integrals. *Journal of Mathematical Physics*, **10**, 494–502.

Bloxham, M. J. W., Olive, D. I., and Polkinghorne, J. C. (1969b). S-matrix singularity structure in the physical region. II. Unitarity integrals. *Journal of Mathematical Physics*, **10**, 545–52.

Bloxham, M. J. W., Olive, D. I., and Polkinghorne, J. C. (1969c). S-matrix singularity structure in the physical region. III. General discussion of simple Landau singularities. *Journal of Mathematical Physics*, **10**, 553–61.

Blum, W., Dürr, H.-P., and Rechenberg, H. (eds.) (1989). *Werner Heisenberg: Gesammelte Werke: Collected Works*, vol. A II. Berlin: Springer–Verlag.

Bogoliubov, N. N., Medvedev, B. V., and Polivanov, M. K. (1958). Probleme der Theorie der Dispersionsbeziehungen. *Fortschritte der Physik*, **6**, 169–245.

Bogoliubov, N. N., and Shirkov, D. V. (1959). *Introduction to the Theory of Quantized Fields*. London: Interscience Publishers, Inc.

Bohm, D. (1952). A suggested interpretation of quantum theory in terms of 'hidden' variables I & II. *Physical Review*, **85**, 166–93.

Bohm, D. (1953). Proof that probability density approaches $|\psi|^2$ in causal interpretation of the quantum theory. *Physical Review*, **89**, 458–66.

Bohm, D. (1980). *Wholeness and the Implicate Order*. London: Routledge and Kegan Paul.

Bohm, D. (1987). Hidden variables and implicate order. In Hiley and Peat, pp. 33–45.

Bohm, D., and Hiley, B. J. (1982). The de Broglie pilot wave theory and the further development of new insights arising out of it. *Foundations of Physics*, **12**, 1001–16.

Bohm, D., Hiley, B. J., and Kaloyerou, P. N. (1987). An ontological basis for the quantum theory. *Physics Reports*, **144** (6), 321–75.

Bohm, D., and Vigier, J.-P. (1954). Model of the causal interpretation of quantum theory in terms of a fluid with irregular fluctuations. *Physical Review*, **96**, 208–16.

Bohm, D., and Vigier, J.-P. (1958). Relativistic hydrodynamics of rotating fluid masses. *Physical Review*, **109**, 1882–91.

Bohr, N. (1936). Neutron capture and nuclear constitution. *Nature*, **137**, 344–8.

Bohr, N., and Kalckar, F. (1937). On the transmutation of atomic nuclei by impact of material particles. *Kongelige Danske Videnskabernes Selskab, Matematisk-fysiske Meddelelser*, **14**, No. 10.

Bonamy, P., *et al.* (1966). $\pi^- p \to \pi^0 n$ polarization at 5.9 and 11.2 GeV/c. *Physics Letters*, **23**, 501–5.

Breit, G. (1932). Quantum theory of dispersion. *Reviews of Modern Physics*, **4**, 504–76.

Breit, G. (1933). Quantum theory of dispersion (continued), parts VI and VII. *Reviews of Modern Physics*, **5**, 91–140.

Breit, G. (1940a). The interpretation of resonances in nuclear reactions. *Physical Review*, **58**, 506–37.

Breit, G. (1940b). Scattering matrix of radioactive states. *Physical Review*, **58**, 1068–74.

Breit, G., and Wigner, E. (1936). Capture of slow neutrons. *Physical Review*, **49**, 519–31.

Bremermann, H., Oehme, R., and Taylor, J. G. (1958). Proof of dispersion relations in quantized field theories. *Physical Review*, **109**, 2178–90.

Brenig, W., and Haag, R. (1959). Allgemeine Quantentheorie der Stossprozesse. *Fortschritte der Physik*, **7**, 183–242.

Brink, L., Schwarz, J. H., and Scherk, J. (1977). Supersymmetric Yang–Mills Theories. *Nuclear Physics*, **B121**, 77–92.

Brinkley, A. (1984). Writing the history of contemporary America: dilemmas and challenges. *Daedalus*, **113**, No. 3, 121–41.

Bros, J., Epstein, H., and Glaser, V. (1965). A proof of the crossing property for two-particle amplitudes in general quantum field theory. *Communications in Mathematical Physics*, **1**, 240–64.

Bros, J., Glaser, V., and Epstein, H. (1972). Local analyticity properties of the *n*-particle scattering amplitude. *Helvetica Physica Acta*, **45**, 149–81.

Bros, J., and Lassalle, M. (1975). Analyticity properties and many-particle structure in general quantum field theory. II. *Communications in Mathematical Physics*, **43**, 279–309.

Brown, H., and Harré, R. (eds.) (1988). *Philosophical Foundations of Quantum Field Theory*. Oxford: Oxford University Press.

Brown, L. M., Dresden, M., and Hoddeson, L. (1989). *Pions to Quarks: Particle Physics in the 1950's*. Cambridge: Cambridge University Press.

Brown, L. M., and Feynmann, R. P. (1952). Radiative corrections to Compton scattering. *Physical Review*, **85**, 231–44.

Brown, L. M., and Hoddeson, L. (1983). *The Birth of Particle Physics*. Cambridge: Cambridge University Press.

Brown, L. M., and Rechenberg, H. (1987). Paul Dirac and Werner Heisenberg – A partnership in science. In Kursunoglu and Wigner, pp. 117–62.

Buchdahl, G. (1970). History of science and criteria of choice. In Stuewer, pp. 204–5.

Burhop, E. H. S. (ed.) (1972). *High Energy Physics*. New York: Academic Press.

Burckhardt, J. (1963). *Briefe, Vol. V*. Basel/Stuttgart: Schwabe & Co. Verlag.

Butts, R. E., and Hintikka, J. (eds.) (1977). *Foundational Problems in the Special Sciences*. Dordrecht: D. Reidel.

Cao, Tian Yu (1988). Attempts at reconciling quantum field theory with

S-matrix theory: the Reggeization program: 1962–1982. Harvard University preprint.

Capella, A., Sukhatme, U., Tan, C.-I. and Tran Thanh Van, J. (1985). The Pomeron story. In De Tar *et al.*, pp. 79–87.

Capps, R. H. (1963). Prediction of an interaction symmetry from dispersion relations. *Physical Review Letters*, **10**, 312–14.

Capps, R. H. (1964). Many-particle self-consistent model. *Physical Review*, **134**, B1396–B1406.

Capps, R. H. (1965). SU(6) in a baryon bootstrap model, *Physical Review Letters*, **14**, 31–3.

Capps, R. H. (1966). SU(6)$_w$-symmetric meson bootstrap model. *Physical Review*, **148**, 1332–40.

Capps, R. H. (1967). SU(6)$_w$ and the Bootstrap Hypothesis. In Takeda and Fuji, pp. 136–60.

Capps, R. H., and Takeda, G. (1956). Dispersion relations for finite momentum-transfer pion–nucleon scattering. *Physical Review*, **103**, 1877–96.

Capra, F. J. (1977). Incorporation of baryons into the topological expansion. *Physics Letters*, **68B**, 93–5.

Capra, F. (1979). Quark physics without quarks: a review of recent developments in S-matrix theory. *American Journal of Physics*, **47**, 11–23.

Capra, F. (1984). Bootstrap theory of particles. *Surveys in High Energy Physics*, **4**, 127–202.

Capra, F. (1985). Bootstrap physics: a conversation with Geoffrey Chew. In De Tar *et al.*, pp. 247–86.

Cartwright, N. (1983). *How the Laws of Physics Lie*. Oxford: Oxford University Press.

Cassidy, D. C. (1981). Cosmic ray showers, high energy physics, and quantum field theories: programmatic interactions in the 1930's. *Historical Studies in the Physical Sciences*, **12**, 1–39.

Castillejo, L., Dalitz, R. H., and Dyson, F. J. (1956). Low's scattering equation for the charged and neutral scalar theories. *Physical Review*, **101**, 453–8.

Chamberlain, O. (1985). Interactions with Geoff Chew. In De Tar *et al.*, pp. 11–13.

Chan, H.-M. (1969). A generalized Veneziano model for the N-point function. *Physics Letters*, **28B**, 425–8.

Chan, H.-M., Paton, J. E., and Tsou, S. T. (1975). Diffractive scattering in the dual model. *Nuclear Physics*, **B86**, 479–525.

Chan, H.-M., Paton, J. E., Tsou, S. T. and Ng, S. W. (1975). Regge parameters from duality and unitarity. *Nuclear Physics*, **B92**, 13–36.

Chand, R. (ed.) (1970). *Symmetries and Quark Models*. New York: Gordon and Breach.

Chandler, C. (1968). Causality in S-matrix theory. *Physical Review*, **174**, 1749–58.

Chandler, C., and Stapp, H. P. (1969). Macroscopic causality and properties of scattering amplitudes. *Journal of Mathematical Physics*, **10**, 826–59.

Charap, J. M., and Fubini, S. (1959). The field theoretic definition of the nuclear potential-I. *Il Nuovo Cimento*, **14**, 540–59.

Chew, G. F. (1948). Elastic scattering of high energy nucleons by deuterons. *Physical Review*, **74**, 809–16.

Chew, G. F. (1950a). The inelastic scattering of high energy neutrons by deuterons and the neutron–neutron interaction. *Physical Review*, **79**, 219.

Chew, G. F. (1950b). The inelastic scattering of high energy neutrons by deuterons according to the impulse approximation. *Physical Review*, **80**, 196–202.

Chew, G. F. (1951a). High energy elastic proton–deuteron scattering. *Physical Review*, **84**, 1057–8.

Chew, G. F. (1951b). A theoretical calculation of the inelastic scattering of 90-MeV neutrons by deuterons. *Physical Review*, **84**, 710–16.

Chew, G. F. (1953). Pion–nucleon scattering when the coupling is weak and extended. *Physical Review*, **89**, 591–3.

Chew, G. F. (1954a). Renormalization of meson theory with a fixed extended source, *Physical Review*, **94**, 1748–54.

Chew, G. F. (1954b). Method of approximation for the meson–nucleon problem when the interaction is fixed and extended. *Physical Review*, **94**, 1755–9.

Chew, G. F. (1954c). Improved calculation of the p-wave pion–nucleon scattering phase shifts in the cut-off theory. *Physical Review*, **95**, 285–6.

Chew, G. F. (1954d). Comparison of the cut-off meson theory with experiment. *Physical Review*, **95**, 1669–75.

Chew, G. F. (1956). Report on theoretical pion physics at Illinois. *Supplemento del Nuovo Cimento*, **4**, 761–5.

Chew, G. F. (1958a). The nucleon and its interaction with pions, photons, and nucleons and antinucleons I. In Ferretti, pp. 93–103.

Chew, G. F. (1958b). Proposal for determining the pion–nucleon coupling constant from the angular distribution for nucleon–nucleon scattering. *Physical Review*, **112**, 1380–3.

Chew, G. F. (1960a). Theory of strong coupling of ordinary particles. In *Proceedings of the Ninth International Annual Conference on High-Energy Physics*. vol. I, pp. 313–52. Moscow.

Chew, G. F. (1960b). The pion–pion interaction. In Sudarshan *et al.*, pp. 273–7.

Chew, G. F. (1961a). Double dispersion relations and unitarity as the basis for a dynamical theory of strong interactions. In Screaton, pp. 167–226, and in DeWitt 1960, pp. 455–514.

Chew, G. F. (1961b). A unified dynamical approach to high and low energy strong interactions. *Reviews of Modern Physics*, **33**, 467–70.

Chew, G. F. (1961c). *S-Matrix Theory of Strong Interactions.* New York: W. A. Benjamin.

Chew, G. F. (1961d). The *S*-matrix theory of strong interactions. Lawrence Radiation Laboratory Preprint, UCRL-9701, May 15, 1961.

Chew, G. F. (1962a). *S*-matrix theory of strong interactions without elementary particles. *Reviews of Modern Physics,* **34**, 394–401.

Chew, G. F. (1962b). Reciprocal bootstrap relationship of the nucleon and the (3,3) resonance. *Physical Review Letters,* **9**, 233–5.

Chew, G. F. (1962c). Strong interaction theory without elementary particles. In Prentki, pp. 525–30.

Chew, G. F. (1963a). Strong-interaction *S*-matrix theory without elementary particles. In *1962 Cargèse Lectures in Theoretical Physics.* pp. XI-1–37. New York: W. A. Benjamin.

Chew, G. F. (1963b). Self-consistent *S*-matrix with Regge asymptotic behavior. *Physical Review,* **129**, 2363–70.

Chew, G. F. (1963c). The dubious role of the space-time continuum in microscopic physics. *Science Progress,* **51**, 529–39.

Chew, G. F. (1964a). Elementary particles. *Physics Today,* **17**, (No. 4), 30–4.

Chew, G. F. (1964b). What is the nucleon? In Hofstadter and Schiff, pp. 3–10.

Chew, G. F. (1964c). Nuclear democracy and bootstrap dynamics. In Jacob and Chew, pp. 103–52.

Chew, G. F. (1965). The analytic *S*-matrix: a theory for strong interactions. In DeWitt and Jacob, pp. 187–250.

Chew, G. F. (1966a). Analyticity as a fundamental principle in physics. *Supplemento del Nuovo Cimento,* **4**, 369–83.

Chew, G. F. (1966b). Nuclear democracy, Regge poles and the analytic *S*-matrix. In Takeda, pp. 1–8.

Chew, G. F. (1966c). Bootstrapping with the Regge boundary condition. In Takeda, pp. 9–16.

Chew, G. F. (1966d). *The Analytic S Matrix: A Basis for Nuclear Democracy.* New York: W. A. Benjamin.

Chew, G. F. (1966e). Crisis for the elementary-particle concept, University of California Berkeley, preprint UCRL-17137.

Chew, G. F. (1967a). Regge-pole daughters and reactions involving unequal-mass particles. *Comments on Nuclear and Particle Physics,* **1**, 17–19.

Chew, G. F. (1967b). Zeros in the Regge formula. *Comments on Nuclear and Particle Physics,* **1**, 58–62.

Chew, G. F. (1967c). The pion mystery. *Comments on Nuclear and Particle Physics,* **1**, 187–90.

Chew, G. F. (1967d). The Pomeranchuk trajectory: actuality or mirage? *Comments on Nuclear and Particle Physics,* **1**, 121–4.

Chew, G. F. (1968a). *S*-matrix theory with Regge poles. In *Proceedings of the Fourteenth Conference on Physics at the University of Brussels.* pp. 65–95. London: Interscience Publishers.

Chew, G. F. (1968b). Horn–Schmid duality. *Comments on Nuclear and Particle Physics*, **2**, 74–7.

Chew, G. F. (1968c). Nuclear and particle physics: two different subjects? *Comments on Nuclear and Particle Physics*, **2**, 107–110.

Chew, G. F. (1968d). Multiperipheralism and the bootstrap, *Comments on Nuclear and Particle Physics*, **2**, 163–8.

Chew, G. F. (1968e). 'Bootstrap'; a scientific idea? *Science*, **161**, 762–5.

Chew, G. F. (1970). Hadron bootstrap: triumph or frustration? *Physics Today*, **23**, (No. 10), pp. 23–8.

Chew, G. F. (1971). Hadron bootstrap hypothesis. *Physical Review*, **D4**, 2330–5.

Chew, G. F. (1974). Impasse for the elementary-particle concept. In *The Great Ideas Today*, pp. 367–99. Chicago: Encyclopaedia Britannica.

Chew, G. F. (1979a). Bootstrap theory of quarks. *Nuclear Physics*, **B151**, 237–46.

Chew, G. F. (1979b). Topological colour and disc mating unitarity as the source of zero triality. *Physics Letters*, **82B**, 439–41.

Chew, G. F. (1983a). Bootstrapping the photon. *Foundations of Physics*, **13**, 217–46.

Chew, G. F. (1983b). The topological bootstrap. In Guth *et al.*, pp. 49–69.

Chew, G. F. (1985). Gentle quantum events as the source of explicate order. *Zygon*, **20**, 159–64; see also, Hiley and Peat (1987), pp. 249–54.

Chew, G. F. (1989). Particles as S-matrix poles: hadron democracy. In Brown, Dresden and Hoddeson, pp. 600–7.

Chew, G. F., and Finkelstein, J. (1983). Topological theory and the standard electroweak model. *Physical Review Letters*, **50**, 795–8.

Chew, G. F., and Frautschi, S. C. (1960). Unified approach to high- and low-energy strong interactions on the basis of the Mandelstam representation. *Physical Review Letters*, **5**, 580–3.

Chew, G. F., and Frautschi, S. C. (1961a). Dynamical theory for strong interactions at low momentum transfer but arbitrary energies. *Physical Review*, **123**, 1478–86.

Chew, G. F., and Frautschi, S. C. (1961b). Potential scattering as opposed to scattering associated with independent particles in the S-matrix theory of strong interactions. *Physical Review*, **124**, 264–8.

Chew, G. F., and Frautschi, S. D. (1961c). Principle of equivalence for all strongly interacting particles within the S-matrix framework. *Physical Review Letters*, **7**, 394–7.

Chew, G. F., and Frautschi, S. C. (1962). Regge trajectories and the principle of maximum strength for strong interactions. *Physical Review Letters*, **8**, 41–4.

Chew, G. F., Frautschi, S. C., and Mandelstam, S. (1962). Regge poles in $\pi\pi$ scattering. *Physical Review*, **126**, 1202–8.

Chew, G. F., and Goldberger, M. L. (1948). High-energy neutron–proton scattering. *Physical Review*, **73**, 1409.

Chew, G. F., and Goldberger, M. L. (1949a). Analysis of low energy proton–proton scattering experiments. *Physical Review*, **75**, 1466.

Chew, G. F., and Goldberger, M. L. (1949b). On the analysis of nucleon–nucleon scattering experiments. *Physical Review*, **75**, 1637–44.

Chew, G. F., and Goldberger, M. L. (1950). The production of fast deuterons in high energy nuclear reactions. *Physical Review*, **77**, 470–5.

Chew, G. F., and Goldberger, M. L. (1952). The scattering of elementary particles by complex nuclei – a generalization of the impulse approximation. *Physical Review*, **87**, 778–82.

Chew, G. F., Goldberger, M. L., and Low, F. E. (1969). An integral equation for scattering amplitudes. *Physical Review Letters*, **22**, 208–12.

Chew, G. F., Goldberger, M. L., Low, F. E., and Nambu, Y. (1957a). Application of dispersion relations to low-energy meson–nucleon scattering. *Physical Review*, **106**, 1337–44.

Chew, G. F., Goldberger, M. L., Low, F. E., and Nambu, Y. (1957b). Relativistic dispersion relation approach to photomeson production. *Physical Review*, **106**, 1345–55.

Chew, G. F., Goldberger, M. L., Steinberger, J. M., and Yang, C. N. (1951). Theoretical analysis of the process $\pi^+ d \Leftrightarrow p + p$. *Physical Review*, **84**, 581–2.

Chew, G. F., Karplus, R., Gasiorowicz, S., and Zachariasen, F. (1958). Electromagnetic structure of the nucleon in local field theory. *Physical Review*, **110**, 265–76.

Chew, G. F., and Lewis, H. W. (1951). A phenomenological treatment of photomeson production from deuterons. *Physical Review*, **84**, 779–85.

Chew, G. F., and Low, F. E. (1956a). Effective-range approach to the low-energy p-wave pion–nucleon interaction. *Physical Review*, **101**, 1570–9.

Chew, G. F., and Low, F. E. (1956b). Theory of photomeson production at low energies. *Physical Review*, **101**, 1579–87.

Chew, G. F., and Low, F. E. (1959). Unstable particles as targets in scattering experiments. *Physical Review*, **113**, 1640–8.

Chew, G. F., and Mandelstam, S. (1960). Theory of low energy pion–pion interaction. *Physical Review*, **119**, 467–77.

Chew, G. F., and Mandelstam, S. (1961). Theory of the low-energy pion–pion interaction II. *Il Nuovo Cimento*, **19**, 752–76.

Chew, G. F., and Moyer, B. J. (1950). High energy accelerators at the University of California Radiation Laboratory. *American Journal of Physics*, **18**, 125–36.

Chew, G. F., and Moyer, B. J. (1951). High-energy nucleon–nucleon scattering experiments at Berkeley. *American Journal of Physics*, **19**, 203–11.

Chew, G. F., and Pignotti, A. (1968). Dolen–Horn–Schmid duality and the Deck effect. *Physical Review Letters*, **20**, 1078–81.

Chew, G. F., and Poénaru, V. (1980). Topological bootstrap prediction of three-'colored', eight-flavored quarks. *Physical Review Letters*, **45**, 229–31.

Chew, G. F., and Poénaru, V. (1981). Topological bootstrap theory of hadrons. *Zeitschrift für Physik*, **C11**, 59–93.

Chew, G. F., and Rosenzweig, C. (1975). Pomeron–Reggeon relationship according to the topological expansion. *Physical Review*, **D12**, 3907–20.

Chew, G. F., and Rosenzweig, C. (1976). Asymptotic planarity: an *S*-matrix basis for the Okubo–Zweig–Iizuka rule. *Nuclear Physics*, **B104**, 290–306.

Chew, G. F., and Rosenzweig, C. (1977). G parity and the breaking of exchange degeneracy. *Physical Review*, **D15**, 3433–40.

Chew, G. F., and Rosenzweig, C. (1978). Dual topological unitarization: an ordered approach to hadron theory. *Physics Reports*, **41**, 263–327.

Chew, G. F., and Snider, D. R. (1970). Multiperipheral mechanism for a schizophrenic Pomeranchon. *Physical Review*, **D1**, 3453–8.

Chew, G. F., and Snider, D. R. (1971). Partial bootstrap of the schizophrenic Pomeranchon. *Physical Review*, **D3**, 420–9.

Chew, G. F., and Stapp, H. P. (1988). Three-space from quantum mechanics. *Foundations of Physics*, **18**, 809–31.

Chew, G. F., and Steinberger, J. L. (1950a). The positive–negative ratio of π^- mesons produced in complex nuclei. *Physical Review*, **78**, 86.

Chew, G. F., and Steinberger, J. L. (1950b). Positive–negative ratio of π^- mesons produced singly in collisions of nucleons with complex nuclei. *Physical Review*, **78**, 497.

Chew, G. F., and Wick, G. C. (1951). A quantitative analysis of the impulse approximation. *Physical Review*, **83**, 239.

Chew, G. F., and Wick, G. C. (1952). The impulse approximation. *Physical Review*, **85**, 636–42.

Chou, T. T., and Dresden, M. (1967). *S*-matrix theory of electromagnetic interactions. *Reviews of Modern Physics*, **39**, 143–66.

Chrétien, M., and Schweber, S. S. (eds.) (1970). *Elementary Particle Physics and Scattering Theory, Vol. 1, Brandeis University Summer Institute in Theoretical Physics, 1967*. New York: Gordon and Breach.

Churchland, P. M., and Hooker, C. A. (eds.) (1985). *Images of Science*. Chicago: University of Chicago Press.

Cini, M. (1980). The history and ideology of dispersion relations. *Fundamenta Scientiae*, **1**, 157–72.

Coccini, G. (1962). High energy physics (experimental). In Prentki, pp. 883–92.

Cohen, R. S., Feyerabend, P. K., and Wartofsky, M. W. (eds.) (1976). *Boston Studies in the Philosophy of Science*. vol. 39. Dordrecht: D. Reidel Publishing Co.

Cohen, R. S., Hooker, C. A., Michalos, A. C., and van Evra, J. W. (eds.) (1976). *Boston Studies in the Philosophy of Science*, vol. 32. Dordrecht: D. Reidel Publishing Co.

Collins, P. D. B., and Squires, E. J. (1968). *Regge Poles in Particle Physics*, *Springer Tracts in Modern Physics*, vol. 45. Berlin: Springer–Verlag.

Coster, J., and Stapp, H. P. (1969). Physical-region discontinuity equations for many-particle scattering amplitudes. I. *Journal of Mathematical Physics*, **10**, 371–96.

Coster, J., and Stapp, H. P. (1970a). Physical-region discontinuity equation. *Journal of Mathematical Physics*, **11**, 2743–63.

Coster, J., and Stapp, H. P. (1970b). Physical-region discontinuity equations for many-particle scattering amplitudes. II. *Journal of Mathematical Physics*, **11**, 1441–63.

Cremmer, E., and Scherk, J. (1972). Factorization of the Pomeron sector and currents in the dual resonance model. *Nuclear Physics*, **B50**, 222–52.

Cushing, J. T. (1966). Internal symmetries in a coupled-channel soluble model with inelasticity. *Physical Review*, **148**, 1558–73.

Cushing, J. T. (1969). Exact static-model bootstrap solutions for arbitrary 2×2 crossing matrices. *Journal of Mathematical Physics*, **10**, 1319–26.

Cushing, J. T. (1971). Internal symmetry propagation in the strong interaction S-matrix. *Physical Review*, **4D**, 1177–84.

Cushing, J. T. (1975). *Applied Analytical Mathematics for Physical Scientists*. New York: John Wiley & Sons.

Cushing, J. T. (1982). Models and methodologies in current theoretical high-energy physics. *Synthese*, **50**, 5–101.

Cushing, J. T. (1983a). Models, high-energy theoretical physics and realism. In Asquith and Nickles, pp. 31–56.

Cushing, J. T. (1983b). A response to Paul Teller. In Asquith and Nickles, pp. 112–13.

Cushing, J. T. (1984). The convergence and content of scientific opinion. In Asquith and Kitcher, pp. 211–23.

Cushing, J. T. (1985). Is there just one possible world? Contingency vs. the bootstrap. *Studies in History and Philosophy of Science*, **16**, 31–48.

Cushing, J. T. (1986a). The importance of Heisenberg's S-matrix program for the theoretical high-energy physics of the 1950's. *Centaurus*, **29**, 110–49.

Cushing, J. T. (1986b). Review of A. Pickering's *Constructing Quarks*. *American Journal of Physics*, **54**, 381–3.

Cushing, J. T. (1986c). Causality as an overarching principle in physics. In Fine and Machamer, vol. 1, pp. 3–11.

Cushing, J. T. (1987a). Representations, reduction and realism: reflections on Schweber's phenomenological quantum field theories. Talk given at the 1987 Meeting of the Eastern Division of the American Philosophical Association, New York City, December 27–30, 1987.

Cushing, J. T. (1987b). Review of Nancy J. Nersessian's *Faraday to Einstein: Constructing Meaning in Scientific Theories*. *Foundations of Physics*, **17**, 101–6.

Cushing, J. T. (1988), Foundational problems in and methodological lessons from quantum field theory. In Brown and Harré, pp. 25–39.

Cushing, J. T. (1989a). The *S* matrix and its relation to quantum field theory. In Dresden (in press).

Cushing, J. T. (1989b). Causal quantum theory: why a nonstarter? (forthcoming).

Cushing, J. T. (1989c). Quantum theory and explanatory discourse: endgame for understanding? *Philosophy of Science* (in press).

Cushing, J. T. (1989d). Is scientific methodology interestingly atemporal? *British Journal for the Philosophy of Science*, **40** (in press).

Cushing, J. T. (1989e). Review of Peter Galison's *How Experiments End*. *Foundations of Physics*, **19**, 625–7.

Cushing, J. T. (1989f), The justification and selection of scientific theories. *Synthese*, **78**, 1–24.

Cushing, J. T., Delaney, C. F., and Gutting, G. (eds.) (1984), *Science and Reality*. Notre Dame, IN: University of Notre Dame Press.

Cushing, J. T., and McMullin, E. (eds.) (1989). *Philosophical Consequences of Quantum Theory*. Notre Dame, IN: University of Notre Dame Press.

Cutkosky, R. E. (1960). Singularities and discontinuities of Feynman amplitudes. *Journal of Mathematical Physics*, **1**, 429–33.

Cutkosky, R. E. (1963a). A model of baryon states. *Annals of Physics*, **23**, 415–38.

Cutkosky, R. E. (1963b). A mechanism for the induction of symmetries among the strong interactions. *Physical Review*, **131**, 1888–90.

Cutkosky, R. E., and Tarjanne, P. (1963). Self-consistent deviations from unitary symmetry. *Physical Review*, **132**, 1354–61.

Cziffra, P., MacGregor, M. H., Moravcsik, M. J., and Stapp, H. P. (1959). Modified analysis of nucleon–nucleon scattering. I. Theory and p–p scattering at 310 MeV. *Physical Review*, **114**, 880–6.

Darrigol, O. (1984). La genèse du concept de champ quantique. *Annales de Physique*, **9**, 433–501.

Darrigol, O. (1986). The origin of quantized matter waves. *Historical Studies in the Physical and Biological Sciences*, **16**, 197–253.

Davidon, W. C., and Goldberger, M. L. (1956). Comparison of spin-flip dispersion relations with pion–nucleon scattering data. *Physical Review*, **104**, 1119–21.

de Alfaro, V., Fubini, S., Furlan, G., and Rossetti, G. (1966). Sum rules for strong interactions. *Physics Letters*, **21**, 576–9.

de Boer, J., Dal, E., and Ulfbeck, O. (eds.) (1986). *The Lesson of Quantum Theory*. Amsterdam: North-Holland Publishing Co.

de Broglie, L. (1927). La mécanique ondulatoire et la structure atomique de la matiére et du rayonnement. *Journal de Physique*, **8**, 225–41.

de Broglie, L. (1928). Nouvelle dynamique des quanta. In *Electrons et Photons, Rapports et Discussions du Cinquiéme Conseil de Physique*, pp. 105–32. Paris: Gauthier-Villars.

de Broglie, L. (1930). *An Introduction to the Study of Wave Mechanics*, (translated from the French by H. T. Flint). London: Methuen & Co. Ltd.

de Broglie, L. (1960). *Non-Linear Wave Mechanics: A Causal Interpretation*, (translated from the French by A. J. Knobel and J. C. Miller). Amsterdam: Elsevier Publishing Co.

de Broglie, L. (1962). *New Perspectives in Physics*, (translated from the French by A. J. Pomerans). Edinburgh: Oliver and Boyd.

de Broglie, L. (1970). The reinterpretation of wave mechanics. *Foundations of Physics*, **1**, 5–15.

de Broglie, L. (1973). The beginnings of wave mechanics. In Price *et al.*, pp. 12–18.

De Tar, C., Finkelstein, J., and Tan, C. I. (1985). *A Passion for Physics: Essays in Honor of Geoffrey Chew*. Singapore: World Scientific Publishing Co.

De Witt, B. S., and Stora, R. (eds.) (1984). *Relativity, Groups and Topology II*. Amsterdam: North-Holland Publishing Co.

De Witt, C. (ed.) (1960). *Relations de Dispersion et Particules Élémentaires*. Paris: Herman.

De Witt, C., and Jacob, M. (eds.) (1965). *High Energy Physics*. New York: Gordon and Breach.

Dicke, R. H. (1957). Gravitation without a principle of equivalence. *Reviews of Modern Physics*, **29**, 363–76.

Diddens, A. N., *et al.* (1962a). High-energy proton–proton diffraction scattering. *Physical Review Letters*, **9**, 108–11.

Diddens, A. N., *et al.* (1962b). High-energy proton–proton scattering. *Physical Review Letters*, **9**, 111–14.

Dirac, P. A. M. (1927). The quantum theory of dispersion. *Proceedings of the Royal Society*, **A114**, 710–28.

Dirac, P. A. M. (1931). Quantized singularities in the electromagnetic field. *Proceedings of the Royal Society*, **A133**, 60–72.

Dirac, P. A. M. (1935). *The Principles of Quantum Mechanics*. London: Oxford University Press.

Dolen, R., Horn, D., and Schmid, C. (1967). Prediction of Regge parameters of ρ poles from low-energy πN data. *Physical Review Letters*, **19**, 402–7.

Dolen, R., Horn, D., and Schmid, C. (1968). Finite-energy sum rules and their applications to πN charge exchange. *Physical Review*, **166**, 1768–81.

Doncel, M. G., Hermann, A., Michel, L., and Pais, A. (eds.) (1987). *Symmetries in Physics (1600–1980), Proceedings of the First International Meeting on the History of Scientific Ideas*. Barcelona: Bellaterra.

Donovan, A., Laudan, L., and Laudan, R. (1988). *Scrutinizing Science*. Dordrecht: Kluwer Academic Publishers.

Drell, S. D. (1962). High-energy physics: theoretical. In Prentki, pp. 897–913.

Drell, S. D., and Zachariasen, F. (1961). *Electromagnetic Structure of Nucleons*. Oxford: Oxford University Press.

Dresden, M. (ed.) (1989). *Field Theory and General Relativity*. Singapore: World Scientific Publishing Co.

Drude, P. (1893). Über die Beziehung der Dielectricitätsconstanten zum optischen Brechunzsexponenten. *Annalen der Physik und Chemie*, **48**, 536–45.

Drude, P. (1900). Zur Geschichte der elektromagnetischen Dispersionsgleichungen. *Annalen der Physik*, **1**, 437–40.

Dyson, F. J. (1949a). The radiation theories of Tomonaga, Schwinger, and Feynman. *Physical Review*, **75**, 486–502.

Dyson, F. J. (1949b). The S matrix in quantum electrodynamics. *Physical Review*, **75**, 1736–55.

Dyson, F. J. (1952). Divergence of perturbation theory in quantum electrodynamics. *Physical Review*, **85**, 631–2.

Dyson, F. J. (1958). Integral representations of causal commutators. *Physical Review*, **110**, 1460–4.

Earman, J. (ed.) (1983). *Testing Scientific Theories*. Minneapolis, MN: University of Minnesota Press.

Eden, R. J. (1948). Analytic behavior of Heisenberg's S-matrix. *Physical Review*, **74**, 982.

Eden, R. J. (1949a). Heisenberg's S matrix for a system of many particles, *Proceedings of Royal Society*, **A198**, 540–59.

Eden, R. J. (1949b). The analytic behavior of Heisenberg's S matrix. *Proceedings of the Royal Society*, **A199**, 256–71.

Eden, R. J. (1952). Threshold behavior in quantum field theory. *Proceedings of the Royal Society*, **A210**, 388–404.

Eden, R. J. (1960a). The use of perturbation methods in dispersion theory. In Sudarshan *et al.*, pp. 219–29.

Eden, R. J. (1960b), Analytic structure of collision amplitudes in perturbation theory. *Physical Review*, **119**, 1763–83.

Eden, R. J. (1960c). Proof of the Mandelstam representation in perturbation theory. *Physical Review Letters*, **5**, 213–15.

Eden, R. J., Landshoff, P. V., Olive, D. I., and Polkinghorne, J. C. (1966). *The Analytic S-Matrix*. Cambridge: Cambridge University Press.

Eden, R. J., Landshoff, P. V., Polkinghorne, J. C., and Taylor, J. C. (1961). Acnodes and cusps on Landau curves. *Journal of Mathematical Physics*, **2**, 656–63.

Edwards, S. F. (1953). A nonperturbative approach to quantum electrodynamics. *Physical Review*, **90**, 284–91.

Einstein, A. (1970). Reply to criticisms. In Schilpp, pp. 663–88.

Electrons et Photons, Rapports et Discussions du Cinquiéme Conseil de Physique, (1928). Paris: Gauthier-Villars.

Erwin, A. R., March, R., Walker, W. D., and West, E. (1961). Evidence for a π–π resonance in the $I = 1$, $J = 1$ state. *Physical Review Letters*, **6**, 628–30.

Fairlie, D. B., Landshoff, P. V. Nuttall, J., and Polkinghorne, J. C. (1962). Singularities of the second type. *Journal of Mathematical Physics*, **3**, 594–602.

Faxén, H., and Holtsmark, J. (1927). Beitrag zur Theorie des Durchganges langsamer Elektronen durch Gase. *Zeitschrift für Physik*, **45**, 307–24.

Federbush, P., Goldberger, M. L., Treiman, S. B. (1958). Electromagnetic structure of the nucleon. *Physical Review*, **112**, 642–65.

Feenberg, E. (1932). The scattering of slow electrons by neutral atoms. *Physical Review*, **40**, 40–54.

Feenberg, E. (1937). A note on the Thomas–Fermi statistical method, *Physical Review*, **52**, 758–60.

Fermi, E. (1936). Sul moto dei neutroni nelle sostanze idrogenate. *La Ricerca Scientifica*, **7**, 2, 13–52.

Fermi, E. (1962). *Collected Papers*, vol. I. Chicago: University of Chicago Press.

Ferretti, B. (1958), *1958 Annual International Conference on High Energy Physics at CERN*. Geneva: CERN.

Feynman, R. P. (1949a). The theory of positrons. *Physical Review*, **76**, 749–59.

Feynman, R. P. (1949b). Space-time approach to quantum electrodynamics. *Physical Review*, **76**, 769–89.

Fierz, M. (1950). Über die Bedeutung der Funktion D_c in der Quantentheorie der Wellenfelder. *Helvetica Physica Acta*, **23**, 731–9.

Fine, A. (1982). Antinomies of entanglement: the puzzling case of the tangled statistics. *The Journal of Philosophy*, **79**, 733–47.

Fine, A. (1984). The natural ontological attitude. In Leplin, pp. 83–107.

Fine, A. (1986). Unnatural attitudes: realist and instrumentalist attachments to science. *Mind*, *XCV*, 149–79.

Fine, A., and Machamer, P. (eds.) (1986–87). *PSA 1986, Proceedings of the 1986 Biennial Meeting of the Philosophy of Science Association*, vols. 1 and 2. East Lansing, MI: Philosophy of Science Association.

Finkelstein, J. (1985). Topological theory of strong and electroweak interactions. In De Tar *et al.*, pp. 184–8.

Finkler, P., and Jones, C. E. (1985). Self-consistent phases in topological particle theory. *Physical Review*, **D31**, 1393–403.

Foley, K. J. *et al.* (1963). 7-to 20-Bev/c π^-p and p+p elastic scattering and Regge pole predictions. *Physical Review Letters*, **10**, 376–81.

Forman, P. (1971). Weimar culture, causality, and quantum theory, 1918–27: adaptation by German physicists and mathematicians to a hostile intellectual environment. *Historical Studies in the Physical Sciences*, **3**, 1–115.

Forman, P. (1979). The reception of an acausal quantum mechanics in Germany and Britain. In Mauskopf, pp. 11–50.

Foucault, M. (1966). *Les mots et les choses*. Paris: Gallimard. [Translated as *The Order of Things* (1970). London: Tavistock Publications.]

Fox, G. C. (1969). Veni, vidi, vici Regge theory. *Comments on Nuclear and Particle Physics*, **3**, 190–7.

Frampton, P. H. (1974), *Dual Resonance Models*. New York: W. A. Benjamin.

Frank, P. (1957). *Philosophy of Science*. Westport, CN: Greenwood Press.

Franklin, A. (1986). *The Neglect of Experiment*. Cambridge: Cambridge University Press.

Frautschi, S. C. (1963). *Regge Poles and S-Matrix Theory*. New York: W. A. Benjamin.

Frautschi, S. C. (1985). My experiences with the *S*-matrix program. In De Tar *et al.*, pp. 44–8.

Frautschi, S. C., Gell-Mann, M., and Zachariasen, F. (1962). Experimental consequences of the hypothesis of Regge poles. *Physical Review*, **126**, 2204–18.

Frazer, W. R. (1961). The electromagnetic structure of pions and nucleons. In Screaton, pp. 227–58.

Frazer, W. R. (1966). *Elementary Particles*. Englewood Cliffs, NJ: Prentice–Hall, Inc.

Frazer, W. R. (1985). The analytic and unitary *S*-matrix. In De Tar *et all.*, pp. 1–8.

Frazer, W. R., and Fulco, J. R. (1959). Effect of a pion–pion scattering resonance on nucleon structure. *Physical Review Letters*, **2**, 365–8.

Frazer, W. R., and Fulco, J. R. (1960a). Partial-wave dispersion relations for the process $\pi + \pi \rightarrow N + \bar{N}$. *Physical Review*, **117**, 1603–8.

Frazer, W. R., and Fulco, J. R. (1960b). Effect of a pion–pion scattering resonance on nucleon structure. II. *Physical Review*, **117**, 1609–15.

Frazer, W. R., and Fulco, J. R. (1960c). Partial-wave dispersion relations for pion–nucleon scattering. *Physical Review*, **119**, 1420–6.

Freund, P. G. O. (1968). Finite-energy sum rules and bootstraps. *Physical Review Letters*, **20**, 235–7.

Freund, P. G. O. (1969). Model for the Pomeranchuk term. *Physical Review Letters*, **22**, 565–8.

Freund, P. G. O., Waltz, R., and Rosner, J. L. (1969). Quark model selection rules for hadron couplings. *Nuclear Physics*, **B13**, 237–45.

Freundlich, Y. (1980). Theory evaluation and the bootstrap hypothesis. *Studies in History and Philosophy of Science*, **11**, 267–77.

Friedman, M. (1988). Philosophy and the exact sciences: logical positivism as a case study. University of Illinois at Chicago preprint.

Froissart, M. (1961). Asymptotic Behavior and Subtractions in the Mandelstam Representation. *Physical Review*, **123**, 1053–7.

Froissart, M., and Taylor, J. R. (1967). Cluster decomposition and the spin-statistics theorem in *S*-matrix theory. *Physical Review*, **153**, 1636–48.

Fröberg, C. E. (1947). Calculation of the interaction between two particles from the asymptotic phase. *Physical Review*, **72**, 519–20.

Fröberg, C. E. (1948a). Calculation of the potential from the asymptotic phase. *Arkiv för Matematik, Astronomi och Fysik*, **34A**, No. 28, 1–16.

Fröberg, C. E. (1948b). Calculation of the potential from the asymptotic phase, II. *Arkiv för Matematik, Astronomi och Fysik*, **36A**, No. 11, 1–55.

Fubini, S.(1970). Multiperipheral model. *Comments on Nuclear and Particle Physics*, **4**, 102–6.

Fubini, S., Gordon, D., Veneziano, G. (1969). A general treatment of factorization in dual resonance models. *Physical Review*, **29B**, 679–82.

Fubini, S., and Veneziano, G. (1969). Level structure of dual-resonance models. *Il Nuovo Cimento*, **64A**, 811–40.

Fubini, S., and Veneziano, G. (1970). Duality in operator formalism, *Il Nuovo Cimento*, **67A**, 29–47.

Gale, G. (1981). The anthropic principle. *Scientific American*, **245**, (No. 6), 114–22.

Gale, G. (1984). Science and the philosophers. *Nature*, **312**, 491–5.

Gale, G. (1986). An attempt to explain two contemporary divergences from mainstream philosophy of physics. In Weingartner and Dorn, pp. 3–17.

Gale, G. (1987). A revised design: teleology and big questions in contemporary cosmology. *Biology and Philosophy*, **2**, 475–91.

Galison, P. (1983a). The discovery of the muon and the failed revolution against quantum electrodynamics. *Centaurus*, **26**, 262–316.

Galison, P. (1983b). How the first neutral-current experiments ended. *Reviews of Modern Physics*, **55**, 477–509.

Galsion, P. (1987). *How Experiments End*. Chicago: University of Chicago Press.

Galison, P. (1988). History, philosophy, and the central metaphor. *Science in Context*, **2**, 197–212.

Gasiorowicz, S. (1966). *Elementary Particle Physics*. New York: John Wiley & Sons.

Gauron, P., Nicolescu, B., and Ouvry, S. (1981). Topological supersymmetric structure of hadron cross sections. *Physical Review*, **D24**, 2501–17.

Gel'fand, I. M., and Levitan, B. M. (1951a). On the determination of a differential equation by its spectral function. *Doklady Akademii Nauk SSSR* **77**, 557–60.

Gel'fand, I. M., and Levitan, B. M. (1951b). On the determination of a differential equation from its spectral function. *Izvestiia Akademii Nauk SSSR* **15**, 309–60. (*American Mathematical Translation*, **1**, 253–304.)

Gell-Mann, M. (1956). Dispersion relations in pion–pion and photon–nucleon scattering. In Ballam *et al.*, Section III, pp. 30–6.

Gell-Mann, M. (1962a). Applications of Regge poles. In Prentki, pp. 533–42.

Gell-Mann, M. (1962b). Factorization of coupling to Regge poles. *Physical Review Letters*, **8**, 263–4.

Gell-Mann, M. (1967). Current topics in particle physics. In *Proceedings of the*

XIIth International Conference on High-Energy Physics, pp. 3–9. Berkeley: University of California Press.

Gell-Mann, M. (1987). Particle theory from S-matrix to quarks. In Doncel *et al.*, pp. 479–97.

Gell-Mann, M., and Goldberger, M. L. (1953). The formal theory of scattering. *Physical Review*, **91**, 398–408.

Gell-Mann, M., and Goldberger, M. L. (1954). The scattering of low energy photons by particles of spin 1/2. *Physical Review*, **96**, 1433–8.

Gell-Mann, M., and Goldberger, M. L. (1962). Elementary particles of conventional field theory as Regge poles. *Physical Review Letters*, **9**, 275–7. [Erratum, *PRL10* (1963), 39–40.]

Gell-Mann, M. Goldberger, M. L., and Low, F. E. (1964). The vacuum trajectory in conventional field theory. *Reviews of Modern Physics*, **36**, 640–9.

Gell-Mann, M. Goldberger, M. L., Low, F. E., Marx, E., and Zacharaisen, F. (1964). Elementary particles of conventional field theory as Regge poles. III. *Physical Review*, **133**, B145–60.

Gell-Mann, M. Goldberger, M. L., Low, F. E., Singh, V., and Zachariasen, F. (1964). Elementary particles of conventional field theory as Regge poles. IV. *Physical Review*, **133**, B161–4.

Gell-Mann, M. Goldberger, M. L., Low, F. E., Zachariasen, F. (1963). Elementary particles of conventional field theory as Regge poles. II. *Physics Letters*, **4**, 265–7.

Gell-Mann, M. Goldberger, M. L., and Thirring, W. E. (1954). Use of causality conditions in quantum theory. *Physical Review*, **95**, 1612–27.

Giere, R. N. (1983). Testing theoretical hypotheses. In Earman, pp. 269–98.

Gliozzi, F., Scherk, J., and Olive, D. (1976). Supergravity and the spinor dual model. *Physics Letters*, **65B**, 282–6.

Gliozzi, F., Scherk, J., and Olive, D. (1977). Supersymmetry, supergravity theories and the dual spinor model. *Nuclear Physics*, **B122**, 253–90.

Glymour, C. (1980). *Theory and Evidence*. Princeton: Princeton University Press.

Goddard, P., Goldstone, J., Rebbi, C., and Thorn, C. B. (1973). Quantum dynamics of a massless relativistic string. *Nuclear Physics*, **B56**, 109–35.

Goebel, C. J., and Sakita, B. (1969). Extension of the Veneziano form to N-particle amplitudes. *Physical Review Letters*, **22**, 257–60.

Goldberger, M. L. (1954). Validity of pseudoscalar meson theory with pseudoscalar coupling. In Noyes *et al.*, pp. 27–39.

Goldberger, M. L. (1955a). Use of causality conditions in quantum theory. *Physical Review*, **97**, 508–10.

Goldberger, M. L. (1955b). Causality conditions and dispersion relations I. Boson fields. *Physical Review*, **99**, 979–85.

Goldberger, M. L. (1960). Introduction to the theory and application of dispersion relations. In De Witt, pp. 15–157.

Goldberger, M. L. (1961). Theory and applications of single variable dispersion relations. In Stoops, pp. 179–208.

Goldberger, M. L. (1969a). The S-matrix. In *Contemporary Physics; Trieste Symposium 1968*, pp. 85–6. Vienna: International Atomic Energy Agency.

Goldberger, M. L. (1969b). Multiperipheral dynamics. In Gudenhus *et al.*, pp. 142–56.

Goldberger, M. L. (1970). Fifteen years in the life of dispersion theory. In Zichichi, pp. 684–93.

Goldberger, M. L. (1985). A passion for physics. In De Tar *et al.*, pp. 241–5.

Goldberger, M. L., and Low, F. E. (1968). Photon scattering from bound atomic systems at very high energy. *Physical Review*, **176**, 1778–81.

Goldberger, M. L., Miyazawa, H., and Oehme, R. (1955). Application of dispersion relations to pion–nucleon scattering. *Physical Review*, **99**, 986–8.

Goldberger, M. L., Nambu, Y., and Oehme, R. (1957). Dispersion relations for nucleon–nucleon scattering. *Annals of Physics*, **2**, 226–82.

Goldberger, M. L., and Watson, K. (1964). *Collision Theory*. New York: John Wiley & Sons.

Gooding, D., Pinch, T. J., and Schaffer, S. (eds.) (1989). *The Uses of Experiment: Studies of Experimentation in the Natural Sciences*. Cambridge: Cambridge University Press.

Gould, S. J. (1986). A triumph of historical excavation. *The New York Review of Books*, **XXXIII**, No. 3, Feb. 27, 9–12.

Green, M. B. (1985). Unification of forces and particles in superstring theories. *Nature*, **314**, 409–14.

Green, M. B., and Schwarz, J. H. (1981). Supersymmetrical dual string theory, *Nuclear Physics*, **B181**, 503–30.

Green, M. B., and Schwarz, J. H. (1984a). Covariant description of super-strings. *Physics Letters*, **136B**, 367–70.

Green, M. B., and Schwarz, J. H. (1984b). Superstring field theory. *Nuclear Physics*, **B243**, 475–536.

Green, M. B., and Schwarz, J. H. (1984c). Anomaly cancellations in supersymmetric $D = 10$ gauge theory and superstring theory. *Physics Letters*, **149B**, 117–22.

Green, M. B., and Schwarz, J. H. (1985). Infinity cancellations in SO(32) superstring theory. *Physics Letters*, **151B**, 21–5.

Gribov, V. N. (1962). Partial waves with complex angular momenta and their moving singularities. In Prentki, pp. 515–21.

Gribov, V. N. (1968). A Reggeon diagram technique. *Soviet Physics JETP*, **26**, 414–23.

Gribov, V. N., and Pomeranchuk, I. Ya. (1962). Limitation on the rate of decrease of amplitudes for various processes. *Soviet Physics JETP*, **16**, 1098–100.

Gribov, V. N., Pomeranchuk, I. Ya., and Ter-Martorosyan, K. A. (1965). Moving branch points in j plane and Regge-pole unitarity conditions. *Physical Review*, **139**, B184–202.

Gross, D. J. (1985). On the uniqueness of physical theories. In De Tar *et al.*, 128–36.

Grythe, I. (1982). Some remarks on the early S-matrix. *Centaurus*, **26**, 198–203.

Grythe, I. N., and Haug, A. (1983). Sorting out the redundant zeroes. Report 83–43 of University of Oslo.

Gudenhus, T., Kaiser, G., and Perlmutter, A. (eds.) (1969). *Coral Gables Conference on Fundamental Interactions at High Energy*. New York: Gordon and Breach.

Gunson, J. (1965a). Unitarity and on-mass-shell analyticity as a basis for S-matrix theories. I. *Journal of Mathematical Physics*, **6**, 827–44.

Gunson, J. (1965b). Unitarity and on-mass-shell analyticity as a basis for S-matrix theories. II. *Journal of Mathematical Physics*, **6**, 845–58.

Guth, A. H., Huang, K., and Jaffe, R. L. (eds.) (1983). *Asymptotic Realms of Physics*. Cambridge, MA: MIT Press.

Hacking, I. (1975). *The Emergence of Probability*. London: Cambridge University Press.

Hacking, I. (1983). *Representing and Intervening*. Cambridge: Cambridge University Press.

Harari, H. (1969). Duality diagrams. *Physical Review Letters*, **22**, 562–5.

Heilbron, J. L. (1982). *Elements of Early Modern Physics*. Berkeley: University of California Press.

Heisenberg, W. (1936). Zur Theorie der 'Schauer' in der Höhenstrahlung. *Zeitschrift für Physik*, **101**, 533–40.

Heisenberg, W. (1938a). Der Durchgang sehr energiereicher Korpuskeln durch den Atomkern. *Il Nuovo Cimento*, **15**, 31–4.

Heisenberg, W. (1938b). Über die in der Theorie der Elementarteilchen auftretende universelle Länge. *Annalen der Physik*, **32**, 20–33.

Heisenberg, W. (1943a). Die 'beobachtbaren Grössen' in der Theorie der Elementarteilchen. *Zeitschrift für Physik*, **120**, 513–38.

Heisenberg, W. (1943b). Die beobachtbaren Grössen in der Theorie der Elementarteilchen. II. *Zeitschrift für Physik*, **120**, 673–702.

Heisenberg, W. (1944). Die beobachtbaren Grössen in der Theorie der Elementarteilchen. III. *Zeitschrift für Physik*, **123**, 93–112.

Heisenberg, W. (1946). Der mathematische Rahmen der Quantentheorie der Wellenfelder. *Zeitschrift für Naturforschung*, **1**, 608–22.

Heisenberg, W. (1947–48). Zur statistischen Theorie der Turbulenz. *Zeitschrift für Physik*, **124**, 628–57.

Heisenberg, W. (1949a). Über die Entstehung von Mesonen in Vielfachprozessen. *Zeitschrift für Physik*, **126**, 569–82.

Heisenberg, W. (1949b). The present situation in the theory of elementary particles. In *Two Lectures by W. Heisenberg*, pp. 9–25. Cambridge: Cambridge University Press.

Heisenberg, W. (1949c). The Electron Theory of Superconductivity. In *Two Lectures by W. Heisenberg*, pp. 27–52. Cambridge: Cambridge University Press.

Heisenberg, W. (1950a). Zur Quantentheorie der Elementarteilchen. *Zeitschrift für Naturforschung*, **5a**, 251–9.

Heisenberg, W. (1950b). Stationäre Zustände in der relativistischen Quantentheorie der Wellenfelder. *Zeitschrift für Naturforschung*, **5a**, 367–73.

Heisenberg, W. (1951). Zur Frage der Kausalität in der Quantentheorie der Elementarteilchen. *Zeitschrift für Naturforschung*, **6a**, 281–4.

Heisenberg, W. (1952). Mesonenerzeugung als Stosswellenproblem. *Zeitschrift für Physik*, **133**, 65–79.

Heisenberg, W. (1953). Zur Quantisierung nichtlinearer Gleichungen. *Nachrichten der Akademie der Wissenschaften in Göttingen aus dem Jahr 1953, Mathematish-Physikalische Klasse*, 109–27.

Heisenberg, W. (1954). Zur Quantentheorie nichtrenormierbarer Wellengleichungen. *Zeitschrift für Naturforschung*, **9a**, 292–303.

Heisenberg, W. (1957). Quantum theory of fields and elementary particles. *Reviews of Modern Physics*, **29**, 269–78.

Heisenberg, W. (1958). Research on the non-linear spinor theory with indefinite metric in Hilbert space. In Ferretti, pp. 119–26.

Heisenberg, W. (1966). *Introduction to the Unified Field Theory of Elementary Particles*. New York: Interscience Publishers.

Heisenberg, W. (1967). Nonlinear problems in physics. *Physics Today*, **20**, No. 5, 27–33.

Heisenberg, W., Kortel, F., and Mitter, H. (1955). Zur Quantentheorie nichtlinearer Wellengleichungen III. *Zeitschrift für Naturforschung*, **10a**, 425–46.

Heitler, W. (1936). *The Quantum Theory of Radiation*. London: Oxford University Press.

Heitler, W. (1941). The influence of radiation damping on the scattering of light and mesons by free particles. I. *Proceedings of the Cambridge Philosophical Society*, **37**, 291–300.

Heitler, W. (1947). The quantum theory of damping as a proposal for Heisenberg's S-matrix. In *Report of an International Conference on Fundamental Particles and Low Temperatures*, Vol. I, pp. 189–94. London: The Physical Society.

Heitler, W., and Hu, N. (1947). Proton isobars in the theory of radiation damping. *Proceedings of the Royal Irish Academy*, **51A**, 123–40.

Heitler, W., and Peng, H. W. (1942). The influence of radiation damping on the scattering of mesons. II. multiple processes. *Proceedings of the Cambridge Philosophical Society*, **38**, 296–312.

Heitler, W., and Peng, H. W. (1943). On the production of mesons by proton–proton collisions. *Proceedings of the Royal Irish Academy*, **49A**, 101–33.

Hempel, C. G. (1945). Studies in the logic of confirmation. *Mind*, **54**, 1–26 and 97–121.

Hendry, J. (1980). Weimar culture and quantum causality. *History of Science*, **18**, 155–80.

Hepp, K. (1965). On the connection between the LSZ and Wightman quantum field theory. *Communications in Mathematical Physics*, **1**, 95–111.

Hermann, A. (1971). *The Genesis of Quantum Theory*, (*1899–1913*). Cambridge, MA: MIT Press.

Hiley, B. J., and Peat, F. D. (eds.) (1987). *Quantum Implications*. London: Routledge & Kegan Paul.

Hofstadter, R., and Schiff, L. I. (eds.) (1964), *Nucleon Structure*. Stanford: Stanford University Press.

Holmberg, B. (1952). A remark on the uniqueness of the potential determined from the asymptotic phase. *Il Nuovo Cimento*, **9**, 597–604.

Hones, M. J. (1987). The neutral-weak-current experiments: a philosophical perspective. *Studies in History and Philosophy of Science*, **18**, 221–51.

Howson, C. (ed.) (1976). *Method and Appraisal in the Physical Sciences*. Cambridge: Cambridge University Press.

Hu, N. (1948a). On the application of Heisenberg's theory of S-matrix to the problems of resonance scattering and reactions in nuclear physics. *Physical Review*, **74**, 131–40.

Hu, N. (1948b). Further investigations on Heisenberg's characteristic matrix. *Proceedings of the Royal Irish Academy*, **52A**, 51–68.

Huang, K., and Low, F. E. (1965). Exact bootstrap solutions in some static models of meson–baryon scattering. *Journal of Mathematical Physics*, **6**, 795–816.

Hwa, R. (ed.) (1986). *Proceedings of the Oregon Meeting*. Singapore: World Scientific Publishing Co.

Hylleraas, E. A. (1958). Calculation of a perturbing central field of force from the elastic scattering phase shift. *Physical Review*, **74**, 48–51.

Iagolnitzer, D. (1969). S-matrix and classical description of interactions. *Journal of Mathematical Physics*, **10**, 1241–64.

Iagolnitzer, D. (1978). *The S-Matrix*. Amsterdam: North-Holland Publishing Co.

Iagolnitzer, D. (1980). Macrocausality, unitarity and physical-region structure of the multiparticle S matrix. *Communications in Mathematical Physics*, **77**, 251–67.

Iagolnitzer, D., and Stapp, H. P. (1969). Macroscopic causality and physical region analyticity in the S-matrix theory. *Communications in Mathematical Physics*, **14**, 15–55.

Iagolnitzer, D., and Stapp, H. P. (1977). The pole-factorization theorem in S-matrix theory. *Communications in Mathematical Physics*, **57**, 1–30.

Igi, K. (1962). π–N scattering length and singularities in the complex J plane. *Physical Review Letters*, **9**, 76–9.

Igi, K. (1963). Two vacuum poles and pion–nucleon scattering. *Physical Review*, **130**, 820–7.

Igi, K., and Matsuda, S. (1967). New sum rules and singularities in the complex J plane. *Physical Review Letters*, **18**, 625–7.

Iizuka, J. (1966). A systematics and phenomenology of meson family. *Supplement to Progress in Theoretical Physics*, **37–38**, 21–34..

Jackson, J. D. (1958). *The Physics of Elementry Particles*. Princeton: Princeton University Press.

Jackson, J. D. (1961). Introduction to dispersion relation techniques. In Screaton, pp. 1–63.

Jackson, J. D. (1975). *Classical Electrodynamics*, 2nd edn. New York: John Wiley & Sons.

Jacob, M. (1974). *Dual Theory*. Amsterdam: North-Holland Publishing Co.

Jacob, M., and Chew, G. F. (1964). *Strong Interaction Physics*. New York: W. A. Benjamin.

Jammer, M. (1974). *The Philosophy of Quantum Mechanics*. New York: McGraw–Hill.

Jardine, N. (1986). *The Fortunes of Inquiry*. Oxford: Clarendon Press.

Jauch, J. M., and Rohrlich, F. (1955). *The Theory of Photons and Electrons*. Reading, MA: Addison–Wesley.

Jones, C. E. (1985). Deducing T, C, and P invariance for strong interactions in topological particle theory. In De Tar *et al.*, pp. 189–94.

Jones, C. E., and Finkler, P. (1985). Derivation of discrete invariances (T, C and P) and the connection between spin and statistics in topological particle theory. *Physical Review*, **D31**, 1404–10.

Jones, C. E., and Uschershon, J. (1980a). Simultaneous bootstrap of mesonic and baryonic trajectories in DTU. *Physics Letters*, **89B**, 409–12.

Jones, C. E., and Uschershon, J. (1980b). Quadratic mass relations in topological bootstrap theory. *Physical Review Letters*, **45**, 1901–4.

Jost, R. (1946). Bemerkungen zu der Vorstehenden Arbeit. *Physica*, **12**, 509–10.

Jost, R. (1947). Über die falschen Nullstellen der Eigenverte der S-Matrix. *Helvetica Physica Acta*, **20**, 256–66.

Jost, R. (1963). TCP-Invarianz der Streumatrix und interpolierende Felder. *Helvetica Physica Acta*, **35**, 77–82.

Jost, R. (1965). *The General Theory of Quantized Fields*. Providence, RI: American Mathematical Society.

Jost, R. (1984). Erinnerungen: Erlesenes und Erlebtes. *Physikalische Blätter*, **7**, 178–81.

Jost, R., and Kohn, W. (1952a). Construction of a potential from a phase shift. *Physical Review*, **87**, 977–92.

Jost, R., and Kohn, W. (1952b). Equivalent Potentials. *Physical Review*, **88**, 382–5.

Jost, R., and Kohn, W. (1953). On the relation between phase shift energy levels and the potential. *Kongelige Danske Videnskabernes Selskab, Matermatisk-fysiske Meddelelser*, **27**, No. 9.

Jost, R., Luttinger, J. M., and Slotnick, M. (1950). Distribution of recoil nucleus in pair production by photons. *Physical Review*, **80**, 189–96.

Källen, G. (1950). Formal integration of the equations of quantum theory in the Heisenberg representation. *Arkiv för Fysik*, **2**, 371–410.

Källen, G. (1952). On the definition of the renormalization constants in quantum electrodynamics. *Helvetica Physica Acta*, **25**, 417–34.

Källen, G. (1964). *Elementary Particle Physics*. Reading: Addison–Wesley.

Kaluza, Th. (1921). Zum Unitätsproblem der Physik. *Sitzungsberichte der Preussischen Akademie der Wissenschaften zu Berlin*, 966–72. [English translation as On the unification problem in physics, by T. Muta (1984) in Lee, pp. 1–9.]

Kaplunovsky, V., and Weinstein, M. (1985). Space-time: arena or illusion? *Physical Review*, **D31**, 1879–98.

Kapur, P. L., and Peierls, R. (1938). The dispersion formula for nuclear reactions. *Proceedings of the Royal Society*, **A166**, 277–95.

Karplus, R., and Ruderman, M. A. (1955). Applications of causality to scattering. *Physical Review*, **98**, 771–4.

Karplus, R., Sommerfield, C. M., and Wichmann, E. H. (1958). Spectral representations in perturbation theory. I. Vertex function. *Physical Review*, **111**, 1187–90.

Keynes, J. M. (1951). *Essays in Biography*. New York: Norton.

Klein, O. (1926). Quantentheorie und fünfdimensionale Relativätstheorie. *Zeitschrift für Physik*, **37**, 895–6. [English translation as Quantum theory and five-dimensional relativity, by T. Muta (1984) in Lee, pp. 10–22.]

Klein, O. (1927). Elektrodynamik und Wellenmechanik vom Standpunkt des Korrespondenzprinzips. *Zeitschrift für Physik*, **41**, 407–42.

Koba, Z., and Nielsen, H. B. (1969). Reaction amplitude for n-mesons: a generalization of the Veneziano–Bardakci–Ruegg–Virasoro model. *Nuclear Physics*, **B10**, 633–55.

Kockel, B. (1949). Prozesse zwischen leichten Teilchen nach der Diracschen Theorie. *Annalen der Physik*, **4**, 279–302.

Kolb, E. W., Seckel, D., and Turner, M. S. (1985). The shadow world of superstring theories. *Nature*, **314**, 415–19.

Kourany, J. A. (1982). Towards an empirically adequate theory of science. *Philosophy of Science*, **49**, 526–48.

Kramers, H. A. (1924). The law of dispersion and Bohr's theory of spectra. *Nature*, **113**, 673–4.

Kramers, H. A. (1927). La diffusion de la lumière par les atomes. *Atti del Congresso Internationale di Fisici, Como*, **2**, 545–57.

Kramers, H. A. (1929). Die Dispersion und Absorption von Rontgenstrahlen. *Physikalische Zeitschrift*, **30**, 522–3.

Kramers, H. A. (1938). *Die Grundlagen der Quantentheorie: Quantentheorie des Elektrons und der Strahlung: Hand-und Jahrbuch der Chemischen Physik*. Leipzig: Akademische Verlagsgesellschaft M. B. H.

Kramers, H. A. (1944). Fundamental difficulties of a theory of particles. *Nederlands tijdschrift voor natuurkunde*, **11**, July–Aug. 1944, 134–47.

Kramers, H. A. (1956). *Collected Scientific Papers*. Amsterdam: North-Holland Publishing Co.

Kramers, H. A., and Heisenberg, W. (1925), Über die Streuung von Strahlen durch Atome. *Zeitschrift für Physik*, **31**, 681–708.

Kroll, N., and Ruderman, M. A. (1954). A theorem on photomeson production near threshold and the suppression of pairs in pseudoscalar meson theory. *Physical Review*, **93**, 233–8.

Kronig, R. (1926). On the theory of dispersion of X-rays. *Journal of the Optical Society of America*, **12**, 547–57.

Kronig, R. (1946). A supplementary condition in Heisenberg's theory of elementary particles. *Physica*, **12**, 543–4.

Kuhn, T. S. (1970). *The Structure of Scientific Revolutions*, 2nd edn. Chicago: University of Chicago Press.

Kursunoglu, B. N., and Wigner, E. P. (1987). *Reminiscences About a Great Physicist: Paul Adrian Maurice Dirac*. Cambridge: Cambridge University Press.

Ladenberg, R. (1921). Die quantentheoretische Deutung der Zahl der Dispersionselektronen. *Zeitschrift für Physik*, **4**, 451–68.

Lakatos, I. (1970). Falsification and the methodology of scientific research programmes. In Lakatos and Musgrave, pp. 91–196.

Lakatos, I. (1976). History of science and its rational reconstructions. In Howson, pp. 1–39.

Lakatos, I., and Musgrave, A. (eds.) (1970). *Criticism and the Growth of Knowledge*. London: Cambridge University Press.

Landau, L. D. (1955). On the quantum theory of fields. In Pauli, pp. 52–69.

Landau, L. D. (1959). On analytic properties of vertex parts in quantum field theory. *Nuclear Physics*, **13**, 181–92.

Landau, L. D. (1960). On analytical properties of vertex parts in quantum field theory. In *Proceedings of the Ninth International Annual Conference on High Energy Physics*, vol. II, pp. 95–101. Moscow.

Landau, L. D. (1965). *Collected Papers*. Oxford: Pergamon Press.

Landau, L. D., Abrikosov, A. A., and Khalatnikov, I. M. (1954a). The removal of infinities in Quantum electrodynamics. *Doklady Akademii Nauk SSSR*, **95**, 497–9.

Landau, L. D., Abrikosov, A. A., and Khalatnikov, I. M. (1954b). An asymptotic expression for the electron Green function in quantum electrodynamics. *Doklady Akademii Nauk SSSR*, **95**, 773–6.

Landau, L. D., Abrikosov, A. A., and Khalatnikov, I. M. (1954c). An asymptotic expression for the photon Green function in quantum electrodynamics. *Doklady Akademii Nauk SSSR*, **95**, 1117–20.

Landau, L. D., Abrikosov, A. A., and Khalatnikov, I. M. (1954d). The electron mass in quantum electrodynamics. *Doklady Akademii Nauk SSSR*, **96**, 261–63.

Landau, L. D., and Pomeranchuk, I. (1955). On point interactions in quantum electrodynamics. *Doklady Akademii Nauk SSSR*, **102**, 489–91.

Landshoff, P. V., Polkinghorne, J. C., and Taylor, J. C. (1961). A proof of the Mandelstam representation in perturbation theory. *Il Nuovo Cimento*, **19**, 939–52.

Lassalle, M. (1974). Analyticity properties implied by the many-particle structure of the n-point function in general quantum field theory. I. *Communications in Mathematical Physics*, **36**, 185–226.

Laudan, L. (1977). *Progress and Its Problems*. Berkeley: University of California Press.

Laudan, L. (1981). *Science and Hypothesis*. Dordrecht: D. Reidel.

Laudan, L. (1984a). *Science and Values*. Berkeley: University of California Press.

Laudan, L. (1984b). Explaining the success of science: Beyond epistemic realism and relativism. In Cushing *et al.*, 83–105.

Laudan, L., Donovan, A., Laudan, R., Barker, P., Brown, H., Leplin, J., Thagard, P. Wykstra, S. (1986). Scientific change: philosophical models and historical research. *Synthese*, **69**, 141–223.

Lax, M. (1950). On a well-known cross-section theorem. *Physical Review*, **78**, 306–7.

Lax, M., Feshbach, H., Chew, G. F., and Lewis, H. W. (1951). A phenomenological treatment of photo-meson production from deuterons. *Physical Review*, **82**, 324.

Lee, H. (1973). How to generate the Pomeranchukon from the background in a dual multiplet model. *Physical Review Letters*, **30**, 719–22.

Lee, H. C. (ed.) (1984). *An Introduction to Kaluza-Klein Theories*. Singapore: World Scientific Publishing Co.

Lehmann, H. (1958). Analytic properties of scattering amplitudes as functions of momentum transfer. *Il Nuovo Cimento*, **10**, 579–89.

Lehmann, H. (1959). Scattering matrix and field operators. *Supplemento del Nuovo Cimento*, **14**, 153–76.

Lehmann, H., Symanzik, K., and Zimmermann, W. (1955). Zur Formulierung quantisierter Feldtheorien. *Il Nuovo Cimento*, **1**, 205–25.

Lehmann, H., Symanzik, K., and Zimmermann, W. (1957). On the formulation of quantized field theories II. *Il Nuovo Cimento*, **6**, 319–33.

Lehr, W. J., and Park, J. L. (1977). A stochastic derivation of the Klein–Gordon equation. *Journal of Mathematical Physics*, **18**, 1235–40.

Leplin, J. (1984). *Scientific Realism*. Berkeley: University of California Press.

Levinson, N. (1940a). Determination of the potential from the asymptotic phase. *Physical Review*, **75**, 1445.

Levinson, N. (1949b). On the uniqueness of the potential in a Schrödinger equation for a given asymptotic phase. *Kongelige Danske Videnskabernes Selskab, Matematisk-fysiske Meddelelser*, **25**, No. 9.

Lichtenberg, D. B., and Rosen, S. P. (1980). *Developments in the Quark Theory of Hadrons, a Reprint Collection*, Vol. I: 1964–1978. Nonantum, MA: Hadronic Press.

Lindenbaum, S. J. (1964). High-energy elastic scattering of $\pi^{\pm}$, p, $\bar{p}$ and $K^{\pm}$ by protons, and Regge pole predictions. In Hofstadter and Schiff, pp. 105–30.

Lindenbaum, S. J. (1969). Asymptotic energies. In *Contemporary Physics; Trieste Symposium 1968*, pp. 123–43. Vienna: International Atomic Energy Agency.

Lippmann, B. A., and Schwinger, J. (1950). Variational principles for scattering processes. I. *Physical Review*, **79**, 469–80.

Losee, J. (1987). *Philosophy of Science and Historical Inquiry*. Oxford: Clarendon Press.

Lovelace, C. (1962). Diffraction scattering and Mandelstam representation. *Il Nuovo Cimento*, **25**, 730–55.

Lovelace, C. (1968). A novel prediction of Regge trajectories. *Physics Letters*, **28B**, 264–8.

Lovelace, C. (1971). Pomeron form factors and dual Regge cuts. *Physics Letters*, **34B**, 500–6.

Lovelace, C. (1972). Regge cut theories. In Tenner and Veltman, pp. 141–51.

Low, F. E. (1954). Scattering of light of very low frequency by systems of spin 1/2. *Physical Review*, **96**, 1428–32.

Low, F. E. (1955). Boson–Fermion scattering in the Heisenberg representation. *Physical Review*, **97**, 1392–8.

Low, F. E. (1959). The quantum theory of scattering. In *Brandeis University 1959 Summer Institute in Theoretical Physics* (no publisher listed), pp. 3–79.

Low, F. E. (1967). Dynamics of strong interactions. In *Proceedings of the XIIth International Conference on High-Energy Physics*, pp. 241–9. Berkeley: University of California Press.

Low, F. E. (1970a). High-energy scattering. I. *Comments on Nuclear and Particle Physics*, **4**, 154–8.

Low, F. E. (1970b). High-energy scattering. II. *Comments on Nuclear and Particle Physics*, **4**, 193–7.

Low, F. E. (1970c). High-energy behavior of scattering amplitudes. In Chrétien and Schweber, pp. 137–208.

Low, F. E. (1985). Complete sets of wave-packets. In De Tar *et al.*, pp. 17–22.

Lu, E. Y. C., and Olive, D. I. (1966). Spin and statistics in S-matrix theory. *Il Nuovo Cimento*, **45A**, 205–18.

Ma, S. T. (1946). Redundant zeros in the discrete energy spectra in Heisenberg's theory of characteristic matrix. *Physical Review*, **69**, 668.

Ma, S. T. (1947). On a general condition of Heisenberg for the S matrix. *Physical Review*, **71**, 195–200.

MacGregor, M. H., Moravcsik, M. J., and Stapp, H. P. (1959). Modified analysis of nucleon–nucleon scattering. II. Completed analysis of p–p scattering at 310 MeV. *Physical Review*, **116**, 1248–56.

Machleidt, R., Holinde, K., and Elster, Ch. (1987). The Bonn meson-exchange model for the nucleon–nucleon interaction. *Physics Reports*, **149**, 1–89.

MacKinnon, E. M. (1982). *Scientific Explanation and Atomic Physics*. Chicago: University of Chicago Press.

Madelung, E. (1926). Quantentheorie in hydrodynamischer Form. *Zeitschrift für Physik*, **40**, 322–6.

Mandelbrot, B. (1983). *The Fractal Geometry of Nature*. New York: W. H. Freeman.

Mandelstam, S. (1958). Determination of the pion–nucleon scattering amplitude from dispersion relations and unitarity. General theory. *Physical Review*, **112**, 1344–60.

Mandelstam, S. (1959a). Analytic properties of transition amplitudes in perturbation theory. *Physical Review*, **115**, 1741–51.

Mandelstam, S. (1959b). Construction of the perturbation series for transition amplitudes from their analyticity and unitarity properties. *Physical Review*, **115**, 1752–62.

Mandelstam, S. (1961). Two-dimensional representations of scattering amplitudes and their applications. In Stoops, pp. 209–33.

Mandelstam, S. (1963a). The Regge formalism for relativistic particles with spin. *Il Nuovo Cimento*, **30**, 1113–26.

Mandelstam, S. (1963b). Cuts in the angular-momentum plane-I. *Il Nuovo Cimento*, **30**, 1127–47.

Mandelstam, S. (1963c). Cuts in the angular-momentum plane-II. *Il Nuovo Cimento*, **30**, 1148–1162.

Mandelstam, S. (1965). Non-Regge terms in the vector-spinor theory. *Physical Review*, , **137**, B949–54.

Mandelstam, S. (1967). Some recent works on Regge poles and the three-body problem. In Takeda and Fuji, pp. 1–21.

Mandelstam, S. (1969a). Rising Regge trajectories and dynamical calculations. *Comments on Nuclear and Particle Physics*, **3**, 65–72.

Mandelstam, S. (1969b). Rising Regge trajectories and dynamical calculations, II. *Comments on Nuclear and Particle Physics*, **3**, 147–54.

Mandelstam, S. (1973a). Interacting-string picture of dual-resonance models. *Nuclear Physics*, **B64**, 205–35.

Mandelstam, S. (1973b). Manifestly dual formulation of the Ramond-model. *Physics Letters*, **46B**, 447–51.

Mandelstam, S. (1974a). Interacting-string picture of the Neveu–Schwarz–Ramond model. *Nuclear Physics*, **B69**, 77–106.

Mandelstam, S. (1974b). Dual-resonance models. *Physics Reports*, **13**, 259–353.

Mandelstam, S. (1985). Composite vector meson and string models. In De Tar et al., pp. 97–105.

Mandl, F. (1959). *Introduction to Quantum Field Theory*. New York: Interscience Publishers.

Martin, A. W., and McGlinn, W. D. (1964). Crossing relations and the prediction of symmetries in a soluble model. *Physical Review*, **136B**, 1515–22.

Mauskopf, S. H. (ed.) (1979). *The Reception of Unconventional Science, AAAS Selected Symposium 25*. Boulder, CO: Westview Press.

Maxwell, J. C. (1890). *Scientific Papers of James Clerk Maxwell*, vol. II, ed. W. D. Niven. Cambridge: Cambridge University Press.

McMullin, E. (1976a). History and philosophy of science: A marriage of convenience? In Cohen, Hooker, Michalos and van Evra, pp. 585–601.

McMullin, E. (1976b). The fertility of theory and the unit for appraisal in science. In Cohen, Feyerabend and Wartofsky, pp. 395–432.

McMullin, E. (1984). A case for scientific realism. In Leplin, pp. 8–40.

McMullin, E. (1988). The shaping of scientific rationality. In McMullin, pp. 1–41.

McMullin, E. (ed.) (1988). *Construction and Constraint: The Shaping of Scientific Rationality*. Notre Dame, IN: University of Notre Dame Press.

Medawar, P. B. (1967). *The Art of the Soluble*. New York: Barnes and Noble.

Medawar, P. B. (1979). *Advice to a Young Scientist*. New York: Harper & Row.

Mehra, J., and Rechenberg, H. (1982). *The Historical Development of Quantum Theory*, vol. 1, part 2. New York: Springer–Verlag.

Meixner, J. (1948). Über den Zusammenhang der Eignewerte der Heisenbergschen *S*-Matrix mit den stationären Zuständen. *Zeitschrift für Naturforschung*, **3a**, 73–8.

Messiah, A. (1965). *Quantum Mechanics*, vol. I. Amsterdam: North-Holland Publishing Co.

Meyerson, É. (1908). *Identité et Réalité*, Paris: Labrairies Félix Alcon et Guillamin Réunés. [Appeared in translation by K. Loewenberg, 1930, as *Identity and Reality*. London: George Allen & Unwin, Ltd.]

Misner, C. W. (1977). Cosmology and theology. In Yourgrau and Breck, pp. 75–100.

Møller, C. (1945). General properties of the characteristic matrix in the theory of elementary particles I. *Kongelige Danske Videnskabernes Selskab, Matematisk-fysiske Meddelelser*, **23**, No. 1.

Møller, C. (1946a). General properties of the characteristic matrix in the theory of elementary particles II. *Kongelige Danske Videnskabernes Selskab, Matematisk-fysiske Meddelelser*, **22**, No. 19.

Møller, C. (1946b). New developments in relativistic quantum theory. *Nature*, **158**, 403–6.

Møller, C. (1947). On the theory of the characteristic matrix. *Report of an International Conference on Fundamental Particles and Low Temperatures*, vol. I, pp. 194–9. London: The Physical Society.

Moravcsik, M. J. (ed.) (1966). *Recent Developments in Particle Physics*. New York: Gordon and Breach.

Moravcsik, M. J. (1985). Thirty years of one-particle exchange. In De Tar *et al.*, pp. 26–34.

Moshe, M. (1978). Recent developments in Reggeon field theory. *Physics Reports*, **37**, 255–345.

Mott, N. F., and Massey, H. S. W. (1933). *The Theory of Atomic Collisions*. London: Oxford University Press.

Moyer, B. J., and Chew, G. F. (1951). Collisions of high-energy nuclear particles with nuclei. *American Journal of Physics*, **19**, 17–26.

Nambu, Y. (1955). Structure of Green's functions in quantum field theory. *Physical Review*, **100**, 394–411.

Nambu, Y. (1957). Parametric representations of general Green's functions. *Il Nuovo Cimento*, **6**, 1064–83.

Nambu, Y. (1970). Quark model and the factorization of the Veneziano amplitude. In Chand, pp. 269–78.

Ne'eman, Y. (1980). Open discussion. In H. Woolf, p. 267.

Nelson, E. (1966). Derivation of the Schrödinger equation from Newtonian mechanics. *Physical Review*, **150**, 1079–85.

Nelson, E. (1967). *Dynamical Theories of Brownian Motion*. Princeton: Princeton University Press.

Nersessian, N. J. (1984). *Faraday to Einstein: Constructing Meaning in Scientific Theories*. Dordrecht: Martinus Nijhoff Publishers.

Nersessian, N. J. (1987). *The Process of Science*. Dordrecht: Martinus Nijhoff Publishers.

Neveu, A. (1987). Superstrings. *Helvetica Physica Acta*, **60**, 1–8.

Neveu, A., and Schwarz, J. H. (1971a). Factorizable dual model of pions. *Nuclear Physics*, **B31**, 86–112.

Neveu, A., and Schwarz, J. H. (1971b). Quark model of dual pions. *Physical Review*, **D4**, 1109–11.

Newton, R. G. (1966). *Scattering Theory of Waves and Particles*. New York: McGraw-Hill.

Nickles, T. (1987). 'Twixt method and madness. In Nersessian, pp. 41–67.

Nicolescu, B. (1981). Topological interpretation of multiquark states. In Arenhövel and Sauris, pp. 223–33.

Nicolescu, B. (1985), *Nous, la particule et le monde*. France: Editions le Mail.

Nicolescu, B., and Poénaru, V. (1985). From baryonium to hexons. In De Tar *et al.*, pp. 195–221.

Nicolescu, B., Richard, J. M., and Vinh Mau, R. (eds.) (1979), *Proceedings of the Workshop on Baryonium and Other Unusual Hadron States*. Orsay: Institut de Physique Nucleàire.

Noyes, H. P., Hafner, E. M., Klarmann, J., and Woodruff, A. E. (eds.) (1954). *Proceedings of the Fourth Annual Rochester Conference on High Energy Nuclear Physics*. Rochester: University of Rochester.

Oehme, R. (1989). Theory of the scattering matrix (1942–1946): An annotation. In Blum *et al.*, pp. 605–10.

Okubo, S. (1963). ϕ-meson and unitary symmetry model. *Physics Letters*, **5**, 165–8.

Okubo, S. (1977). Consequences of quark-line (Okubo–Zweig–Iizuka) rule. *Physical Review*, **D16**, 2336–52.

Okun', . B., and Pomeranchuk, I. Ia. (1956). The conservation of isotopic spin and the cross section of the interaction of high-energy π-mesons and nucleons with nucleons. *Soviet Physics JETP*, **3**, 307–8.

Oldershaw, R. L. (1988). The new physics – physical or mathematical science? *American Journal of Physics*, **56**, 1075–81.

Olive, D. I. (1964). Exploration of S-matrix theory. *Physical Review*, **135B**, B745–60.

Olive, D., and Scherk, J. (1973a). No-ghost theorem for the Pomeron sector of the dual model. *Physics Letters*, **44B**, 296–300.

Olive, D., and Scherk, J. (1973b). Towards satisfactory scattering amplitudes for dual fermions. *Nuclear Physics*, **B64**, 334–48.

Omnès, R. (1958). On the solution of certain singular integral equations of quantum field theory. *Il Nuovo Cimento*, **8**, 316–26.

Paton, J. E., and Chan, H.-M. (1969). Generalized Veneziano model with isospin. *Nuclear Physics*, **B10**, 516–20.

Pauli, W. (1928). Discussion générale des idées nouvelles émises. *Electrons et Photons, Rapports et Discussions du Cinquieme Conseilde Physique*, pp. 280–2. Paris: Gauthier-Villars.

Pauli, W. (1946). *Meson Theory of Nuclear Forces*. New York: Interscience Publishers, Inc.

Pauli, W. (1947). Difficulties of field theories and of field quantization. In *Report of an International Conference on Fundamental Particles and Low Temperatures*, vol. I, pp. 5–10. London: The Physical Society.

Pauli, W. (ed.) (1955). *Neils Bohr and the Development of Physics*. London: Pergamon Press.

Peres, A., and Zurek, W. H. (1982). Is quantum theory universally valid? *American Journal of Physics*, **50**, 807–10.

Pham, F. (1967). Singularités des processus de diffusion multiple. *Annales de l'Institut Henri Poincaré*, **6A**, 89–204.

Phillips, R. J. N. (1972). Regge cuts and related topics. In Tenner and Veltman, pp. 110–39.

Phillips, R. J. N., and Ringland, G. A. (1972). Regge phenomenology. In Burhop, pp. 187–247.

Pickering, A. (1984). *Constricting Quarks – A Sociological History of Particle Physics*. Chicago: University of Chicago Press.

Pickering, A. (1987). Against correspondence: A constructivist view of experiment and the real. In Fine and Machamer, vol. 2, pp. 196–206.

Pickering, A. (1989a). From field theory to phenomenology: the history of dispersion relations. In Brown, Dresden and Hoddeson, pp. 579–99.

Pickering, A. (1989b). Living in the material world: on realism and experimental practice. In D. Gooding, *et al.*, pp. 275–97.

362   References

Planck, M. (1949). *Scientific Autobiography and Other Papers*. Westport, CN: Greenwood Press.

Polkinghorne, J. C. (1956). General dispersion relations. *Il Nuovo Cimento*, **4**, 216–30.

Polkinghorne, J. C. (1962a). Analyticity and unitarity. *Il Nuovo Cimento*, **23**, 360–7.

Polkinghorne, J. C. (1962b). Analyticity and unitarity – II. *Il Nuovo Cimento*, **25**, 901–11.

Polkinghorne, J. C. (1963a). High-energy behavior in perturbation theory. *Journal of Mathematical Physics*, **4**, 503–6.

Polkinghorne, J. C. (1963b). Singularities of Regge trajectories and asymptotes to Landau curves. *Journal of Mathematical Physics*, **4**, 1393–5.

Polkinghorne, J. C. (1963c). High-energy behavior of perturbation theory. II. *Journal of Mathematical Physics*, **4**, 1396–400.

Polkinghorne, J. C. (1963d). Cancelling cuts in the Regge plane. *Physics Letters*, **4**, 24.

Polkinghorne, J. C. (1964a). The complete high-energy behavior of ladder diagrams in perturbation theory. *Journal of Mathematical Physics*, **5**, 431–4.

Polkinghorne, J. C.(1964b). Asymptotic behavior of Feynman integrals with spin. *Journal of Mathematical Physics*, **5**, 1491–8.

Polkinghorne, J. C. (1979). *The Particle Play*. Oxford: W. H. Freeman and Company.

Polkinghorne, J. C. (1985). Salesman of ideas. In DeTar *et al.*, pp. 23–5.

Polkinghorne, J. C., and Screaton, G. R. (1960a). The analytic properties of perturbation theory I. *Il Nuovo Cimento*, **15**, 289–300.

Polkinghorne, J. C., and Screaton, G. R. (1960b). The analytic properties of perturbation theory II. *Il Nuovo Cimento*, **15**, 925–31.

Pomeranchuk, I. Ia. (1956a). The conservation of isotopic spin and the scattering of antinucleons by nucleons. *Soviet Physics JETP*, **3**, 306–7.

Pomeranchuk, I. Ia. (1956b). Vanishing of the renormalized charge in electrodynamics and in meson theory. *Il Nuovo Cimento*, **3**, 1186–1203.

Pomeranchuk, I. Ia. (1958). Equality of the nucleon and antinucleon total interaction cross section at high energies. *Soviet Physics JETP*, **34**, 7, 499–501.

Popper, K. (1959). *The Logic of Scientific Discovery*. New York: Harper & Row.

Popper, K. R. (1963). *Conjectures and Refutations*. New York: Harper & Row.

Prentki, J. (ed.), (1962). *1962 International Conference on High Energy Physics at CERN*, Geneva: CERN.

Price, W. C., Chissick, S. S., and Ravendale, T. (eds.) (1973). *Wave Mechanics: The First Fifty Years*. London: Butterworths.

Putnam, H. (1975). *Mathematics, Matter and Method*. Cambridge: Cambridge University Press.

Putnam, H. (1978). *Meaning and the Moral Sciences*. London: Routledge & Kegan Paul.

Raine, D. J., and Heller, M. (1981). *The Science of Space – Time*. Tucson, AZ: Pachart Publishing House.

Rebbi, C. (1974). Dual models and relativistic quantum strings. *Physics Reports*, **12**, 1–73.

Rechenberg, H. (1989). The early S-matrix theory and its propagation (1942–1952). In Brown, Dresden and Hoddeson, pp. 551–78.

Redhead, M. L. G. (1980). Some philosophical aspects of particle physics. *Studies in History and Philosophy of Science*, **11**, 279–304.

Redhead, M. L. G. (1983). Nonlocality and peaceful coexistence. In Swinburne. pp. 151–89.

Regge, T. (1958a). Analytic properties of the scattering matrix. *Il Nuovo Cimento*, **8**, 671–9.

Regge, T. (1958b). On the analytic behavior of the eigenvalue of the S-matrix in the complex plane of the energy. *Il Nuovo Cimento*, **9**, 295–302.

Regge, T. (1959). Introduction to complex orbital momenta. *Il Nuovo Cimento*, **14**, 951–76.

Regge, T. (1960). Bound states, shadow states and Mandelstam representation. *Il Nuovo Cimento*, **18**, 947–56.

Rho, M., and Wilkinson, D. (eds.) (1979). *Mesons in Nuclei*, vol. I. Amsterdam: North-Holland Publishing Co.

Rigden, J. S. (1987). Heisenberg, February 1927, and physics. *American Journal of Physics*, **55**, 107.

Rivier, D. (1984). Ernst Carl Gerlach Stuekelberg von Breidenbach. *Helvetica Physica Acta*, **57**, 577–8.

Rivier, D., and Stückelberg, E. C. G. (1948). A convergent expression for the magnetic moment of the neutron. *Physical Review*, **74**, 218.

Rohrlich, F., and Gluckstern, R. L. (1952). Forward scattering of light by a Coulomb field. *Physical Review*, **86**, 1–9.

Rohrlich, F., and Hardin, L. (1983). Established theories. *Philosophy of Science*, **50**, 603–17.

Rolnik, W. B. (1966). Comment on the application of Cutkosky's rules. *Physical Review Letters*, **16**, 544–5.

Rorty, R. (1988). Is natural science a natural kind? In McMullin, pp. 49–74.

Rosenthal-Schneider, I. (1980). *Reality and Scientific Truth*. Detroit, MI: Wayne State University Press.

Rosenzweig, C., and Chew, G. F. (1975). A systematic lifting of exchange-degeneracy that clarifies the relationship between Pomeron, Reggeons and SU3-symmetry violation. *Physics Letters*, **58B**, 93–6.

Rosenzweig, C., and Veneziano, G. (1974). Regge couplings and intercepts from the planar dual bootstrap. *Physics Letters*, **52B**, 335–40.

Rosner, J. L. (1969). Graphical form of duality. *Physical Review Letters*, **22**, 689–92.

Ruelle, D. (1961). Connection between Wightman functions and Green functions in p-space. *Il Nuovo Cimento*, **19**, 356–76.

Salam, A. (1956). On generalized dispersion relations. *Il Nuovo Cimento*, **3**, 424–9.

Salam, A. (ed.) (1965). *High Energy Physics and Elementary Particles*. Vienna: International Atomic Energy Agency.

Salam, A., and Gilbert, W. (1956). On generalized dispersion relations II. *Il Nuovo Cimento*, **3**, 607–11.

Salmon, W. C. (1984). *Scientific Explanation and the Causal Structure of the World*. Princeton: Princeton University Press.

Salzman, G., and Salzman, F. (1957). Solutions of the static theory integral equations for pion–nucleon scattering in the one-meson approximation. *Physical Review*, **108**, 1619–28.

Scadron, M., and Weinberg, S. (1964). Potential theory calculations by quasiparticle method. *Physical Review*, **133**, B1589–96.

Scadron, M., Weinberg, S., and Wright, J. (1964). Functional analysis and scattering theory. *Physical Review*, **135**, B202–7.

Schaap, M. M., and Veneziano, G. (1975). Self-consistent ρ–P′ trajectory from the planar dual bootstrap. *Lettre al Nuovo Cimento*, **12**, 204–6.

Scherk, J. (1975). An introduction to the theory of dual models and strings. *Reviews of Modern Physics*, **47**, 123–64.

Scherk, J., and Schwarz, J. H. (1974a), Dual models and the geometry of space-time. *Physics Letters*, **52B**, 347–50.

Scherk, J., and Schwarz, J. H. (1974b). Dual models for non-hadrons. *Nuclear Physics*, **B81**, 118–44.

Scherk, J., and Schwarz, J. H. (1975). Dual field theory of quarks and gluons. *Physics Letters*, **57B**, 463–6.

Schilpp, P. A. (1970). *Albert Einstein: Philosopher – scientist*. La Salle, IL: Open Court.

Schmid, C. (1968a). Meson bootstrap with finite-energy sum rules. *Physical Review Letters*, **20**, 628–31.

Schmid, C. (1968b). Direct-channel resonances from Regge-pole exchange. *Physical Review Letters*, **20**, 689–91.

Schmid, C. (1970). Duality and exchange degeneracy. In Zichichi, pp. 108–43.

Schrödinger, E. (1926). Quantisierung als Eigenvertproblem. *Annalen der Physik*, **81**, 109–39.

Schützer, W., and Tiomno, J. (1951). On the connection of the scattering and derivative matrices with causality. *Physical Review*, **83**, 249–51.

Schwarz, J. H. (1967). Superconvergence relations from Regge-pole theory. *Physical Review*, **159**, 1269–71.

Schwarz, J. H. (1973). Dual resonance theory. *Physics Reports*, **8**, 269–335.

Schwarz, J. H. (1975). Dual-resonance models of elementary particles. *Scientific American*, **232**, No. 1, 61–7.

Schwarz, J. H. (1982a). Supersymmetry, strings, and the unification of fundamental interactions. CALT-68-905 DOE Research and Development Report.

Schwarz, J. H.(1982b). Superstring theory. *Physics Reports*, **89**, 223–322.

Schwarz, J. H. (1984). What are superstrings? *Comments on Nuclear and Particle Physics*, **13**, 103–15.

Schwarz, J. H. (1985). From the bootstrap to superstrings. In De Tar *et al.*, pp. 106–13.

Schwarzschild, B. M. (1985). Anomaly cancellation launches superstring bandwagon. *Physics Today*, **38**, 7, 17–20.

Schweber, S. S. (1961). *An Introduction to Relativistic Quantum Field Theory*. Evanston, IL: Row, Peterson and Company.

Schweber, S. S. (1984). Some chapters for a history of quantum field theory; 1938–1952. In De Witt and Stora, pp. 37–220.

Schweber, S. S. (1987). Some philosophical reflections on the history of quantum field theory. Talk given at the 1987 Meeting of the Eastern Division of the American Philosophical Association, New York City, December 27–30, 1987.

Schweber, S. S. (1989). Particle theory in the 50's: An historical assessment. In Brown, Dresden and Hoddeson, pp. 668–93.

Schweber, S. S. (1990). *QED (1946–1950): An American Success Story*. Princeton: Princeton University Press (in press).

Schwimmer, A., and Zachariasen, F. (1975). The radius of the Pomeron and absorptive cuts. *Physics Letters*, **57B**, 357–60.

Schwinger, J. (1948). Quantum electrodynamics. I. A covariant formulation. *Physical Review*, **74**, 1439–61.

Screaton, G. R. (ed.) (1961). *Dispersion Relations*. Edinburgh: Oliver and Boyd.

Shapere, D. (1984). *Reason and the Search for Knowledge*. Dordrecht: D. Reidel Publishing Co.

Shapere, D. (1986). External and internal factors in the development of science. *Science and Technology Studies*, **4**, No. 1, 1–9.

Shimony, A. (1978). Metaphysical problems in the foundations of quantum mechanics. *International Philosophical Quarterly*, **18**, 3–17.

Siegel, H. (1985). What is the question concerning the rationality of science? *Philosophy of Science*, **52**, 517–37.

Siegert, A. J. F. (1939). On the derivation of the dispersion formula for nuclear reactions. *Physical Review*, **56**, 750–2.

Slater, J. C. (1924). Radiation and Atoms. *Nature*, **113**, 307–8.

Slater, J. C. (1973). The development of quantum mechanics in the period 1924–1926. In Price *et al.*, pp. 19–25.

Slater, J. C., and Frank, N. H. (1933). *Introduction to Theoretical Physics*. New York: McGraw-Hill.

Stapp, H. P. (1962a). Derivation of the CPT theorem and the connection between spin and statistics from postulates of the *S*-matrix theory. *Physical Review*, **125**, 2139–62.

Stapp, H. P. (1962b). Axiomatic *S*-matrix theory. *Reviews of Modern Physics*, **34**, 390–4.

Stapp, H. P. (1962c). The decomposition of the S matrix and the connection between spin and statistics. Lawrence Radiation Laboratory Report UCRL-10289 (unpublished).

Stapp, H. P. (1965a). Space and time in S-matrix theory. *Physical Review*, **139**, B257–70.

Stapp, H. P. (1965b). Analytic S-matrix theory. In Salam, pp. 3–54.

Stapp, H. P. (1968). Crossing, hermitian analyticity, and the connection between spin and statistics. *Journal of Mathematical Physics*, **9**, 1548–92.

Stapp, H. P. (1971). S-matrix interpretation of quantum theory. *Physical Review*, **D3**, 1303–20.

Stapp, H. P. (1972a). Foundations of S-matrix theory I: Theory and measurement. Lawrence Berkeley Laboratory Report 759 Rev, UC-34 Physics.

Stapp, H. P. (1972b). The Copenhagen interpretation. *American Journal of Physics*, **40**, 1098–116.

Stapp, H. P. (1977). Topological expansion for baryons and mesons. *Lettre al Nuovo Cimento*, **19**, 622–4.

Steinmann, O. (1960). A rigorous formulation of LSZ field theory. *Communications in Mathematical Physics*, **10**, 245–68.

Stern, A. W. (1964). The third revolution in 20th Century physics. *Physics Today*, **17**, 42–5.

Stirling, A. V., *et al.* (1965). Small-angle change exchange of π^- mesons between 6 and 8 GeV/c. *Physical Review Letters*, **14**, 763–7.

Stöckler, M. (1984). *Philosophische Probleme der relativistischen Quantenmechanik, Erfahrung und Denken Band 65*. Berlin: Duncker & Humbolt.

Stoops, R. (ed.) (1961). *La Theorie Quantique des Champs*, (Proceedings of the Twelfth Solvay Conference on Physics). New York: Interscience Publishers.

Streater, R. F., and Wightman, A. S. (1964). *PCT, Spin and Statistics, and All That*. New York: W. A. Benjamin, Inc.

Stückelberg, E. C. G. (1938). Die Wechselwirkungskräfte in der Elektrodynamik und in der Feldtheorie der Kernkräft. (Teil I). *Helvetica Physica Acta*, **11**, 225–44; (Teil II und III), 299–328.

Stückelberg, E. C. G. (1939). A new model of the point charge electron and other elementary particles. *Nature*, **144**, 118.

Stückelberg, E. C. G. (1941). Un nouveau modéle de l'électron ponctuel en théorie classique. *Helvetica Physica Acta*, **14**, 51–80.

Stückelberg, E. C. G.(1942). La mécanique du point matériel en théorie de relativité et en théorie des quanta. *Helvetica Physica Acta*, **15**, 23–37.

Stückelberg, E. C. G. (1944). Un modéle de l'électron ponctuel II. *Helvetica Physica Acta*, **17**, 3–26.

Stückelberg, E. C. G. (1945). Mécanique fonctionnelle. *Helvetica Physica Acta*, **18**, 195–220.

Stückelberg, E. C. G. (1946). Une propriété de l'opréteur S en mécanique asymptotique. *Helvetica Physica Acta*, **19**, 242–3.

Stückelberg, E. C. G., and Green, T. A. (1951). Elimination des constantes arbitraires dans la théorie relativiste des quanta. *Helvetica Physica Acta*, **24**, 153–74.

Stückelberg, E. C. G., and Rivier, D. (1950a). Causalité et structure de la matrice *S*. *Helvetica Physica Acta*, **23**, 215–22.

Stückelberg, E. C. G., and Rivier, D. (1950b). A propos des divergences en théorie des champs quantifiés. *Helvetica Physica Acta*, **23**, Supplement III, 236–9.

Stuewer, R. H. (ed.) (1970). *Historical and Philosophical Perceptives of Science*, (*Minnesota Studies in the Philosophy of Science*, vol. 5.). Minneapolis, MN: University of Minnesota Press.

Sudarshan, E. C. G., Tinlot, J. H., Melissinos, A. C. (1960). *Proceedings of the 1960 Annual International Conference on High Energy Physics at Rochester*. New York: Interscience Publishers.

Swinburne, R. (ed.) (1983). *Space, Time and Causality*. Dordrecht: D. Reidel Publishing Co.

Symanzik, K. (1957). Derivation of dispersion relations for forward scattering. *Physical Review*, **105**, 743–9.

Takeda, G. (ed.) (1966). *High Energy Physics, Part II*. Tokyo: Syokabo Publishing Co.

Takeda, G., and Fuji, A. (eds.) (1967). *Elementary Particle Physics, Part II*. Tokyo: Syokabo Publishing Co.

Tarski, J. (1960). Analyticity of the fourth order scattering amplitude with two complex invariants. *Journal of Mathematical Physics*, **1**, 149–63.

Taylor, T. B., and Chew, G. F. (1950). Energy and angular distribution of π^- mesons produced by 350-MeV-protons. *Physical Review*, **78**, 86.

Teller, P. (1988). Three problems of renormalization. In Brown and Harré, pp. 73–89.

Teller, P. (1989a). Infinite renormalization. *Philosophy of Science*, **56**, 238–57.

Teller, P. (1989b). Relativity, relational holism and the the Bell inequalities. In Cushing and McMullin, pp. 208–23.

Tenner, A. G., and Veltman, M. J. G. (eds.) (1972). *Proceedings of the Amsterdam International Conference on Elementary Particles*. Amsterdam: North-Holland Publishing Co.

ter Haar, D. (1946). On the redundant zeros in the theory of the Heisenberg matrix. *Physica*, **12**, 501–8.

Thorn, C. B. (1971). Embryonic dual model for pions and fermions. *Physical Review*, **D4**, 1112–16.

Titchmarsh, E. C. (1937). *Introduction to the Theory of Fourier Integrals*. London: Oxford University Press.

Toll, J. S. (1952). The dispersion relation for light and its application to problems involving electron pairs, unpublished Ph.D. dissertation, Princeton University.

Toll, J. S. (1956). Causality and the dispersion relation: logical foundations, *Physical Review*, **104**, 1760–70.

Toll, J. S., and Wheeler, J. A. (1951). Some pair-theoretic applications of the dispersion relation. *Physical Review*, **81**, 654–5.

Toulmin, S. (1961). *Forsight and Understanding*. London: Hutchinson.

Touschek, B. (1948). Zum analytischen Verhalten Schrödinger'scher Wellenfunktionen. *Zeitschrift für Physik*, **125**, 293–7.

Toynbee, A. (1935–1961). *A Study of History*. London: Oxford University Press.

Udgaonkar, B. M. (1962). High-energy cross sections, Pomeranchuk theorems, and Regge poles, *Physical Review Letters*, **8**, 142–5.

Udgaonkar, B. M., and Gell-Mann, M. (1962). High-energy nuclear scattering and Regge poles. *Physical Review Letters*, **8**, 346–9.

van der Waerden, B. L. (1968). *Sources of Quantum Mechanics*. New York: Dover Publications.

van Fraassen, B. C. (1980). *The Scientific Image*. Oxford: The Clarendon Press.

van Fraassen, B. C. (1985). Empiricism in the philosophy of science. In Churchland and Hooker, pp. 245–308.

van Hove, L. (1967). Pion–nucleon charge-exchange scattering and high-energy dynamics, I. The differential cross section. *Comments on Nuclear and Particle Physics*, **1**, 191–5.

van Hove, L. (1968). Pion–Nucleon charge-exchange scattering and high-energy dynamics, II. Polarization effects. *Comments on Nuclear and Particle Physics*, **2**, 10–14.

van Kampen, N. G. (1951). Note on the analytic continuation of the S-matrix. *Philosophical Magazine*, **42**, 851–5.

van Kampen, N. G. (1953a). S-matrix and causality condition. I. Maxwell field. *Physical Review*, **89**, 1072–9.

van Kampen, N. G. (1953b). S-matrix and causality condition. II. Nonrelativistic particles. *Physical Review*, **91**, 1267–76.

vann Woodward, C. (1986). Gone with the wind. *The New York Review of Books XXXIII*, No. 12 (July 17), pp. 3–6.

Veneziano, G. (1968). Construction of a crossing-symmetric, Regge-behaved amplitude for linearly rising trajectories. *Il Nuovo Cimento*, **57A**, 190–7.

Veneziano, G. (1974a). Regge intercepts and unitarity in planar dual models. *Nuclear Physics*, **B74**, 365–77.

Veneziano, G. (1974b). Large N expansion in dual models. *Physics Letters*, **52B**, 220–2.

Veneziano, G. (1974c). An introduction to dual models of strong interactions and their physical motivations. *Physics Reports*, **9**, 199–242.

Vinh Mau, R. (1979). The Paris nucleon–nucleon potential. In Rho and Wilkinson, pp. 151–96.

Vinh Mau, R. (1985). The S-matrix theory of nuclear forces. In De Tar *et al.*, pp. 49–63.

Virasoro, M. A. (1969a). Alternative constructions of crossing-symmetric amplitudes with Regge behavior. *Physical Review*, **177**, 2309–11.

Virasoro, M. A. (1969b). Generalization of Veneziano's formula for the five-point function. *Physical Review Letters*, **22**, 37–9.

Virasoro, M. A. (1970). Subsidiary condition and ghosts in dual resonance models. *Physical Review*, **D1**, 2933–6.

von Neumann, J. (1955). *Mathematical Foundations of Quantum Mechanics.* Princeton: Princeton University Press.

Waller, I. (1927). On the scattering of radiation from atoms. *Philosophical Magazine*, **4**, 1228–37.

Waller, I. (1928). Über ein verallgemeinerte Streuungsformel. *Zeitschrift für Physik*, **51**, 213–31.

Watkins, J. W. N. (1984). *Science and Skepticism.* Princeton: Princeton University Press.

Weinberg, S. (1963a). Elementary particle theory of composite particles. *Physical Review*, **130**, 776–83.

Weinberg, S. (1963b). Quasiparticles and the Born series. *Physical Review*, **131**, 440–60.

Weinberg, S. (1964a). Systematic solution of multiparticle scattering problems, *Physical Review*, **133**, B232–56.

Weinberg, S. (1964b). Feynman rules for any spin. *Physical Review*, **133**, B1318–32.

Weinberg, S. (1964c). Feynman rules for any spin. II. Massless particles. *Physical Review*, **134**, B882–96.

Weinberg, S. (1964d). Derivation of gauge invariance and the equivalence principle from Lorentz invariance of the S-matrix. *Physics Letters*, **9**, 357–9.

Weinberg, S. (1964e). Photons and gravitons in S-matrix theory: derivation of charge conservation and equality of gravitational and inertial mass. *Physical Review*, **135**, B1049–56.

Weinberg, S. (1965a). Evidence that the deuteron is not an elementary particle. *Physical Review*, **137**, B672–8.

Weinberg, S. (1965b). Photons and gravitons in perturbation theory: derivation of Maxwell's and Einstein's equations. *Physical Review*, **138**, B988–1002.

Weinberg, S. (1965c). Comments on relativistic supermultiplet theories. *Physical Review*, **139**, B597–601.

Weinberg, S. (1965d). Infrared photons and gravitons. *Physical Review*, **140**, B516–24.

Weinberg, S. (1977). The search for unity: Notes for a history of quantum field theory. *Daedalus*, **106**, No. 4, 17–35.

Weinberg, S. (1985). The ultimate structure of matter. In De Tar *et al.*, pp. 114–27.

Weinberg, S. (1986a). Particles, fields, and now strings. In de Boer, Dal and Ulfbeck, pp. 227–39.

Weinberg, S. (1986b). Particle physics: past and future. In Hwa, pp. 186–203.

Weingartner, P., and Dorn, G. (eds.) (1986). *Foundations of Physics*. Vienna: Holder–Pichter–Tempsky.

Weisskopf, V. (1931). Zur Theorie der Resonanzfloureszenz. *Annalen der Physik*, **9**, 23–66.

Weisskopf, V., and Wigner, E. (1930). Berechnung der natürlichen Linienbreite auf Grund der Diracschen Lichttheorie. *Zeitschrift für Physik*, **63**, 54–73.

Wentzel, G. (1943). *Einführung in die Quantentheorie der Wellenfelder*. Wien: Franz Deuticke. [Translated as *Quantum Theory of Fields*, by C. Houtermann and J. M. Jauch (1949). New York: Interscience Publishers.]

Wentzel, G. (1947). Recent research in meson theory. *Reviews of Modern Physics*, **19**, 1–18.

Wess, J., and Zumino, B. (1974). Supergauge transformations in four dimensions. *Nuclear Physics*, **B70**, 39–50.

Wheeler, J. A. (1937a). Molecular viewpoints in nuclear structure. *Physical Review*, **52**, 1083–106.

Wheeler, J. A. (1937b). On the mathematical description of light nuclei by the method of resonating group structure. *Physical Review*, **52**, 1107–22.

Wheeler, J. A. (1977). Genesis and observership. In Butts and Hintikka, pp. 3–33.

Whewell, W. (1857). *History of the Inductive Sciences*, Three volumes. London: John W. Parker and Son. (Reprinted in 1967. London: Frank Cass and Co. Ltd.)

Wick, G. C. (1955). Introduction to some recent work in meson theory. *Reviews of Modern Physics*, **27**, 339–62.

Wick, G. C., Wightman, A. S., and Wigner, E. P. (1952). The intrinsic parity of elementary particles. *Physical Review*, **88**, 101–5.

Wightman, A. S. (1956). Quantum field theory in terms of vacuum expectation values. *Physical Review*, **101**, 860–6.

Wightman, A. S. (1989). The general theory of quantized fields in the 1950's. In Brown, Dresden and Hoddeson, pp. 608–29.

Wigner, E. P. (1946a). Resonance reactions and anomalous scattering. *Physical Review*, **70**, 15–33.

Wigner, E. P. (1946b). Resonance reactions. *Physical Review*, **70**, 606–18.

Wigner, E. P. (1951). On a class of analytic functions from the quantum theory of collisions. *Annals of Mathematics*, **53**, 36–67.

Wigner, E. P. (1955a). Lower limit for the energy derivative of the scattering phase shift. *Physical Review*, **98**, 145–7.

Wigner, E. P. (1955b). On the development of the compound nucleus model. *American Journal of Physics*, **23**, 371–80.

Wigner, E. P., and Eisenbud, L. (1947). Higher angular momenta and long range interaction in resonance reacions. *Physical Review*, **72**, 29–41.

Wigner, E. P., and von Neumann, J. (1954). Significance of Loewner's theorem in the quantum theory of collisions. *Annals of Mathematics*, **59**, 418–33.

Wildermuth, K. (1949a). Der analytische Zusammenhang zwischen den Streumatrix-Elementen und den diskreten stationären Zuständen in der Heisenbergschen S-Matrix-Theorie. *Zeitschrift für Physik*, **127**, 85–91.

Wildermuth, K. (1949b). Das analytische Verhalten der asymptotischen Wellenfunktion und die S-bzw. η-Matrix für mehrere Teilchen, Teil I. *Zeitschrift für Physik*, **127**, 92–121.

Wildermuth, K. (1949c). Das analytische Verhalten der asymptotischen Wellenfunktion und die zugehörige S-Matrix für mehrer Teilchen, Teil II. *Zeitschrift für Physik*, **127**, 122–152.

Wilson, P. A. (1989). The anthropic cosmological principle, unpublished Ph.D. dissertation, University of Notre Dame.

Woolf, H. (ed.) (1980). *Some Strangeness in the Proportion*. Reading, MA: Addison–Wesley.

Yang, C. N. (1980) Einstein and the physics of the future. In Woolf, pp. 501–2.

Yang, C. N., and Feldman, D. (1950). The S-matrix in the Heisenberg representation. *Physical Review*, **79**, 972–8.

Yang, C. N., and Mills, R. L. (1954). Conservation of isotopic spin and isotopic gauge invariance. *Physical Review*, **96**, 191–5.

Yourgrau, W., and Breck, A. D. (eds.) (1977). *Cosmology, History and Theology*. New York: Plenum Press.

Zachariasen, F.(1961). Self-consistent calculation of the mass and width of the $J=1$, $T=1\pi\pi$ Resonance. *Physical Review Letters*, **7**, 112–3 [Erratum, *PRL* **7** (1961), 268].

Zachariasen, F. (1963). The theory and application of Regge poles. In *1962 Cargèse Lectures in Theoretical Physics*, pp. X-1–X-42. New York: W. A. Benjamin.

Zachariasen, F. (1965). What, if anything, is the bootstrap? In Salam, pp. 823–55.

Zachariasen, F. (1966). Lectures on bootstraps. In Moravcsik, pp. 86–151.

Zachariasen, F. (1971). Theory and practice of complex Regge poles. *Acta Physica Austriaca, Supplement* **VIII**, 50–90.

Zachariasen, F., and Zemach, C. (1962). Pion resonances. *Physical Review*, **128**, 849–58.

Zahar, E. (1976). Why did Einstein's programme supersede Lorentz's? In Howson, pp. 211–75.

Zichichi, A. (ed.) (1970). *Subnuclear Phenomena*, vol. 1. New York: Academic Press.

Zweig, G. (1964). An SU$_3$ model for strong interaction symmetry and its breaking. II. CERN Report 8419/TH.142. Reprinted in Lichtenberg and Rosen (1980), pp. 22–101.

Zweig, G. (1980). Origins of the quark model. Invited talk, Baryon 1980 Conference, CALTECH preprint, CALT-68-805.

Glossary of technical terms (from physics and from philosophy)

We give here brief definitions of technical terms that appear in the text. These are intended to provide an easy reference for the reader, not to furnish exhaustive and highly precise definitions. Many of the informal comments made below are meant to relate these terms to the context (or 'spirit') of the discussion in the case study. They ought not be taken in all cases as *literally* accurate.

adiabatic invariant: a ratio of quantities (such as the energy E divided by the frequency v for a harmonic oscillator) that remains constant as external conditions on a mechanical system are varied in such a way that neither quantity itself remains constant. (A classic example is E/v for a simple pendulum as the length of the pendulum is slowly shortened.)

Aharonov–Bohm effect: a quantum-mechanical effect in which a particle (such as an electron) can be affected by a potential even when the corresponding physical field (such as a magnetic one) vanishes identically in the entire region accessible to the particle.

analytic continuation: the process by which an analytic function can be determined *everywhere* in its domain of analyticity by its values in any region *however small*. Loosely speaking, once we know an analytic function anywhere, we know it (in principle) everywhere.

analytic function: one that is defined and has a unique derivative everywhere in a region of the complex plane. Analytic functions are very 'smooth' and tightly constrained.

angular distribution: the variation in intensity of a scattered beam (of particles or of light) as a function of the angle (usually measured relative to the incident beam direction).

anomalous magnetic moment: departures of the actual magnetic moment of a particle (such as a proton) from the value predicted by the (single-particle) Dirac equation.

anomaly cancellation: arranging to have terms, which make a theory mathematically inconsistent (in the sense of $0 = \infty$), cancel each other and, hence, disappear from (the predictions of) the theory.

anthropic principle: a generic term, referring to any one of several conjectures, for attempts to use the existence of life in the universe as an *explanation* for the evolution of the universe to its present state.

anticommutator: defined as $\{A, B\} \equiv AB + BA$ for two operators A and B.

asymptotic behavior: the behavior (or functional form) of a function (such as a scattering amplitude) as an independent variable approaches infinity.

asymptotic freedom: a property of some gauge field theories for which, at asymptotically high energies and momentum transfers, the constituents (e.g., the quarks) of a hadron scatter as free particles would.

asymptotic series: a series which, although not convergent, can nevertheless be used (when truncated) to obtain accurate asymptotic values for a function.

autonomous S-matrix program: an attempt to set up a complete and consistent S-matrix theory independent of any connections with quantum field theory.

auxiliary hypotheses: in Lakatos' methodology of scientific research programs, those ('modifiable') parts of a theory (or program) that are more likely to be falsified and changed when anomalies (or difficulties) arise. These hypotheses are subordinate to the theory under test and are implicitly assumed by the proponents of the theory. When anomaly arises, falsification may *logically* be directed at either the theory or the auxiliary hypotheses. But, by methodological agreement, criticism will more likely be aimed at the latter, which are often referred to also as auxiliary assumptions in the text.

axiomatic quantum field theory: an attempt to construct a mathematically rigorous quantum field theory on the basis of a precise and well-defined set of axioms, thus avoiding the mathematically objectionable features of ordinary (perturbative) Lagrangian field theory.

baryon: a half-integral spin (in units of $\hbar = h/2\pi$) particle (such as a nucleon) that undergoes strong interactions.

baryon number (B): an integer assigned to the baryons, typically, $+1$ for a baryon (e.g., a proton, p) and -1 for an antibaryon (e.g., an antiproton, $\bar{p}$).

baryonium: 'exotic' states having quantum numbers that would be associated (in quark language) with a system $qq\bar{q}\bar{q}$ (i.e., two quarks and two antiquarks).

Bev: billion electron volts (10^9 ev), a unit of energy used in nuclear and elementary-particle physics. Modern accelerator energies (e.g., Fermilab, CERN) lie in the hundreds of Bev range.

bifurcation point: a point (or parameter value) at which a trajectory or solution splits into two branches or parts. In this monograph the term is used to denote a critical, unstable set of circumstances at which developments may go in either of two radically different directions.

bootstrap condition: a mathematical constraint that must be fulfilled if a solution is to be a bootstrap one. (This could, for example, be a requirement that certain values of input and output parameters be equal to each other, as in the $\pi\pi$ bootstrap of the ρ meson.)

Born amplitude: the amplitude generated by the exchange of a single (virtual) particle (as, for example, in Figure 1.3).

boson: a particle having integer spin (in units of $\hbar = h/2\pi$) (i.e., $0, 1, 2, \ldots$). Such particles obey Einstein–Bose statistics.

boundary conditions: certain values or asymptotic behavior that functions (such as scattering amplitudes) are required to take on.

bound state: a configuration of a system in which the system is localized in space at any time. For a quantum-mechanical system, the corresponding wave function vanishes rapidly at large distances.

branch point: a point at which two or more branches (or 'pieces') of a multivalued (analytic) function of a complex variable become (numerically) equal. A simple example is $z = 0$ for the function $f(z) = \sqrt{z}$ (or, more loosely, $f(z) = \pm\sqrt{z}$).

Breit–Wigner formula: a mathematical form for the scattering amplitude, peaked at a resonance energy.

causal interpretation of quantum mechanics: an interpretation of the Schrödinger equation, associated largely with David Bohm (but having its origins with Louis de Broglie), in which particles follow definite trajectories and in which a nonlocal quantum potential (produced by the wave function) is responsible for peculiarly quantum effects.

causal S-matrix: a scattering matrix associated with causal field operators (i.e., those satisfying the microcausality condition).

causality: a requirement on a quantum field theory that the field operators commute at spacelike separations.

channel: a possible reaction (or scattering) process, such as $\pi^- + p \rightarrow \pi^0 + n$ (or, more generally, $a + b \rightarrow c + d$).

charge conjugation: the (mathematical) operation of changing particles into antiparticles (roughly, changing the sign of the charge of a particle, as in $e^- \rightarrow e^+$, the replacement of an electron with a positron).

charge exchange reaction: a process in which two interacting particles exchange charge (subject, of course, to overall conservation of charge) as in $\pi^- p \rightarrow \pi^0 n$.

charge independence: an (approximate) invariance of nuclear interactions under the replacement of a neutron by a proton, and *vice versa* (other quantum numbers remaining unchanged).

charge renormalization: the modification of the 'bare' charge of a particle due to interactions with the rest of the universe (including the quantum-field-theory 'vacuum').

Chew-Low effective range formula: a form for the low-energy behavior of the pion–nucleon phase shift.

Chew-Low extrapolation procedure: a mathematical recipe for extrapolating production cross sections (such as that for $\pi N \rightarrow \pi\pi N$) to elastic cross sections (such as that for $\pi\pi \rightarrow \pi\pi$).

Chew model: a particular form of a model for the interaction between pions and (static) nucleons.

cluster decomposition: an expression of the physical requirement that observations sufficiently far apart in space or time should be independent. Formally this is imposed as a factorization condition on transition amplitudes into products of partial transition amplitudes ('clusters') in the limit that the subgroups of particles are infinitely displaced from each other.

commutator: defined as $[A, B] \equiv AB - BA$ for two operators A and B.

compactification: a process whereby a space (or certain dimensions) reduce (or 'curl up') so as to occupy a very small volume.

complementarity: a basic quantum-mechanical principle according to which certain mutually exclusive pairs of properties (e.g., wave–particle) apply to a system under different circumstances.

completeness: a mathematical property possessed by any set of functions (or basis states) that is large enough (i.e., 'complete') to allow any function (or state) to be expressed as a linear combination of them. A familiar example is to set $\sin nx$, $\cos nx$, $n = 0, 1, 2, \ldots$, which can be used to expand 'any' function $f(x)$ on the range $(0, 2\pi)$ (Fourier's theorem).

complex angular momentum: the mathematical extension of the angular momentum index l from its physical, integral values $(0, 1, 2, \ldots)$ to arbitrary complex values.

Compton scattering: the scattering of a photon (γ) by a free electron (e).

conservation of probability: the requirement that the probabilities for the transition from a *given* initial state to all possible final states must sum to unity.

constructive empiricism: the philosophical position that the only those components of a theory to be believed are those that can be directly observed.

contour integral: an integral taken around a closed curve (usually in the complex plane).

Copenhagen interpretation of quantum mechanics: the (standard) interpretation of quantum mechanics according to which the formalism of quantum mechanics is taken as providing a literally true and (in principle) complete description of physical phenomena.

correspondence principle: a limiting procedure in which quantum-mechanical quantities pass over into their classical analogues.

cosmological anthropic principle: see entry under 'anthropic principle'.

Coulomb field: the electric field generated by a charge.

Coulomb scattering: the scattering of two charged particles *via* their electromagnetic interaction.

Coulomb scattering amplitude: the quantum-mechanical scattering amplitude for the purely electromagnetic scattering of two charged particles (neglecting any other possible interactions, such as the strong one).

coupling constant: a number that is a measure of the strength with which one particle (such as an electron e) interacts with another (such as a photon γ) at a vertex (such as eeγ).

CPT theorem: a general quantum field theory result that requires a theory to be invariant under the combined operations of charge conjugation (C), parity inversion (P) and time reversal (T).

creation and annihilation operators: operators $a^\dagger$ and a, such that $[a, a^\dagger] = 1$, having the property (on certain state vectors ψ) that the state $a^\dagger\psi$ has one more quantum than does the state ψ, while $a\psi$ has one fewer.

cross section: a measure of the rate (or strength) of a given reaction. The differential cross section $d\sigma/d\Omega$ measures the reaction as a function of the scattering angle, while the total cross section σ is integrated over all angles (and depends upon the energy only).

crossed channel: a reaction obtained by 'crossing' (or interchanging) two particles in a reaction. Thus, $a + c \rightarrow b + d$ is a crossed channel of the (direct) reaction $a + b \rightarrow c + d$. (We have neglected niceties about particles and antiparticles here.)

crossing: the operation of interchanging (or 'crossing') the particles in a reaction (or the 'legs' in the corresponding diagram).

current commutator: the commutation relation satisfied by the current operators (such as that for charged particles) in a quantum field theory.

cut: a line joining two branch points.

Cutkosky rules: a mathematical formula for calculating the contribution to the Mandelstam double spectral function in terms of Feynman diagrams.

cutoff: a mathematical procedure for replacing an infinite upper limit in a divergent integral to produce a finite result.

de Broglie wavelength: the wavelength λ given by the de Broglie relation $\lambda = h/p$, where p is the momentum of the particle. This assigns a wavelength λ to any system having a momentum p.

degenerating research program: in Lakatos' methodology of scientific research programs, a program that has lost its ability to suggest new, fruitful avenues for research and that must make repeated *ad hoc* adjustments to keep up with events.

Delbrück scattering: the scattering of light (photons, γ) by the Coulomb field of a (heavy) nucleus.

δ_{33} *resonance*: the resonance in the pion nucleon phase shift (around 200 MeV for the kinetic energy of the pion in the laboratory frame) in the J (angular momentum) $= \frac{3}{2}$, T (isospin) $= \frac{3}{2}$ state. This shows up as a peak (or 'bump') in the cross section.

demarcation problem: the question of how one distinguishes, by criteria internal to the discipline, science from other rational knowledge-generating enterprises.

diffraction peak: a sharp peak in the differential cross section in the forward direction characteristic of (but not exclusively restricted to) elastic scattering processes.

(Dirac) δ function: a 'function' (actually, a distribution or generalized function in the mathematical sense) such that $\delta(x) \equiv 0$, $x \neq 0$, $\int_{-\infty}^{\infty} \delta(x)\,dx = 1$.

Dirac hole theory: an interpretation of the negative-energy solutions of the relativistic (single-particle) Dirac equation in which the 'vacuum' consists of all of these negative energy states being filled with electrons. When an electron is removed from this 'sea', a 'hole' is left and this hole behaves as a positively charged, physical particle (a positron).

direct channel: the physical reaction, say, $a + b \rightarrow c + d$, for which the invariant $(p_a + p_b)^2$ plays the role of the energy variable.

dispersion relation: a relation between the real and imaginary parts of an analytic function of a complex variable. Typically, Re $f(z)$ is expressed in terms of an integral of Im $f(z)$ taken along the real axis.

dispersion theory: the program of employing dispersion relations to correlate experimentally measured quantities.

divergences: the infinite terms that arise in quantum field theory (perturbation) calculations.

double dispersion relation: a dispersion relation in both relativistically invariant, independent kinematic variables (usually taken to be the 'energy' s and 'momentum transfer' t).

duality: the property that a scattering amplitude can be considered either as a sum of its direct-channel resonances or as a sum of its cross-channel Regge exchanges.

Duhem–Quine thesis: the observation that any theory in science is always underdetermined by the (finite) empirical base that supports it. That is, in principle, many theories can exist to explain any given set of data or observations.

effective range formula: a low-energy expansion (or approximation) for the phase shift $\delta(k)$:

$$k \cot \delta(k) = -\frac{1}{a_0} + \frac{1}{2} r_{\text{eff}} k^2$$

Here a_0 is termed the scattering length and r_{eff} the effective range.

eightfold way: the strong-interaction symmetry scheme proposed by Murray Gell-Mann based on the mathematical group SU(3), which has eight generators. Also, octets of particles play a central role in that theory.

elastic scattering: a scattering process in which the same (type of) particles appear in the initial and final states (as $a+b \rightarrow a+b$). Such processes are distinguished from inelastic ones, $a+b \rightarrow c+d$ (where c and d are different from a and b).

elastic unitarity: the unitarity relation that holds above the two-body threshold, but below the threshold for the production of more (or of other) particles.

electromagnetic interaction: the interaction of charged particles due to their electric charge only.

electroweak interactions: the generic term applied to the interactions of particles *via* the electromagnetic or weak forces. The Salam–Weinberg 'standard' model combines both forces into a single, fundamental interaction (much as Maxwell's electromagnetic theory unified electric and magnetic interactions).

empiricism: a tenet that all (scientific) knowledge must be based on observation (or 'data') alone. This is the position of logical positivism.

equivalence principle of strong interactions: the postulate that *all* strongly interacting particles are associated with Regge poles. Chew and Frautschi proposed this as a concrete formulation of Feynman's criterion that a correct theory should not allow a decision as to which particles are elementary.

essential singularity: any singularity of a *single-valued* (analytic) function other than a pole (which is a singularity of the form $(z-z_0)^{-k}$, where k is a positive integer).

Euler–Lagrange equations: the equations that must be satisfied by a set of functions $\phi^\alpha(x)$ that will render an integral $I = \int \mathscr{L}(\phi^\alpha, \phi^\alpha{}_{,\nu}) \, dx$ an extrenum. Those equations are $\dfrac{\partial \mathscr{L}}{\partial \phi^\alpha} - \dfrac{\partial}{\partial x_\nu} \left(\dfrac{\partial \mathscr{L}}{\partial \phi^\alpha{}_{,\nu}} \right) = 0$. Here $\mathscr{L}$ is the Lagrangian density and ϕ^α a set of fields. These variational formulations apply to particle mechanics and to field theories.

exchange degeneracy: the fact that Regge trajectories of opposite signature coincide.

exotic mesons: meson states having quantum numbers characteristic of multiquark states (such as $qq\bar{q}\bar{q}$), as opposed to the usual quark–antiquark ($q\bar{q}$) states.

factorization: the property that the residue of the scattering amplitude at a pole (in, say, the energy variable) reduces to a product of factors, one of which is labeled only by the quantum numbers of the entrance channel and the other only by those of the exit channel.

Fermi theory of β decay: the original quantum field theory of the neutron decay process $n \rightarrow p + e^- + \nu$. In this theory the basic interaction is the four-point vertex with all four particles (n, p, e, v) meeting at a single point. It is not renormalizable.

fermion: a particle having half-odd integer spin (in units of $\hbar = h/2$) (i.e., $\frac{1}{2}, \frac{3}{2}, \frac{5}{2}$, ...). Such particles obey Fermi–Dirac statistics.

Feynman diagram: a pictoral representation (in terms of the exchange of virtual particles) of the terms in the perturbation expansion of quantum field theory.

fine structure constant: the parameter $\alpha = e^2/\hbar c (\cong 1/137)$. This is the coupling constant of the eeγ vertex and serves as the expansion parameter in the perturbation series of quantum electrodynamics.

finite-energy sum rule (FESR): a sum rule at finite (but variable) energy that relates Regge trajectory parameters to an integral over total cross sections.

fixed-source theory: quantum field theory in which the nucleon is treated in static approximation (i.e., it does not move when mesons are scattered from it).

form factor: the function that gives the behavior of a vertex function as a function of the momentum transfer variable (as, for instance, for $p\bar{p} \rightarrow \gamma$). These factors can be related (in a nonrelativistic regime) to the Fourier transform of the charge (or current) distribution of the particle (e.g., a proton).

Forman thesis: a claim about the influence of the post WWI cultural milieu in Germany upon the rejection of causality and the formulation and rapid acceptance of quantum theory. In this book the term refers to the general influence of social and sociological factors upon the construction and acceptance of scientific theories. This broader thesis is actually more central to the claims made by writers like Bloor and other members of the 'Edinburgh School' of the sociology of knowledge.

foundationist: a term referring to an approach to the philosophy of science based on *a priori*, 'obvious' or necessary principles of rationality. That is how this term is employed in the text. In the philosophy of science the term is often applied to a position that science itself rests on statements (either axioms or observation statements) that are themselves not open to challenge. In this latter school would be Aristotle, Descartes and the logical positivists. (See similar comments made under the term 'rationalist'.)

Fourier decomposition: the expansion of a function (or of a field operator) in terms of the (Fourier) basis functions e^{ikx}.

fractal: a 'dimension' assigned to a set of points when the dimension is not an integer (e.g., $n=1$, line; $n=2$, surface; $n=3$, volume). Loosely speaking, a badly broken curve may have a dimension between one and two. In technical terms, a fractal is a set for which the Hausdorff Besicovitch dimension exceeds the topological dimension (Mandelbrot, 1983, p. 15).

Fredholm determinant: a determinant (typically infinite dimensional) whose roots are the eigenvalues of the corresponding linear Fredholm integral equation.

Froissart bound: a theoretical upper bound (based on the Mandelstam representation and unitarity) on the asymptotic growth of a scattering amplitude.

fundamental interactions: the four basic interactions: strong, electromagnetic, weak and gravitational. The relative strengths of these interactions can be characterized with the following scale:

strong	1
electromagnetic	10^{-2}
weak	10^{-14}
gravitational	10^{-38}

gamma function $[\Gamma(z)]$: the generalization of the factorial, $n! = 1 \cdot 2 \cdot 3 \cdots (n-1)n$, to continuous, complex values of z. $\Gamma(z = n + 1) = n!$

gauge field theory: a quantum field theory that is invariant under a general set of gauge transformations (e.g., $\phi(x) \to e^{i\omega(x)}\phi(x)$).

gauge invariance: the invariance of a theory under (local or global) gauge transformations.

Gev: giga eleltron volts (10^9 ev). See entry under Bev.

ghost states: (spurious) states having negative norm (or probability) or, in Regge theory, negative (mass)^2.

global gauge principle: the invariance of a theory under global gauge transformations (e.g., $\phi(x) \to e^{i\alpha}\phi(x)$), where the gauge parameter α is a *constant*. That is, the parameter has a ('global') value independent of the space-time variables.

Goldstone bosons: the massless, spinless particles (or excitations) associated with spontaneous symmetry breaking in a theory.

G parity: a quantum number associated with a particular combination of charge conjugation (C) and isospin rotation ($G = C\,e^{i\pi T_2}$).

grand unified theory (GUT): a gauge theory that unifies (or encompasses) the strong, electromagnetic and weak interactions (in a manner similar to electroweak unification).

Green's function: the solution to a field equation for a point source (i.e., typically for a Dirac delta function as the source term). The use of Green's functions is that they allow one to build up (by direct superposition) the solution for an arbitrary source term. Thus, for Laplace's equation, $\nabla^2\phi = \rho$, the Green's function would be the solution to $\nabla^2 G = \delta(\mathbf{x} - \mathbf{x}')$.

Gribov singularity: a complicated type of singularity that can be present in the complex angular momentum plane.

group: a mathematical structure, having a set of elements and an operation, and satisfying a specific set of properties with respect to that operation. A simple but important example of this is the rotation group in ordinary three-dimensional space.

hadron: any subatomic particle that takes part in the strong interactions.

Hamiltonian formalism: a formulation (of classical particle mechanics or (quantum) field theory) in which the equations of motion are given in terms of a Hamiltonian, $H = T + V$, where T is a kinetic energy and V is the potential energy. The dynamical equations for the canonical variables q ('position') and p ('momentum') are $\dot{q} = \partial H/\partial p$, $\dot{p} = -\partial H/\partial q$.

hard core: in Lakatos' methodology of scientific research programs, those assumptions (or hypotheses) of the research program that are kept unfalsifiable (or unquestioned) by methodological decision. Other components of a theory (or research program) are modified as necessary to make required adjustments with observations. Thus, for classical mechanics, Newton's equation of motion would be the hard core.

hard-scattering events: processes in which there is a large momentum transfer transverse to the beam direction.

harmony of random choices: the Leibnizian doctrine that certain conceptually independent events (such as a decision to kick a stone and the movement of my foot) are correlated through a pre-established harmony (rather than by any type of direct cause and effect).

Heisenberg field operators: quantum field operators that satisfy the equations of motion in the form.

$$i\frac{\partial \psi}{\partial t} = [\psi, H].$$

helicity: the projection of the angular momentum $\mathbf{J}$ of a particle along the direction of its momentum $\mathbf{p}$. This is basically the projection of the (intrinsic) spin of the particle along the direction of $\mathbf{p}$.

hermitian analyticity: a requirement about the type of analytic continuations that can be made beyond the physical region. As an example, for a scattering amplitude with a simple, normal threshold cut structure, this property would imply that, below the lowest threshold, the amplitude can be chosen to have the property $A(s + i\varepsilon) = A(s - i\varepsilon)$.

heuristic: in Lakatos' methodology of scientific research programs, a set of suggestions (or general guidelines) for modifying the auxiliary assumptions (or hypotheses) of a theory (or program). For example, in the old quantum theory, the correspondence principle would be the heuristic. Lakatos uses the term 'positive heuristic', rather than simply 'heuristic' as we do in the text.

Higgs boson: the quantum of the field responsible for spontaneous symmetry breaking in the standard electroweak gauge theory.

hypothetico-deductive method: the method of hypothesis in which the test of an assumption lies in its successful predictions.

implicate order: David Bohm's (1980) term for an enfolded, deeper-level order or structure not contained in the regular, explicit order of a system. Such implicate order is a hallmark of the undivided wholeness of, say, quantum phenomena in which certain events cannot be analyzed into distinct and well-defined parts.

impulse approximation: a theoretical approximation for the scattering of a projectile from a compound target in which the constituents of the target are treated as essentially free.

in and out fields: the asymptotic form an interacting field takes long before ('in', $t \to -\infty$) or long after ('out', $t \to +\infty$) the interaction acts. These are essentially 'free' fields before or after the interaction.

instrumentalism: the philosophical view that the only function of a scientific theory is to provide successful predictions. This is typically opposed to scientific realism which also seeks true explanations and 'pictures' of the world from a successful physical theory.

intermediate state: a (virtual) state or process *via* which a reaction may proceed (such as, for example, $e^+e^- \to \gamma \to e^+e^-$ for positron–electron scattering).

internal history: a historical account of an episode in the history of science (e.g., in physics) in which the intellectual developments are explained and judged solely on the basis of activities and criteria that characterize (or are internal to) the discipline itself.

internal symmetry: a symmetry or invariance operation involving only internal (or intrinsic) properties of a system (such as charge or baryon number), but not space-time properties.

invariance group: a group of transformations that leaves a theory invariant under the operation of its elements.

isotopic spin (isospin): a mathematical spin (similar to angular momentum) assigned to a particle and linked to its charge. The simplest (and historically first) example is that of the nucleon (N) which has an isospin of $\frac{1}{2}$, the $+\frac{1}{2}$ state being a proton (p) and the $-\frac{1}{2}$ one a neutron (n), so that Q (charge) $= \frac{1}{2} + T_3$ (z-component of isospin).

Jost function: the incoming plane-wave solution to the Schrödinger equation, evaluated at the origin. That is, if $f(k,r) \xrightarrow[r \to \infty]{} e^{-ikr}$, then the Jost function, $f(k)$, is defined as $f(k) = f(k, r=0)$. The S matrix is given in terms of it as $S(k) = f(k)/f(-k)$.

Kaluza–Klein theory: field theories formulated in space-time geometries of more than four dimensions with the aim of making the additional dimensions compactify (or essentially 'disappear' from the theory).

Kaluza (1921) attempted a five-dimensional theory, as an extension of general relativity, in which gravity and electromagnetism would determine the structure of space.

Klein–Nishina formula: the relativistic form of the differential cross section for Compton (i.e., photon–electron) scattering.

Kramers–Kronig dispersion relation: a dispersion relation derived in (classical) electromagnetic theory, relating the real and the imaginary parts of the dielectric constant (which is directly related to the index of refraction for light).

Lagrangian: a function (typically the difference between the kinetic and potential energies, $L = T - V$) that satisfies Hamilton's variational principle to yield the (Euler–Lagrange) equations of motion, $\dfrac{\mathrm{d}}{\mathrm{d}t}\left(\dfrac{\partial L}{\partial \dot{q}_j}\right) - \dfrac{\partial L}{\partial q_j} = 0$. These are essentially equivalent to Newton's second law, $\mathbf{F} = m\mathbf{a}$. This formulation can also be extended to fields.

Lagrangian formalism: the formulation of dynamical equations (for particles or for fields) in terms of a variational principle.

Lamb shift: the splitting of the $2s_{\frac{1}{2}}$ and $2p_{\frac{1}{2}}$ levels in the hydrogen spectrum. Single-particle Dirac theory had predicted that these levels should be degenerate. A correct account of this experimental observation was an early accomplishment of (renormalized) quantum electrodynamics.

Landau rules: a set of rules for locating the singularities of a Feynman diagram occurring (only) when the values of the external variables (e.g., the invariants s, t and u discussed in the text) allow all internal momenta to be simultaneously on their own mass shells (i.e., $k_j^2 = m_j^2$).

Landau surfaces: those (in general, higher dimensional) surfaces containing the Landau singularities.

Lehmann–Symanzik–Zimmermann (LSZ) formalism: a (fairly abstract) axiomatic formulation of quantum field theory, based on the existence of a Hilbert space for the field operators, certain transformation properties of the field operators, and locality, in addition to the imposition of asymptotic conditions on the fields.

leptons: light fermions (e.g., the neutrino, ν, the electron, e, and the muon, μ), as well as their antiparticles.

local field: a field $\psi(x)$ satisfying the locality condition, $[\psi(x), \psi(y)] = 0$, whenever the space-time points x and y are spacelike separated (i.e., when $(x - y)^2 \equiv (x_0 - y_0)^2 - (\mathbf{x} - \mathbf{y})^2 < 0$).

local gauge principle: the invariance of a theory under local gauge transformations (e.g., $\phi(x) \to e^{i\omega(x)}\phi(x)$), where the gauge parameter $\omega(x)$ is a function of the space-time variable x.

logic of induction: a formal theory of the degree of confirmation of an hypothesis h based on a definition of the (quantitative) probability of h relative to evidence e. This program is often associated with Rudolf Carnap.

logical positivism: a philosophical position which holds that all meaningful statements are either analytic (roughly, necessarily true) or else confirmable by observation (and experiment) and which holds metaphysical theories to be meaningless.

Lorentz invariance: the invariance of a theory under the group of Lorentz transformations (i.e., the invariance requirement of special relativity which imposes the formal equivalence of all inertial frames).

Low equation: a type of extended unitarity equation for the matrix elements of the transition operator.

macroscopic causality: a form of causality based on the assumed short-range nature (i.e., exponential fall-off with distance) of the interactions among (strongly-interacting) particles. It is a relaxation of the microscopic causality requirement which forbade *any* influence (even over microscopic distances) at spacelike separations.

magnetic moment: an intrinsic property of a particle that describes the interaction of the particle with a magnetic field (similar to the way the electric charge of a particle accounts for the basic interaction of a charged particle with an electric field).

Mandelstam representation: a double dispersion relation, in both independent variables (say, s and t) for a relativistic scattering amplitude.

mass renormalization: the modification of the 'bare' mass of a particle due to interactions with the rest of the universe (including the quantum-field-theory vacuum).

mass shell: the surface (or 'shell') in (four-dimensional) momentum space defined by the constraint $p^2 \equiv p_0^2 - \mathbf{p}^2 = m^2$, where m is the mass of the particle.

matrix mechanics: a formulation (due to Heisenberg) of the dynamical equations of quantum mechanics in terms of (infinite-dimensional) matrices (rather than in terms of functions and differential operators as in Schrödinger's version).

maximal analyticity: the maximum analyticity (or 'smoothness') for a scattering amplitude allowed by the constraint of unitarity.

Maxwell's equations: the dynamical equations for the (classical) electromagnetic field.

mesons: the strongly interacting bosons. (Notice that the muon, μ, sometimes in older literature referred to as a mu meson, is now classified as a lepton).

meta-level: in the philosophy of science, a term used in descriptions of scientific methodology to refer to the ('higher') levels of the methods (or criteria) for judging theories and of the goals of science. These meta-levels are usually contrasted with the ('lower') level of (day-to-day) scientific practice. The criteria for the evaluation of theories are said to be at a 'meta' (or higher) level, logically speaking, than the theories themselves.

Mev: million electron volts (10^6 ev), a unit of energy used in nuclear and elementary-particle physics. For example, the rest-mass (energy) of a proton is 938 Mev.

microscopic causality: the form of causality that forbids any influences between spacelike separated events, even over microscopic distances.

Møller scattering: electron–electron scattering (including positron–positron and electron–positron scattering).

momentum transfer: the momentum transferred from one particle to another during a scattering process. In the center of momentum frame, this would be given as ($\mathbf{k}_f - \mathbf{k}_i$), where $\mathbf{k}_f$ and $\mathbf{k}_i$ are, respectively, the final and initial three-momenta of the particle.

monopole: an isolated unit of electric or magnetic 'charge'. Typically, the term is used to refer to the (hypothetical) magnetic monopole.

multiperipheral model: a model of the high-energy collision between two incident particles (to produce many outgoing particles) in which the process is pictured as taking place through a succession of one-particle, low momentum-transfer exchanges.

naturalized philosophy of science: a program of using the methods and norms of science itself to do the philosophy of science. This usually implies a heavy emphasis on historical case studies.

N/D equations: a set of linear, coupled integral equations for a scattering amplitude A (which is decomposed as $A = N/D$).

nearby singularities: those singularities (usually poles) closest to the physical region for a particular scattering process.

Noether's theorem: a formula that gives the conserved quantity (such as energy, momentum, angular momentum, charge, etc.) associated with a transformation that leaves a theory (actually, the Lagrangian) invariant.

non-abelian gauge theory: a gauge theory based on a group whose elements do not commute with one another.

nonlinear spinor theory: Heisenberg's theory of fundamental interactions based on a nonlinear dynamical equation.

N-point amplitude: a scattering amplitude having N external particles (or 'legs').

nuclear democracy: the conjecture that no (strongly interacting) particle is more elementary than any other. This effectively means that there are to be *no* elementary particles since all particles are generated through the exchange of other particles.

nucleon (N): a term used to refer to a 'particle' that can exist in either of two 'states': the proton, p, or the neutron, n.

old quantum theory: the set of semiempirical rules used to treat atomic systems from Planck's discovery until the formulation of quantum mechanics.

optical theorem: a formula that gives the total scattering cross section in terms of the imaginary part of the forward (i.e., $\theta = 0$) scattering amplitude.

order: a topological concept introduced into S-matrix theory. The *order* referred to is the order (or sequence) of the particle types (or legs) on a diagram representing a scattering process.

OZI rule: a semiphenomenological rule (named after Okubo, Zweig and Iizuka) forbidding (or greatly suppressing) certain types of particle decay modes.

parity: the formal operation of space inversion (i.e., $\mathbf{r} \to -\mathbf{r}$) in a theory. It is sometimes referred to as mirror inversion (or reflection).

partial wave: a state of definite (orbital) angular momentum. This term usually refers to the decomposition of a scattering amplitude into angular-momentum eigenstates ($l = 0, 1, 2, \ldots$).

particle exchange: the (virtual) exchange of an intermediate particle between two other ('external' or real) particles in a scattering process.

perturbation theory: the expansion, in quantum-field-theory calculations, of a scattering amplitude in powers of a coupling constant (such as the fine-structure constant α for quantum electrodynamics).

phase shift: the shift (or change) in the phase of the asymptotic form of a scattering wave function compared to what it would be in the absence of any interaction. The phase shift δ_l is a measure (in the lth partial wave) of the effectiveness of the interaction potential $V(r)$ in producing scattering effects.

phase-shift analysis: an analysis (or reduction) of scattering data (usually a cross section) in terms of the phase shifts in various angular momentum states.

phenomenology: the application of theories and of theoretical models directly to the analysis of experimental data. Also, this includes rules and regularities abstracted from the data.

philosophical relativism: the position that there are no absolute or invariant epistemological norms or criteria for judging theories (or knowledge claims in general).

photomeson production: a scattering process (such as $\gamma + p \to \pi^+ + n$) in which a high-energy photon is incident upon a target, producing an outgoing meson.

physical region: the region (of the independent variables) accessible to physical processes. For example, for a two-particle scattering amplitude $A(s, t)$, the physical region for direct-channel scattering would be $s \geqslant 4m^2$, $t \leqslant 0$, where these variables are (essentially) the center-of-momentum (CM) energy and (CM) momentum transfer, respectively.

physical-region analyticity: analyticity of a scattering amplitude in the physical region.

pion–nucleon (πN) coupling constant: a number that is a measure of the strength with which a nucleon (N) interacts with a pion (π) at the vertex πNN.

planar diagram: a diagram (for a reaction) which, when drawn on a plane, has no lines that cross.

Poisson brackets: a operation defined for functions of the canonical variables q_j and p_j of classical mechanics. The equations of motion of classical mechanics can be formulated in terms of Poisson brackets. Dirac showed how to pass directly from these Poisson brackets to the commutator formulation of quantum mechanics.

polarization: in a scattering process between elementary particles, a measure of the flip (or change in direction of the quantization) of the spin of a particle.

pole: a specific type of singularity of an analytic function of a complex variable. A function $f(z)$ has a pole of order $k (k = 1, 2, 3, \ldots)$ at $z = z_0$ if $f(z)$ behaves as $R/(z - z_0)^k$ in the neighborhood of z_0. Here R is the *residue* of $f(z)$ at the pole.

pole factorization: see the entry under 'factorization'.

pole–particle corrspondence: the rule that a scattering amplitude has a (simple; i.e., first-order, or $k = 1$) pole in the external variables (such as s or t) when that variable equals the square of the mass of a particle that can be exchanged in an intermediate state.

Pomeranchuk theorems: restrictions on the asymptotic structure of cross sections based on an assumed asymptotic singularity behaviors for scattering amplitudes. For example, these theorems show that certain cross sections are required to approach common constant values at high energy.

positive definite: the requirement that a term or form be strictly positive (not zero or negative).

potential theory: scattering theory governed by the nonrelativistic Schrödinger equation.

principal value integral: a limiting procedure to assign a value to an integral when the integrand becomes undefined on the path of integration. The (Cauchy) principal value is defined, for an $f(x)$ that is singular at x_0, as the symmetric limit

$$\mathrm{P}\!\!\int_a^b f(x)\,\mathrm{d}x = \lim_{\delta \to 0}\left\{\int_a^{x_0 - \delta} f(x)\,\mathrm{d}x + \int_{x_0 + \delta}^b f(x)\,\mathrm{d}x\right\},$$

if the limit exists.

principle of maximal analyticity: see entry under 'maximal analyticity'.

principle of maximum strength: the (S-matrix) assumption that the strong interactions are as strong as is allowed by unitarity. This becomes a statement about the behavior of certain Regge trajectories.

production threshold: the minimum value of the energy for which an additional particle can be produced.

propagator: the mathematical expression (in a quantum field theory) representing the exchange (or propagation) of an intermediate-state (virtual) particle in a scattering process between two other particles.

pseudoscalar particle: a particle whose state vector (or field) is an eigenstate of the parity operator P with eigenvalue -1.

quantum chromodynamics (QCD): a gauge quantum field theory of strongly interacting particles.

quantum electrodynamics (QED): the quantum field theory of the electromagnetic interactions.

quantum field theory (QFT): a dynamical theory based on field operators that satisfy commutation relations (i.e., 'quantization' conditions) and equations of motion.

quarks: hypothetical (in principle unobservable) elementary particles having fractional charge (in terms of the quantum of charge e).

radiation damping: the loss of energy, through radiation (of photons or of other quanta of energy, such as mesons), by an interacting particle.

radiative corrections: effects resulting from higher-order calculations in quantum field theory.

rationalist: in the philosophy of science, a position that the methodological criteria of science must be justified solely on the basis of *a priori* or of necessary first principles. That is the use of the term in this text. Traditionally, the term has also been applied to the position that the principles of the science itself must be justified on the grounds of reason alone. (See similar comments made under the term 'foundationist'.)

Ravens' Paradox: a logical paradox in confirmation theory, based on the intuitive assumption that the observation of an A that is a B tends to confirm the hypothesis that *all* A's are B's (cf. Hempel, 1945). The paradox arises from the logical equivalence of $A \Rightarrow B$ and $\sim B \Rightarrow \sim A$, plus the intuitive feeling that observation of a $\sim$ B that is a $\sim$ A does not tend to confirm $A \Rightarrow B$. For example, does the observation of a green frog confirm the claim that all ravens are black?

Rayleigh scattering: the incoherent scattering of light by randomly distributed scatterers. Rayleigh scattering is responsible for the blue color of the day-time sky.

Regge cuts: (moving) branch points of the scattering amplitude $A(l, k)$ in the complex angular momentum plane.

Regge poles: (moving) poles of the scattering amplitude $A(l, k)$ in the complex angular momentum plane.

Regge trajectory: the curve (in the complex angular momentum plane) described by the moving Regge pole $l = \alpha(k)$.

renormalization: the formal mathematical procedure for assigning finite, physically observable quantities to divergent terms in the perturbation expansion of a quantum field theory.

research program: in Lakatos' methodology of scientific research programs, a sequence of theories, sharing a common hard core and (positive) heuristic, and generated by progressively modifying auxiliary assumptions (or hypotheses).

resonance: in scattering data, a peak (or 'bump') in the cross section at a particular energy.

rho (ρ) *meson*: a resonance consisting of the bound state of two pions.

Riemann–Lebesgue lemma: a mathematical theorem that states that an integral of the form $\displaystyle\int_a^b e^{i\lambda x} f(x)\,dx$ vanishes as $\lambda \to \infty$ for 'any' function $f(x)$.

Salam–Weinberg model: a gauge quantum field theory of the electroweak interactions.

scalar particle: a particle whose state vector (or field) is an eigenstate of the parity operation P with eigenvalue $+1$.

scattering amplitude: the (complex) function whose absolute value squared yields the probability for a given reaction to occur.

scattering cross section: see the entry under 'cross section'.

scattering length: see the entry under 'effective range formula'.

Schrödinger theory: (nonrelativistic) wave mechanics.

scientific realism: the position that successful scientific theories give us an (essentially) accurate picture of the physical universe. It is the view that the long-term success of a scientific theory gives us good reason to believe that the entities postulated by the theory actually exist.

self-energy correction: quantum-field-theory corrections arising from the emission and reabsorption of a virtual particle by another particle.

separability: the assumption that spatially separated systems always have independently definable properties and existence and that these properties exhaust the description of any system made up of these subsystems. Quantum mechanics is an example of a theory that does *not* respect separability.

shadow matter: a conjectured type of matter in the universe that may be 'transparent' to all of the interactions except gravity. This possibility is fairly naturally allowed by some modern theories of cosmology. It could provide the additional mass needed (but not yet observed) for consistency between the amount of matter actually observed in the universe and that expected on the basis of the observed motion of certain stars near galaxies in the universe.

shrinkage of the diffraction peak: the narrowing (or sharpening) of the forward (diffraction) peak observed in (for example) elastic scattering processes.

signature: the plus or minus sign (positive or negative signature) in the factor $[1 \pm e^{-i\pi\alpha(s)}]$ arising in the analytic continuation of (Regge) partial-wave scattering amplitudes. This factor is necessary to produce amplitudes with suitable mathematical properties. Amplitudes of definite signature have physical recurrences only at alternate integers.

S-matrix theory (SMT): a theory based only on 'directly observable' quantities, such as the scattering matrix, the absolute squares of whose elements give (observable) transition probabilities.

sociological: relating to the study of the development, structure, interaction and collective behavior of organized groups (of scientists, for us). In the philosophy of science, this refers to the influence of such factors upon the form and content of scientific knwledge produced (in, say, physics).

soft-scattering events: those processes in which there is small transverse-momentum transfer perpendicular to the beam direction.

solitons: solitary (or single, highly localizable) waves that retain their shape and speed after interacting (or 'colliding') with each other.

spacelike separation: two space-time points x and y so situated relative to each other that $(x-y)^2 \equiv (x_0 - y_0)^2 - |\mathbf{x} - \mathbf{y}|^2 < 0$. The basic idea is that a light signal could not propagate from the 'first' event to the location of the 'second' before the second one occurred.

spin-statistics theorem: a proof that it is not possible (consistently) to quantize fermions (spin '$\frac{1}{2}$' particles) with commutation relations (Bose–Einstein statistics) nor bosons (integral-spin particles) with anticommutation relations (Fermi–Dirac statistics).

spontaneous symmetry breaking: a situation in which the solutions to a set of dynamical equations possesses a lower degree of symmetry than do the equations themselves. The spontaneous magnetization of a ferromagnetic material (along an arbitrary but specific direction) upon being cooled is a commonly cited example of this.

static model (fixed-source theory): a model of (say, pion–nucleon) scattering in which the recoil of the target is neglected.

strangeness: a quantum number (similar to electric charge or baryon number) that is conserved in certain reactions.

strong interaction: a very short-range interaction (effective only over subatomic or nuclear dimensions) acting, for example, between mesons and nucleons.

strong program: a radical position in the sociology of knowledge according to which not only the form but even the very content of scientific knowledge is to be explained ultimately in sociological terms.

SU(2) symmetry: the symmetry group employed in the isotopic spin formalism. In its mathematical structure it is similar to the ordinary rotation group in three-dimensional space.

SU(3) symmetry: a mathematical extension of the SU(2) group. It has proven useful in classifying the strongly interacting particles.

substitution law: a rule discovered in perturbation theory in quantum electrodynamics. It relates scattering processes in which one particle is interchanged with (or 'substituted' for) another. It is closely related to crossing symmetry.

subtractions: a procedure for producing a convergent dispersion relation for a scattering amplitude that does not vanish sufficiently rapidly at infinity. This is different from the 'subtraction procedure' of quantum field theory.

subtraction procedure: a set of (*ad hoc*) rules for throwing away (or 'subtracting') infinities that arose in early quantum field theory calculations. This was a precursor to the renormalization program.

sum rule: a relation, typically, between an integral over a scattering cross section and a parameter (or parameters) characteristic of that reaction (such as a coupling constant, Regge trajectory intercept, etc.). Sum rules provide useful checks since they involve directly and independently measurable experimental quantities.

superconvergence relations: integrals over certain single spectral functions that must vanish for all values of an external variable, as, for example, $\int_{s_0}^{\infty} \mathrm{Im}\, A(s, t)\, \mathrm{d}s = 0$. Such relations can be obtained by demanding the absence of certain singularities in the complex angular momentum plane.

superdeterminism: an absolute and universal form of determinism in which even the 'choices' made by an experimenter are predetermined. Such superdeterminism would block a derivation of the Bell inequalities since 'free' (or independent) choices of instrument settings are essential there.

superselection rule: any statement (such as charge conservation) that singles out certain state vectors as not physically realizable.

superstrings: a quantum field theory of strings that includes supersymmetry.

supersymmetry: an assumed fundamental symmetry or invariance of a theory under the interchange of fermions and bosons. Such a symmetry must be badly broken in the real world since the observed interactions do not possess this 'blindness' to fermions versus bosons.

tachyon: a (hypothetical) particle that moves at speeds exceeding that of light. Tachyons could never move at subluminal velocities.

Thomson scattering: the scattering of electromagnetic radiation by a free charge.

time reversal: the formal mathematical operation of replacing the time variable t by $-t$. One can also speak of a time reversed reaction that proceeds in the reverse temporal order to another one (e.g., $a + b \rightarrow c + d$ versus $c + d \rightarrow a + b$).

topology: the mathematical study of those properties of geometrical configurations or relations that are independent of the deformation (or 'stretching') of a figure, surface or space.

unified field theory: a field theory that incorporates all of the known types of interactions (strong, electroweak and gravitational) into a single formalism governed by one general principle (such as guage invariance).

unitarity: a mathematical constraint imposed on the scattering operator by the requirement that the norm of a state vector be preserved during the time evolution of the system. This is equivalent to the statement that, given some fixed initial state, the probabilities of the transitions to *all possible* final states must sum to unity.

unphysical region: that region (or those regions) of the complex plane (in the variables of the scattering amplitude) that lies outside the physical region. Such regions are not directly accessible to experiment.

unphysical unitarity relations: the unitarity relation continued beyond the physical region. The validity of such a relation depends upon certain (assumed) analytic properties of the scattering amplitudes.

vacuum polarization: photon self-energy corrections arising from the photon's creating virtual electron–positron pairs (which subsequently annihilate again). An external electromagnetic field will modify the distribution of the virtual charge pair (i.e., 'polarize' the vacuum).

vector particle: a particle of spin one (in units of $\hbar$).

vertex function: the function representing the diagram (actually, the sum of all possible diagrams) with three legs (e.g., $ee\gamma$).

vertex modification: the higher order corrections to the simple vertex in which three lines ('particles') meet at a point (cf. the left $ee\gamma$ vertex in Figure 1.6 versus the basic vertex in the left of Figure 1.4).

wave mechanics: that formulation of quantum mechanics in terms of wave functions and differential operators (basically, the Schrödinger equation).

weak interaction: the interaction among the leptons, a characteristic example of which is that responsible for the decay of a free neutron as $n \rightarrow p + e^- + v$.

Yang–Mills theory: a theory of strongly interacting particles based on the principle of local gauge invariance.

Yukawa potential: a potential of the form $V(r) = \dfrac{\lambda e^{-\mu r}}{r}$. This form arises naturally from a field theory of nuclear interactions in which the nucleons exchange mesons.

Some key figures and their positions

Below I list some scientists and philosophers of science whose research and writing have influenced my own thinking or have been important for the episodes discussed in the text. No claim is made that *every* relevant, influential scientist or philosopher is represented below. Like the preceding glossary, this appendix is intended to serve as a convenient reference for those figures or positions unfamiliar to the reader. The References for this text contain specific citations to the works of nearly all of these people. (Here, the term 'American' signifies that the scientist or philosopher is either native to the United States or has worked there for the major part of his or her professional life.)

Gaston Bachelard (1884–1962) – a French philosopher of science whose ideas in many ways presaged current tendencies in England and in the United States to naturalize the philosophy of science by giving the history and the study of actual scientific practice a central role in the philosophical analysis of science.

David Bloor (1942–   ) – a key figure in the (Edinburgh school) 'strong programme' in the sociology of knowledge, according to which not only the form but even the very content of scientific knowledge are to be explained ultimately by sociological factors.

David Bohm (1917–   ) – an American-born and trained, English theoretical physicist who has been influential in expounding a causal theory of quantum phenomena.

Nancy Cartwright (1944–   ) – an American philosopher of science who has argued that the fundamental laws of physics gain their great simplicity and explanatory power at the expense of their being able to represent directly and accurately the actual (highly complex) phenomena of nature.

Geoffrey Chew (1924–   ) – a theoretical high-energy physicist who was a prime mover in the modern S-matrix program that began in the early 1960s. Chew's early semiphenomenological work led him to reject quantum field theories (because of their large degree of arbitrariness) and to suggest an *S*-matrix theory that would allow just one (i.e., a unique) solution for strong-interaction dynamics.

Freeman Dyson (1923–   ) – a British-trained, American theoretical physicist who was instrumental in formalizing the renormalized quantum electrodynamics program in the late 1940s and early 1950s. Dyson also proved the equivalence of the (more formal or abstract) Schwinger–Tomonaga version of quantum electrodynamics and of the (more intuitive or picturable) one of Feynman.

Richard Eden (1922–   ) – an English theoretical physicist who was one of the first to study the analytic properties of scattering amplitudes, especially within the framework of (Feynman) perturbation theory. Eden became the senior member of the 'English school' of an analytic *S*-matrix program.

Richard Feynman (1918–1988) – an American theoretical physicist who formulated quantum electrodynamics in a version that is extremely powerful for making calculations. Feynman's diagrams (which can be put into a one-to-one correspondence with the perturbation expansion for transition amplitudes) give a pictoral representation of reactions in terms of the exchange of virtual particles.

Arthur Fine (1937–   ) – an American philosopher of science expert in the foundational problems of quantum mechanics. Fine advocates for the philosophy of science a position that is neither that of scientific realism nor that of antirealism. He stresses science as an *historical* entity with its varieties of practices and disciplines.

Peter Galison (1955–   ) – an American historian of modern physics who emphasizes that when experimentalists finally agree that a phenomenon has been reliably observed (or 'seen'), they employ a complex network of overlapping arguments that have many dimensions, some of them sociological. He stresses that neither theory nor experiment *alone* should be taken as *the* paradigm of science.

Murray Gell-Mann (1929–   ) – an American theoretical physicist who was instrumental in initiating the modern (quantum-field) dispersion theory in the mid 1950s. Although Gell-Mann contributed several key ideas to the *S*-matrix program, he is best known for his proposal of the 'eightfold way' (i.e., SU(3) symmetry) for classifying the strongly interacting particles.

Marvin Goldberger (1922–   ) – an American theoretical physicist who was the central figure in the modern dispersion-theory program during the 1950s and 1960s. Goldberger gave a very influential proof of quantum-field-theory dispersion relations based on the (micro-) causality condition.

Ian Hacking (1936–   ) – a Canadian-born, British-educated philosopher of science, working in America and Canada, whose stance is one of scientific realism, but one based on a philosophy of *experimental* (as opposed to *theoretical*) practice as providing compelling reasons for that position.

Walter Heitler (1904–1981) – A German-born theoretical physicist who worked at the Dublin Institute for Advanced Studies during World War II. Heitler wrote an influential book on quantum field theory in the 1930s and was a master of its calculational techniques.

Carl Hempel (1905–   ) – a German-born and educated, American philosopher of science whose deductive-nomological (or 'covering-law') model of scientific explanation has been very influential among philosophers. Hempel's model represents scientific explanation as consisting of an accounting for phenomena by deductively reasoning to them from general laws (or hypotheses).

Rest Jost (1918–   ) – a Swiss theoretical physicist who was a leader in axiomatic quantum field theory in the 1950s and 1960s.

Hendrik Kramers (1894–1952) – a Dutch theoretical physicist whose career bridged the revolution in quantum mechanics and continued into the era of quantum field theory. Kramers proposed some of the earliest dispersion relations for electromagnetic waves and also, much later, suggested how a program of mass and charge renormalization might be carried through for quantum electrodynamics.

Thomas Kuhn (1922–   ) – an American historian of science whose representation of the history of science in terms of periods of normal science punctuated by revolutions in science has been one of the most influential models of science in the last quarter of a century. He is largely responsible for the modern trend of emphasizing the social and sociological dimensions of science.

Imre Lakatos (1922–1974) – a Hungarian-born, English philosopher of science who branched off from Karl Popper's school to suggest that the proper unit for appraisal in studying science is the scientific research program. In Lakatos' methodology of scientific research programs, a hard core of more or less fixed premises is protected against falsification by modifying the auxiliary hypotheses (the 'protective belt') that surround it. In this way, a succession of theories is generated.

Lev Landau (1908–1968) – a Russian theoretical physicist whose professional interests covered several fields of physics. Landau (along with Ilya Lifshits) authored a series of advanced texts on theretical physics, which have influenced generations of students (since they have been translated into many languages). He was among the most vocal proponents of an anti-quantum field theory stance.

Larry Laudan (1941–   ) – an American philosopher of science who has been extremely critical of scientific realism. Laudan has proposed a model of science in which the practice, methodological rules, and goals of science are coupled and evolve, mutually influencing and modifying each other in the process.

Harry Lehmann (1924–   ) – a German theoretical physicist who gave an early rigorous derivation of *nonforward* dispersion relations. Lehmann (along with Kurt Symanzik and Wolfhart Zimmermann) laid the foundations for a rigorous, axiomatic quantum field theory based on the assumption of well-defined asymptotic field operators.

Francis Low (1921–   ) – an American theoretical physicist whose main field of research has been quantum field theory. Although Low collaborated with Chew in the formulation of the Chew–Low model, which was influential for the subsequent *S*-matrix program, he remained skeptical that the *S*-matrix alone would provide the basis for a complete dynamical theory.

Stanley Mandelstam (1928–   ) – a (South African) British-educated, American theoretical physicist who first proposed a double dispersion relation for relativistic scattering amplitudes. Mandelstam is an expert in quantum field theory who has contributed significantly to the dispersion-theory and *S*-matrix theory programs but who never became anti-field theory.

Ernan McMullin (1924–   ) – an Irish-born, American philosopher of science who has been one of the strongest and most consistent advocates of scientific realism in recent times. McMullin has argued for the fertility of a scientific theory as one of the most important hallmarks of successful, reliable theories.

Emil Meyerson (1859–1933) – a French philosopher who proposed describing science in terms of a principle of lawfulness (i.e., relations among phenomena) and of a principle of causality (i.e., our need to understand the persistence of certain observed regularities). Meyerson saw both of these principles as rooted in modes of human thought (which remain essentially fixed).

Christian Møller (1904–1980) – a Danish theoretical physicist who gave an early, pedagogically clear and mathematically precise presentation in the mid 1940s of Heisenberg's original *S*-matrix program.

Nancy Nersessian (1947–   ) – an American philosopher of science who has studied, in the context of specific episodes in physics, how scientists create the language they use and how the meanings of terms evolve through developing scientific practice.

Thomas Nickles (1943–   ) – an American philosopher of science who has emphasized that the discovery and the justification of scientific theories are not disjoint enterprises (either temporarlly or logically). Nickles has shown that an important aspect of justifying a theory already accepted in practice is to show that it is 'demanded' by some other accepted, apparently nearly self-evident principle.

Andy Pickering (1948–   ) – a British educated, American sociologist of science whose position is that the theories that scientists eventually accept are not judged against wholly independent experimental results, but rather against phenomena that have been to some extent already selected in light of a particular theory. For Pickering, there thus is a great deal of flexibility about how things might finally turn out.

John Polkinghorne (1930–   ) – an English theoretical physicist expert in the analytic properties of scattering amplitudes in (quantum-field) perturbation theory. He became the leader of the 'Cambridge school' of the *S*-matrix program in the 1960s and 1970s.

Karl Popper (1902–   ) – an Austrian-born, English philosopher of science who claims the hallmark of a scientific theory to be its potential *falsifiability*. That is, for Popper, scientific theories are bold conjectures whose truth is never guaranteed, since they must always remain subject to stringent tests of observation.

Tullio Regge (1931–   ) – an Italian theoretical physicist who proposed extending the angular momentum variable (l) to continuous, complex values for an analysis of the analytic properties of scattering amplitudes. Regge's original work was in potential scattering theory and he never backed the once-fashionable program of extending (by guess) these ideas to a relativistic context.

Silvan Schweber (1928–   ) – a French-born, American-educated theoretical physicist who specialized in quantum field theory in his early work. After writing an influential textbook on quantum field theory, Schweber began work in the history of science and has since become a leading historian of modern theoretical physics.

Julian Schwinger (1918–   ) – an American theoretical physicist who formulated renormalized quantum electrodynamics in the late 1940s. Schwinger published one of the early impressive results of the renormalized quantum field theory calculational program: the anomalous magnetic moment of the electron.

Dudley Shapere (1928–   ) – an American philosopher of science who has been active in the movement away from logical positivism. Shapere holds the view that *all* aspects of science (e.g., practice, methods, goals, language) are subject to change as we learn more about nature. Still, he argues against a picture of science as an arbitrary, chaotic enterprise incapable of discovering truth.

Ernst Stückelberg (1905–1984) – a Swiss theoretical physicist who was particularly active in quantum field theory, and later in *S*-matrix theory, in the 1930s and 1940s. Stückelberg and his school sought to formulate a quantum field theory free of divergences.

Bas van Fraassen (1941–   ) – a Dutch-born, American philosopher of science who has developed an empiricist position as an alternative to logical positivism and to scientific realism. According to his 'constructive empiricism', science aims to give us theories that are empirically adequate (i.e., that predict correctly), and acceptance of a theory involves as belief only that it *is* empirically adequate. Only those entities are to be accepted which are observable.

John Watkins (1924–   ) – a British philosopher of science of the Popper school. Watkins' goal is to salvage as much of the rationality of science as possible, given the potential falsifiability of the basic theoretical components (or entities) of science.

Steven Weinberg (1933–   ) – an American theoretical physicist who (with Abdus Salam) formulated the unified theory of electroweak interactions. Although Weinberg has made major contributions to modern gauge field theories, he sees value in the *S*-matrix paradigm and has stressed the oscillation in modern theoretical physics between the quantum field theory and *S*-matrix theory paradigms.

Gregor Wentzel (1898–1978) – a German theoretical physicist who worked in Switzerland and later in America. Wentzel wrote a very influential book on quantum field theory (published in Vienna) during World War II. He was among the first to bring news of Heisenberg's *S*-matrix program to the United States in the late 1940s.

John Wheeler (1911–   ) – an American theoretical physicist who first introduced an *S*-matrix as a calculational tool in nuclear theory. Wheeler has been influential not only in nuclear theory but also in general relativity.

William Whewell (1794–1866) – a nineteenth century English historian and philosopher of science who, in today's terminology, adopted a scientific realist stance and argued that scientific theories converged to truth. Whewell's prodigious scholarship laid the foundation for much of modern history and philosophy of science.

Index

(References to glossaries are indicated by italicized numbers.)

Abers, E., 159
Abrikosov, A. A., 18
adiabatic invariant, 219, *372*
aether, 234, 263
Aharonov, Y., 275
Aharonov–Bohm effect, 275, *372*
Aks, S. Ø., 322
Alessandrini, V., 205
Allen, A. D., 181
Amati, D., 143, 154, 161, 205
analogy, 213, 219, 224, 228, 244, 257
analytic continuation, 55, *372*
analytic functions, 55, 75, 176, 303, *372*
analyticity, 30, 55, 76, 90, 135, 229, 303
  hermitian, 184, *381*
  maximal, 173, 174, 185, *384*
  physical region, 184, *386*
Anderson, H. L., 79
angular distribution, 99, *372*
  see also cross section, scattering
anomalous magnetic moment, 19, *372*
anomaly cancellation, 206, *372*
anthropic principle, 221, 259, *373*
anticommutator, 213, *373*
Araki, H., 167
Ascoli, R., 39
asymptotic behavior, 107, 131, 141, 142, 297, *373*
asymptotic freedom, 25, *373*
asymptotic series, 18, *373*
Atkinson, D., 322
Aubert, J. J., 202
Augustin, J-E., 202
auxiliary hypotheses, xii, *373*

Bachelard, G., 283, *393*
Bacon, F., ix
Bacon–Descartes ideal, 282
Balázs, L. A. P., 157, 200, 202, 207, 216, 261
Ball, J., 216
Bardakci, K., 196
Barger, V. D., 155
Bargmann, V., 50, 60, 61

Barrow, J. D., 221, 259, 285
baryon, 197, *373*
  nonexotic, 325
baryon number (B), 173, *373*
baryonium, 201, *373*
Baumann, K., 321
Belinfante, J. G., 159
Bell, J. S., 274
Bethe, H., 13, 14, 60, 69, 256, 321
Bev, *373*
bifurcation point, 256, 267, *373*
Blankenbecler, R., 123, 127, 143, 154, 157, 160
Blatt, J. M., 223
Bloor, D., xiii, 283, *393*
Bloxham, M. J. W., 186, 324
Bogoliubov, N. N., 81, 130, 324
Bohm, D., 231–5, 274, 275, 289, *393*
Bohr, N., 5, 10, 218, 224
Bonamy, P., 153
bootstrap, 8, 116, 127–9, 134, 135, 139–41, 168, 176–9, 190, 212, 227, 240, 258, 265, 268, *374*
  applications of, 153–61
  of internal symmetries, 159
  of rho meson, 157–9
  partial, 168, 179, 181
  and Venziano model, 195
Born amplitude, 123, *374*
boson, *374*
  intermediate vector, 23, 25, 246
bound state, 44, 60, 298, *374*
boundary conditions, 104, 131, *374*
  see also asymptotic behavior
branch point, 55, 153, *374*
Breit, G., 36, 38, 51, 52, 226, 306, 309
Breit–Wigner formula, 30, 36, 52–4, 66, 213, 226, 227, 309–12, 314, *374*
Bremermann, H., 81
Brenig, W., 35
Brink, L., 326
Brinkley, A., 4
Bros, J., 83
Brown, L. M., 122, 273

Brownian motion, 234
Brueckner, K., 142
Buchdahl, G., 272
Burckhardt, J., 4

Cao, T. Y., 324, 325
Capella, A., 164
Capps, R. H., 81, 159, 206
Capra, F., 89, 94, 96, 128, 129, 142–5, 164, 200–3
Cartwright, N., 1, 289, *393*
Cassidy, D. C., 2, 33, 39
Castillejo, L., 108
causal quantum theory, 231–6, 264, *374*
causality, 76, *374*
    and analyticity, 226, 302, 303
    and dispersion relations, 57–63, 265
    and first signal, 30, 70, 211, 214, 223
    Heisenberg and, 220
    macroscopic, 176, 184, *384*
    microscopic, 68–73, 76, 116, 132, 213, 229, 230, 265, *385*
    and phase shifts, 79
    principle, 226–30, 319
    and quantum mechanics, 232
Chamberlain, O., 94
Chan, H-M., 196, 199
Chandler, C., 184
channel, 31, 103, 223, 296, 299, *374*
    crossed, 128, 140, 146, 189, *376*
    direct, 128, 140, 377
Charap, J. M., 127
charge conjugation, 138, 203, *374*
charge exchange reaction, 153, *374*
charge independence, 93, *374*
charge renormalization, 16, 21, *374*
Chew, G. F., 80, 106, 112, 113, 147, 154, 162, 164, 177, 185, 187, 251, 256, 261, 281, *393*
    analyticity, 90, 180
    bootstrap condition, 8, 127–9, 139–41, 155–61, 181, 192
    crossing, 103
    Fermi, 90
    form factors, 83
    graduate work, 67
    Heisenberg program, 28, 64, 65, 144, 266
    impulse approximation, 94–6
    Lagrangian field theory, 96
    La Jolla Conference, 135, 142–5
    Landau, 129
    Low, 90, 97

Chew, G. F. (*continued*)
    Mandelstam, 139–41, 322
    maximal analyticity, 173
    maximum strength, 148, 173
    methodology, 144, 175, 181, 264, 265
    multiperipheral model, 193
    order, 200
    phase shift, 104, 108
    potential theory, 90
    recollections, 89, 320
    rho meson, 128
    S-matrix program, 118, 129–32, 167–71, 173–6
    sociological factors, 90, 91
    space-time, 178
    static model, 97–100, 105
    sufficient reason, 135, 144, 174
    (3, 3) resonance, 109
    topological expansion, 199–203
    see also Chew model; Chew–Low effective range formula; Chew–Low extrapolation; Chew–Low model; pole–particle correspondence
Chew–Frautschi plot, 148–50
Chew–Low effective range formula, 108, 129, *375*
Chew–Low extrapolation, 91, 113, 128, 214, *375*
Chew–Low model, 90, 91, 104–9, 111, 119, 122, 212, 214
Chew model, 97–100, *375*
Chou, T. T., 324
Cini, M., 67, 80, 214, 216, 288
Cline, D. E., 155
cluster decomposition, 184, *375*
Cocconi, G., 151
Coester, F., 318–20
Collins, P. D. B., 148, 152
commutation relation, 70, 71
commutator, *375*
    current, 73, 191, *376*
compactification, 206, *375*
complementarity, 284, *375*
completeness, 44, *375*
complex angular momentum, 126, 212, *375*
compound nucleus model, 223–31
conservation of probability, 32, 176, *375*
consilience of inductions, 111, 247
contingency, 250, 255, 259, 269, 283
contour integral, 126, *375*
convergence of scientific opinion, 243, 247, 251, 271–81

Cook, L. F., 157
Coon, D. D., 160
Copenhagen interpretation, 231, 234, 264, 375
correspondence principle, 39, 172, *375*
cosmic ray 'explosions', 33, 39
cosmological anthropic principle, *375*
  see also anthropic principle
Coster, J., 185
Coulomb amplitude, 78, *376*
Coulomb field, 59, *376*
coupling constant, 18, 97, *376*
  pion–nucleon, 79, 99, 108, 111, *387*
CPT theorem, 182, *376*
creation and annihilation operators, 12, 71, 74, 205, *376*
Cremmer, E., 205
cross section, 31, 32, 37, 223, 294, *376*
  scattering, 60, 98, 147, 294, *389*
crossing, 75, 100–4, 107, 266, *376*
Cushing, J. T., 8, 28, 33, 34, 38, 51, 70, 160, 176, 180, 213, 215, 217–21, 223, 228, 229, 241, 251, 253, 254, 256–8, 261, 264, 265, 270, 273–5, 277–9, 285, 289, 292
cut, 56, 128, *376*
  see also Regge cuts
Cutkosky, R. E., 130, 143, 159, 185
Cutkosky rules, 130, 185, *376*
cutoff, 106, 141, 159, *376*
Cziffra, P., 104, 105

Dalitz, R. H., 108
Darrigol, O., 8
Davidon, W. C., 79, 80
de Alfaro, V. S., 191
de Broglie, L., 231–5
de Broglie wavelength, 94, *376*
de Haas, W. J., 246
demarcation problem, 241, 377
De Tar, C., 89
determinism, 233, 274
Dicke, R. H., 259
Diddens, A. N., 151
diffraction peak, 148, 152, *377*
  shrinkage of, 147, 151, 215, *389*
Dirac, P. A. M., 10, 11, 19, 26, 35, 219, 232, 256, 258, 260, 261, 308
(Dirac) δ function, *377*
Dirac hole theory, 13, *377*
discovery, xi
dispersion relations, 57–63, 116, 211, 303, *377*

dispersion relations (*continued*)
  for form factors, 85
  Kramers–Kronig, 57, 70, 72, 87, 213, 305, 306, *383*
  model example, 108
  for N/D equations, 141
  nucleon–nucleon, 80
  phenomenology of, 77–80
  pion–nucleon, 79
  proof of, 70–7, 81–3
  and superconvergence, 191
  see also Mandelstam representation
dispersion theory, 8, 30, *377*
divergences, 12–14, *377*
Dolen, R., 157, 191, 192
Donovan, A., 249
double dispersion relations, 115, 119, 121, 123–7, *377*
  see also Mandelstam representation
double solution, 231
Drell, S. D., 86, 151
Dresden, M., 324
Drude, P., 213, 224
duality, 26, 189, 191–3, 197, 212, 220, 266, *377*
Duhem–Quine thesis, x, 275, *377*
Dyson, F. J., 15, 18, 47, 48, 56, 82, 108, 130, 167, 256, *394*

Earman, J., 272
Eden, R. J., 55, 56, 69, 119, 123, 143, 184, 267, *394*
Edwards, S. F., 18, 130
effective range formula, 95, 108, 295, *377*
Ehrenfest, P., 219
eightfold way, 143, *377*
Einstein, A., 5, 91, 231, 233, 235, 246, 266
Eisenbud, L., 53, 54, 61, 226
Elster, Ch., 280
empiricism, 1, 242, 243, *378*
  constructive, *375*
epistemology, 241
Epstein, H., 83
equivalence principle of strong interactions, 145, *378*
Erwin, A. R., 128
essential singularity, 55, 123, 155, *378*
Euler–Lagrange equations, 22, *378*
exchange degeneracy, 200, 201, *378*
explanation, 251, 258–60

factorization, *378*
  see also scattering (*S*) matrix,
    factorization
Fairlie, D. B., 131
false zeros, 49, 126
Faxén, H., 294
Feenberg, E., 32, 36, 60, 296, 299
Feldman, D., 73
Fermi, E., 2, 17, 23, 33, 67, 94, 180
Fermi theory of β decay, 17, 23, *378*
Fermion, *379*
Ferretti, B., 40
fertility, 215, 218, 234, 251
Feshbach, H., 86, 96
Feynman, R. P., 15, 20, 122, 256, 267, *394*
Feynman diagrams, 19–21, 26, 47, 101,
    379
Fierz, M., 41
Fine, A., xii, xiii, 250, 251, 254, 255, 275,
    394
fine struture constant, 15, *379*
finite-energy sum rule (FESR), 191, 193,
    379
Finkelstein, J., 89, 203, 216
Finkler, P., 203
fixed-source theory, 97, *379*
Foley, H. M., 14
Foley, K. J., 152
form factor, 85, 86, 128, *379*
Forman, P., 6, 7, 216, 232, 285
Forman thesis, 6, 7, 216, *379*
Foucault, M., 257
foundationist, xi, 240, 242, 282, *379*
Fourier decomposition, 44, 58, 71, *379*
Fox, G. C., 161
fractal, 284, *379*
Frampton, P. H., 205
Frank, N. H., 291
Frank, P., 326
Franklin, A., 246
Frautschi, S. C., 89, 141–5, 147, 148, 154,
    157, 164, 173
Frazer, W. R., 91, 128, 129, 143, 165, 207,
    216, 270
Fredholm determinant, 31, 380
Freund, P. G. O., 162, 192, 198
Freundlich, Y., 177, 258
Friedman, M., x
Fröberg, M., 50, 51
Froissart, M., 142, 143, 184
Froissart bound, 142, 148, *380*
Fubini, S., 127, 143, 154, 161, 191, 204, 205
Fulco, J. R., 128, 129, 216

fundamental length, 33, 48
Furlan, G., 191

G parity, 200, *380*
Gale, G., 221, 241, 259
Galison, P., xii, 2, 5, 33, 217, 227, 238,
    246–8, 252, 272, 277, 285, *394*
gamma function, 194, *380*
Gamow, G., 38
Gasiorowicz, S., 83, 138
gauge field theory, 22–6, 222, 245, *380*
gauge invariance, 15, *380*
  global, 22, *380*
  local, 23, *383*
Gauron, P., 202
Gel'fand, I. M., 51, 126
Gell-Mann, M., 24, 81, 144, 167, 170, 174,
    191, 213, 251, 256, 265, 268, 280, 281,
    394
  bootstrap, 155–6, 158, 274
  crossing, 75, 76, 90, 101–3
  on discovery, 267
  dispersion relations, 65, 68–71, 77, 87,
    109, 119, 123, 229
  Heisenberg program, 28, 64
  La Jolla Conference, 142, 143
  Landau, 129, 130
  mass-shell QFT, 116–18, 145, 316
  nuclear democracy, 145, 156
  quarks, 156
  Regge poles, 145, 148–50, 164
  Reggeized QFT, 157
  scattering theory, 35, 95
  S-matrix program, 115–118, 171
generative potential, 218, 219
Gev, *380*
ghost states, 196, 204, 205, *380*
Gierre, R. N., 254
Gilbert, W., 81
Glaser, V., 83
Gliozzi, F., 206
Gluckstern, R. L., 59, 60, 62, 69, 87, 265
Glymour, C., 227
Goddard, P., 205
Goebel, C. J., 196
Goldberger, M. L., 83, 86, 91, 96, 167, 174,
    185, 251, 256, 265, *394*
  causality, 87
  crossing, 75, 76, 101–3
  dispersion relations, 65, 68–72, 77, 79,
    90, 109, 169–71
  graduate work, 67
  Heisenberg program, 28, 64, 67

impulse approximation, 94, 95
La Jolla Conference, 142, 143
Mandelstam representation, 123, 124, 127
mass-shell QFT, 145
microcausality, 70, 71, 213, 229
multiperipheral model, 161, 162
pragmatic attitude, 80, 81, 115, 214
QFT versus SMT, 169, 170
Regge cuts, 154
Reggeized QFT, 157
scattering theory, 35
Goldstone, J., 24, 205
Goldstone bosons, 24, *380*
Gordon, D., 204
Gould, S. J., 288
grand unified theory (GUT), 25, 26, *380*
Green, M. B., 156, 206, 281
Green, T. A., 47
Green's function, 119, 291, 301, *380*
Gribov, V. N., 155, 164
Gribov singularity, 155, 157, *380*
Gross, D. J., 206
group, 93, *380*
Grythe, I., 28, 38–40, 42, 49, 64
Gunson, J., 185

Haag, R., 35, 167
Haas, A. E., 218
Hacking, I., 1, 257, 276, *394*
Hadron, *381*
hamiltonian formalism, 9, 172, *381*
Harari, H., 196
hard core, xii, *381*
Hardin, L., 277
harmony of random choices, 275, *381*
Heilbron, J. L., 252
Heisenberg, W., 1, 5, 6, 56, 185, 213, 256, 306
   analyticity, 38
   Fermi theory, 23
   fundamental length, 33, 34
   hamiltonian field theory, 33
   matrix mechanics, 10, 11, 26, 219
   nonlinear spinor QFT, 39, 40
   optical theorem, 37
   scattering theory, 35
   S-matrix program, 27, 33–39, 65, 82, 89, 167; causality, 57; Chew on, 143; Gell-Mann on, 143; Heisenberg's later views, 41, 42; Kramers on 38, 42; Møller on, 40, 43, 44; motivation for, 2, 26, 33, 34; Pauli on, 40; Wentzel on, 47

Heisenberg, W. (*continued*)
   superconductivity, 40
   unitarity, 35, 36
   Wheeler, 36, 318
Heisenberg operator, 70, 73, *381*
Heitler, W., 45, 46, 56, 291, 321, *395*
helicity, 191, *381*
Heller, M., 6, 286
Hempel, C. G., 275, *395*
Hendry, J., 7
Hepp, K., 83
Hermann, A., 218
heuristic, xii, *381*
   mathematical, 190, 214, 215, 219, 236, 258, 260, 261, 264, 265
Higgs, P. W., 24
Higgs boson, 25, 222, *381*
Hiley, B. J., 231–4, 289
historicist, xi, *381*
Hoddeson, L., 273
Holinde, K., 280
Holmberg, B., 51
Holtsmark, J., 294
Hones, M. J., 220, 255
Horn, D., 157, 191, 192
Hu, N., 45, 51, 54, 55, 57
Huang, K., 160
Huygens, C., 257
hydrodynamical model, 232, 233
Hylleraas, E. A., 50, 51
hypothetico-deductive method, 190, 258, *382*

Iagolnitzer, D., 182, 184, 185
Igi, K., 154, 191
Iizuka, J., 198
implicate order, 275, *382*
impulse approximation, 94–6, *382*
in and out fields, 73, *382*
indeterminacy, 240
instrumentalism, x, *382*
interaction, *380*
   electromagnetic, 168, 174, 176, 179, 181, 202, *378*
   electroweak, 203, *378*
   gravitational, 206, 208
   strong, 244, *390*
internal history, 3–6, *382*
internal symmetry, 159, *382*
invariance group, 22, *382*
isotopic spin (isospin), 78, 93, 173, *382*

Jackson, J. D., 67, 73, 75
Jacob, M., 205

Jammer, M., 231
Jardine, N., 327
Jauch, J. M., 321
Jones, C. E., 202, 203, 216
Jones, G. L., xv
Jost, R., 40, 49, 51, 58, 59, 80, 82, 117, 126, 130, 182, 183, *395*
Jost function, 49, 50, 126, 392, *395*
justification, xi

Kalckar, F., 38, 224
Källen, G., 18, 73
Kaloyerou, P. N., 289
Kaluza, Th., 206
Kaluza–Klein theory, 206, *382*
Kaplunovsky, V., 284
Kapur, P. L., 38, 226, 309
Karplus, R., 59, 70, 77–9, 83, 122
Kemmer, N., 48, 232
Keynes, J. M., 239
Khalatnikov, I. M., 18
Khuri, N. N., 123, 127
Klein–Nishina formula, 170, *383*
Klein, O., 206, 308
Knudsen, O., 317, 318
Koba, Z., 196, 204
Kockel, B., 321
Kohn, W., 51, 126
Kolb, E. W., 207
Kortel, F., 39
Kourany, J. A., 242
Kramers, H. A., 14, 38, 39, 42, 44, 57, 65, 66, 70, 213, 229, 256, 304–6, *395*
Kramers–Kronig dispersion relation, *see* dispersion relations
Kroll, N., 14, 77, 108
Kronig, R., 57, 58, 60, 65, 66, 70, 213, 229, 304, 305
Kruse, U. E., 79
Kuhn, T. S., xi, xiii, 1, 217, 222, 253, 293, *395*
Kusch, P., 14

ladder of axiom, ix
Ladenberg, R., 213
Lagrangian, 22, 131, 132, *383*
  field theory, 170, 171; effective, 270
  formalism, *383*
LaJolla Conference, 135, 142–5
Lakatos, I., xi, 7, 217, 222, 240, 242, 248, 254, 281, *395*
Lamb, W. E., 12
Lamb shift, 13, 14, 19, *383*

Landau, L. D., 18, 129, 130, 143, 167, 168, 171, 174, 221, *395*
Landau rules, 130, *383*
Landau singularities, 221
Landau surfaces, 184, *383*
Landshoff, P. V., 123, 131, 184, 185
Lassalle, M., 320
Laudan, L., 242, *395*
  antirealism, 264, 278
  Leibnizian ideal, 276
  method of hypotheses, 257, 258
  naturalized philosophy of science, xiii, 249–54
  reticulated model, 243, 249, 257, 272, 275, 278, 284, 285
Laudan, R., 249
Lax, M., 60, 96
Le Bellac, M., 205
Lee, H. C., 199
Lehmann, H., 81–3, 116, 130, 214, *396*
Lehmann–Symanzik–Zimmermann (LSZ) formalism, 81, 131, 214, 270, *383*
Lehr, W. J., 234
Leibniz' principle of sufficient reason, 135, 144, 168, 174, 176, 229, 259
Leptons, 174, *383*
Levinson, N., 50, 51
Levitan, B. M., 51, 126
Lewis, H. W., 95, 96
lifetime, 310
Lindenbaum, S. J., 152
Lippmann, B. A., 35, 90
Lippman–Schwinger equation, 90, 92, 96, 106
local field, 132, 171, *383*
logic of induction, *383*
logical positivism, x, 240, 242, 281, *384*
Lorentz invariance, 15, 25, 37, *384*
Losee, J., 241
Lovelace, C., 164, 196, 205
Low, F. E., 73, 77, 80, 89, 91, 97, 100, 105, 106, 108, 109, 113, 145, 157, 160–62, 164, 170, 171, 174, 328, *396*
Low equation, 100, 129, *384*
Lu, E. Y. C., 184
Luttinger, J. M., 58, 117

Ma, S. T., 44, 49
MacGregor, M. H., 104, 105
MacKinnon, E. M., 231, 257, 258, 288
McGlinn, W. D., 160

McMullin, E., xiv, xv, 1, 218, 228, 229, 274, 289, 327, *396*
Machleidt, R., 280
Madelung, E., 232
magnetic moment, 14, 15, *384*
  anomalous, 19, 99, *372*
Mandelbrot, B., *379*
Mandelstam, S. L., 89, 90, 115, 174, 261, *396*
  bootstrap mechanism, 128, 129, 139, 141, 144
  double-dispersion relation, 109
  duality, 191, 196, 205
  Gribov singularities, 157
  linear Regge trajectories, 151, 194
  mass-shell QFT, 131, 132, 145, 171
  microcausality, 132
  Regge cuts, 154, 155, 164
  Regge poles, 131, 323
  string theory, 281
  see also Mandelstam representation
Mandelstam representation, 115, 118–23, 212, 214, *384*
  proofs of, 123–6
Mandl, F., 73
March, R., 128
Martin, A. W., 160
Marx, E., 157
mass renormalization, 16, *384*
mass shell, 117, 131, 132, 185, *384*
Massey, H. S. W., 291
matrix mechanics, 11, 34, *384*
Matsuda, S., 154, 191
Matthews, P., 17
Maxwell, J. C., 263
Maxwell's equations, 11, *384*
measurement problem, 168, 175, 185, 202, 234
Medawar, P. B., 240, 288
Medvedev, B. V., 81
Mehra, J., 232
Meixner, J., 50
mesons, 197, *384*
  exotic, 201, *378*
  nonexotic, 325
Messiah, A., 295, 328
meta-level, 242, 250, 251, 257, *384*
methodologial rules, 213, 220, 221, 238, 249–62
Mev, *385*
Meyerson, É, 290, *396*
Millikan, R. A., 247
Mills, R. L., 22, 23

Misner, C. W., 277
Mitter, H., 39
Miyazawa, H., 79
models, 203, 213, 220
Møller, C., 40, 42–4, 49, 50, 67, *396*
momentum transfer, 115, *385*
monopole, 260, *385*
Moravcsik, M. J., 104, 105, 279
Moshe, M., 138
Mott, N. F., 291
Moyer, B. J., 96
multiperipheral model, 161, 162, *385*
  Regge form of, 193, 199

N/D equations, 140, 158, *385*
N-point amplitude, 196, 204, *385*
Nambu, Y., 80, 81, 90, 109, 119, 190, 205
Nash, N., xv
naturalized philosophy of science, xiii, 1, 241, 255, 271, 294, *385*
nearby singularities, 104, 129, 142, *385*
Ne'eman, Y., 24, 234
Nelson, E., 234
Nersessian, N. J., xiii, 230, *396*
neutral current, 247
Neveu, A., 205, 207
Newton, R. G., 51
Ng, S. W., 199
Nickles, T., xiii, 218, 226, 227, 243, 254, *396*
Nicolescu, B., 202, 203
Nielsen, H. B., 196, 204
Noether's theorem, 22, 25, *385*
non-abelian gauge theory, 261, *385*
nonlinear spinor theory, 39, 40, *385*
nonlocality, 235
nonseparability, 240
normal science, xi
November Revolution, 202, 244, 245
nuclear democracy, 135, 145, 156, 164, 168, 180, *385*
nucleon (N), 93, *385*
Nuttall, J., 131

Oehme, R., 28, 33, 79–81
Okubo, S., 198
Okun', L. B., 139
Oldershaw, R. L., 327
Olive, D., 182, 184, 186, 205, 206, 268, 281
Omnès, R., 86
ontology, 241
Oppenheimer, J. R., 15, 38, 48

optical theorem, 32, 37, 56, 60, 78, 296, 299, 301, *386*
order, 190, 200, 325, *386*
Ouvry, S., 202
OZI rule, 198, 200, 201, *386*

paradigm, xi
parity, 203, *386*
Park, J. L., 234
partial wave, 78, 92, 294, *386*
  amplitude, 126, 127
particle exchange, 105, 280, *386*
Pascal, B., 257
Paton, J. E., 196, 199
Pauli, W., 5, 40, 45, 50, 56, 98, 231, 232, 256
Peierls, R., 38, 226, 309
Peng, H. W., 45
Peres, A., 289
perturbation theory, 20, *386*
Pham, F., 186
phase shift, 50, 62, 63, 92, 295, 311, *386*
phase-shift analysis, 78–80, 104, *386*
phenomenology, 104, *386*
  degenerating Regge, 152–5
Phillips, R. J. N., 164
philosophical relativism, 241, 248, *386*
photomeson production, 108, *386*
physical region, 146, *386*
Pickering, A., 202, 260, *396*
  analogy, 213
  dispersion theory, 67
  pragmatic attitude, 80, 214
  SMT versus QFT, 165, 207, 269
  sociologial factors, 215, 217, 238, 242–6, 248, 261, 285
  strong program, xiii
  symbiosis, 220, 243, 254, 276
Pignotti, A., 193
pilot wave, 231
planar diagram, 197, *386*
Planck, M., 227
Poénaru, V., 202, 203
Poisson brackets, 11, *387*
polarization, 153, *387*
pole, *387*
  see also pole–particle correspondence;
    Regge poles
pole factorization, 149, 204, 205, *387*
pole–particle correspondence, 90, 107, 109–13, 129, 130, *387*
Politzer, H. D., 25
Polivanov, M. K., 81

Polkinghorne, J., xv, 81, 123, 130, 131, 155, 157, 164, 173, 184–6, 216, *397*
Pomeranchuk, I. Ia., 130, 134, 155, 164
Pomeranchuk theorems, 136–9, 150, *387*
Pomeron, 148, 199
Popper, K., xi, 7, 240, 242, 243, 253, 281, *397*
positive definite, 204, *387*
positivism, xi
potential, 123, 142
potential theory, 33, 90, *387*
pragmatic attitude, 80, 81, 113, 214, 220, 237
prescriptive criteria, 242
principal value integral, 59, *387*
principle of equivalence, 145
principle of maximal analyticity, 173, 174, 185, *387*
principle of maximum strength, 142, 144, 148, 173, *387*
production threshold, 55, *387*
propagator, 84, *387*
pseudoscalar particle, *388*
Putnam, H., 264, 277

quantum chromodynamics (QCD), 25, 26, 222, *388*
quantum electrodynamics (QED), 12–19, *388*
quantum field theory (QFT), 8–12, 33, *388*
  axiomatic, 83, 167, 184, *373*
  Reggeized, 157, 325
  versus SMT, 161, 169–73, 245, 269, 276
  see also gauge field theory
quantum potential, 233
quantum theory (old), 218, *385*
quarks, 25, 156, 190, *388*

radiation damping, 45, 213, 304, *388*
radiative corrections, 14, *388*
Raine, D. J., 6, 286
rational reconstruction, 240
rationalist, xi, *388*
rationality, 219, 237, 240, 260, 271, 272, 274, 281
Ravens' paradox, 275, *388*
Rayleigh, Lord, 219
Rayleigh sattering, 77, 306, *388*
Rebbi, C., 205
Rechenberg, H., 28, 36, 64, 232
recycling of analogies, 244, 255
Redhead, M. L. G., xv, 118, 148
reduction technique, 73–5

Regge, T., 90, 116, 126, 129, 142, 143, *397*
Regge cuts, 153, 155, 200, 201, *388*
Regge phenomenology, 248
  degenerating, 152–5, 164, 165, 248
Regge poles, 126, 127, 131, 135, 141, 142,
  145, 146, 151, 212, 214, 323, *388*
Regge trajectory, *388*
Reggeon calculus, 164, 324
renormalizatin, 12–19, 99, 221, 268, *388*
research program, xii, 218, 249, *388*
  degenerating, 154, 248, *376*
  see also Regge phenomenology
resonance, 29, 30, 55, 63, 98, 99, 224, 305,
  308–15, *389*
  δ_{33}, 94, 99, 109, 141, *376*
  rho (ρ), 128, 129, *389*
Retherford, R. C., 12
reticulated model x, 243, 249, 273, 274,
  277, 287
Revolutionary science, xi
rho (ρ) meson, 128, 129, *389*
Richard, J. M., 202
Riemann–Lebesgue lemma, 62, *389*
Rigden, J. S., 5
Ringland, G. A., 164
Rivier, D., 47
Rohrlich, F., 59, 60, 62, 69, 87, 265, 277,
  321
Rolnik, W. B., 322
Rorty, R., 327, 328
Rosenthal-Schneider, I., 91
Rosenzweig, C., 199–202
Rosner, J. L., 196, 198
Rossetti, G., 191
Roy, S. M., 160
Ruderman, M. A., 77, 79, 108
Ruegg, H., 196
Ruelle, D., 83

S-matrix theory (SMT), 29, *390*
  see also sattering (S) matrix program
Sakita, B., 196
Salam, A., 17, 23–5, 81
Salam–Weinberg model, 25, 26, 203, 260,
  389
Salmon, W. C., 242
Salzman, F., 106, 216
Salzman, G., 106
Scadron, M., 160
scalar particle, 159, *389*
scattering, 19
  amplitude, 32, 57, 291, *389*
  Compton, 20, 170, *375*

scattering (*continued*)
  Coulomb, 78, *376*
  Delbrück, 59, 60, 62, *376*
  diffractive, 199
  elastic, 31, *378*
  hard, 244, *381*
  length, 94, 295, *389*
  Møller, 20, 21, *385*
  soft, 244, *390*
scattering (S) matrix, 17, 31, 74, 295
  defined, 31, 47, 177, 317, 318
  factorization of, 149, 185, 196, 204
  false zeros of, 49, 50, 318, 319
  ordered, 201
  planar, 201
  poles of, 43, 44, 61
  poles versus zeros, 55
  program, 212, 229, 258, 265;
    autonomous, 8, 129–32, 145, 168,
    172, 186, 215, 220, *373*; axiomatic,
    169, 182—5, 188; causal, 82, 270, *374*;
    conceptual framework of, 173–6;
    dual topologial (DTU), 201, 202;
    Heisenberg's, 28, 34, 39, 42, 57, 64,
    65, 67, 69, 89, 105, 115, 117, 118, 143,
    144, 167, 168, 210, 220, 266;
    topological (TSMT), 190. 200–3, 207;
    versus QFT, 161, 169–73, 245, 269,
    276
  Wheeler, 31
  zeros of, 38, 49
  see also pole–particle correspondence
Schaap, M. M., 199
Scherk, J., 205, 206
Schmid, C., 157, 191, 191–3
Schrödinger, E., 5, 10, 11, 219, 231, 308
Schrödinger equation, 11, 22, 31, 35, 42,
  124, 223, 226, 233, 234, 292, 299, 307,
  309
Schrödinger theory, *389*
Schützer, W., 60, 61, 63, 68–70, 226
Schwarz, J. H., 156, 191, 195, 202, 205,
  206, 215, 216, 281, 325
Schwarzschild, B. M., 206, 207
Schweber, S. S., 3, 8, 9, 214, 248, 256, 270,
  285, 288, *397*
Schwimmer, A., 324
Schwinger, J., 15, 18, 35, 47, 90, 256, *397*
scientific realism, xiii, 230, 264, *389*
Screaton, G. R., 123, 130
Seckel, D., 207
self-energy correction, 13, 14, *389*
separability, 229, *389*

Serber, R., 38
shadow matter, 207, *389*
Shapere, D., xiii, 227, 241, 242, 250, 251, 254, 255, 275, 284, *397*
Shimony, A., 228
Shirkov, D. V., 324
Siegel, H., 241
Siegert, A. J. F., 38, 52, 226, 309, 310
signature, 195, 325, *389*
simple model of science, x, 254
Singh, V., 157
singularity, *see* cut; essential singularity; pole
Slater, J. C., 213, 232, 291
Slotnick, M., 58, 117
Snider, D. R., 162
sociologial, *390*
    factors, 217, 238, 243–9, 284
solitons, 261, *390*
Solvay Congress, 14, 41, 131, 161, 169, 231, 232, 235
Sommerfeld, A., 220
Sommerfield, C. M., 122
spacelike separation, 69, 229, *390*
space-time continuum, 168, 175, 178, 240, 265
spectral functions, 124, 125
spin-statistics theorem, 37, 182, 183, *390*
spontaneous symmetry breaking, 24, 25, 206, *390*
Squires, E. J., 148, 152
Stack, J., 216
Stanghellini, A., 143, 154, 161
Stapp, H. P., 104, 105, 178, 182–6, 200, 284
static model (fixed-source theory), 97, *390*
Steinberger, J. L., 95, 96
Steinmann, O., 83
Stern, A. W., 175
Stirling, A. V., 153
stochastic mechanics, 234
Stöckler, M., 8
Stoops, R., 41
strangeness, 159, 173, *390*
Streater, R. F., 320
string theories, 26, 190, 203–7, 213, 266
strong program, 283, *390*
Stückelberg, E. C. G., 45–7, *397*
SU(2) symmetry, 93, *390*
SU(3) symmetry, 24, 25, 159, 196, *390*
substitution law, 101, 321, *390*
subtraction procedure, *391*
    see also renormalization

subtractions, 14, 15, 127, 131, *391*
Sukatme, U., 164
sum rule, 77, 189, *391*
superconvergence relations, 154, 191, *391*
superdeterminism, 289, *391*
superselection rule, 183, *391*
superstrings, 26, 190, 207, 208, 220, 221, 266, *391*
supersymmetry, 206, 326, *391*
Sursock, J., 216
Symanzik, K., 81, 82, 116, 126, 214

tachyon, 195, 205, *391*
Takeda, G., 81
Tan, C-I., 89, 164
Tarjanne, P., 159
Tarski, J., 123
Taylor, J. C., 123
Taylor, J. G., 81
Taylor, J. R., 184
Taylor, T. B., 96
Teller, P., 221, 228
ter Haar, D., 49
Ter-Martorosyan, K. A., 164
theory construction and selection, 223, 230, 231, 237, 246, 255, 270
Thirring, W. E., 67, 70, 77, 87, 213, 229, 265
Thomas–Kuhn sum rule, 77
Thomson scattering, 77, 306, *391*
't Hooft, G., 25
Thorn, C. B., 205
(3, 3) resonance, *see* resonance, δ_{33}
time reverwal, *391*
Tiomno, J., 60, 61, 63, 68–70, 226
Tipler, F. J., 221, 259, 285
Titchmarsh, E. C., 328
TOE, 191
Toll, J. S., 59, 62, 68–70
Tomonaga, S., 15, 256
topological expansion, 190, 196–200
topology, *391*
Toulmin, S., 327
Touschek, B., 50
Toynbee, A., 250
Tran Thanh Van, J., 164
Treiman, S. B., 86, 123, 127
triadic model, *see* reticulated model
Tsou, S. T., 199
Turner, M. S., 207

Udgaonkar, B. M., 148, 150
unified field theory, *391*

unitarity, 23, 35–37, 58, 59, 87, 107, 176, 177, *392*
  elastic, 92, 140, *378*
  unphysical region, 185, 186, *392*
unphysical region, 117, *392*
Uschersohn, J., 202

vacuum, 263
  polarization, 14, *392*
van der Waerden, B. L., 213
van Fraassen, B. C., 1, 254, *397*
van Hove, L., 153
van Kampen, N. G., 50, 60–3, 69, 70, 76
vann Woodward, C., 269
Varnak, S. J., xv
vector particle, 159, *392*
Veneziano, G., 157, 189, 199, 201, 204, 205, 251
Veneziano model, 189, 193–6, 203, 212, 214
vertex function, 84, 117, *392*
vertex modification, 21, *392*
Vigier, J-P., 232, 233
Vinh Mau, R., 165, 202, 280
Virasoro, M. A., 196, 205
virtual process, 14
von Neumann, J., 126, 232

W boson, *see* boson, intermediate vector
Walker, W. D., 128
Waller, I., 308
Waltz, R., 198
Watkins, J. W. N., xi, xiv, 282, *398*
Watson, K. M., 94
wave mechanics, 11, *392*
weak interaction, 168, 174, 185, 202, *392*
Weinberg, S., 24, 160, 161, 164, 165, 207, 229, 270, *398*
  folk theorems, 207, 270

Weinstein, M., 284
Weisskopf, V. F., 45, 52, 223, 224, 226, 256, 306, 308
Wentzel, G., 28, 45, 47, 48, 67, 89, 145, *398*
Wess, J., 326
West, E., 128
Wheeler, J. A., 27, 30, 31, 36, 59, 62, 65, 68, 69, 259, *398*
Whewell, W., ix, 4, *398*
Wichmann, E. H., 122
Wick, G. C., 36, 95, 96, 104–6, 183
Wightman, A. S., 83, 167, 183, 320
Wigner, E. P., 38, 68, 126
  causality, 60
  compound nucleus model, 223, 228
  phase-shift bound, 62, 63, 313
  radiation damping, 45, 224, 306
  resonance formula, 51–4, 61, 66, 226, 227, 309
  superselection rule, 183
Wildermuth, K., 50
Wilson, P. A., 221, 259, 260
Wong, D. Y., 143
Wouthuysen, S. A., 38
Wright, J., 160

Yang, C. N., 22, 23, 73, 95, 261
Yang–Mills theory, 23–5, 185, 222, *392*
Yukawa, H., 2, 256, 280
Yukawa potential, 97, 105, 127, *392*

Zachariasen, F., 84, 86, 148, 150, 151, 154, 157, 159, 160, 164, 178
Zahar, E., 215, 248
Zemach, C., 157, 159, 178
zeros, *see* scattering (S) matrix
Zimmermann, W., 81, 82, 116, 214
Zumino, B., 326
Zurek, W. H., 289
Zweig, G., 198, 215, 280